INTERNET POLICIES AND ISSUES

BROADBAND AND RURAL AMERICA

INTERNET POLICIES AND ISSUES

Additional books in this series can be found on Nova's website at:

https://www.novapublishers.com/catalog/index.php?cPath=23_29&seriesp=Internet +Policies+and+Issues

Additional e-books in this series can be found on Nova's website at:

https://www.novapublishers.com/catalog/index.php?cPath=23_29&seriespe=Internet +Policies+and+Issues

INTERNET POLICIES AND ISSUES

BROADBAND AND RURAL AMERICA

BETHANY J. LENZ
EDITOR

Nova Science Publishers, Inc.

New York

For permission to use material from this book please contact us:
Telephone 631-231-7269; Fax 631-231-8175
Web Site: http://www.novapublishers.com

LIBRARY OF CONGRESS CATALOGING-IN-PUBLICATION DATA

>Broadband and rural America / editors, Bethany J. Lenz.
> p. cm.
> Includes bibliographical references and index.
> ISBN 978-1-61668-637-6 (hardcover)
> 1. Rural telecommunication--Government policy--United States. I. Lenz,
>Bethany J.
>
> TK5102.3.U6B73 2010
> 384.3'3--dc22
> 2010016684

Published by Nova Science Publishers, Inc. † New York

CONTENTS

PREFACE

This book explores broadband Internet use and rural America. For many Americans, a world without broadband is unimaginable. However, large parts of rural America have languished on the sidelines of the digital revolution. As many of their fellow citizens in more densely populated parts of the country go online for work, education, entertainment, healthcare, civic participation, and much more, too many rural Americans are being left behind. Rural governments and businesses are missing opportunities to function more efficiently and effectively. At a time when access to affordable, robust broadband services is a fundamental part of efforts to restore America's economic well-being in both rural and urban areas, we must ensure that this capability is available to open the doors of opportunity for everyone.

Chapter 1 - 1. For many Americans, a world without broadband is unimaginable. For them, broadband Internet access has transformed the way they live their lives. But we have not succeeded in bringing broadband to everyone. For years, large parts of rural America have languished on the sidelines of the digital revolution. Home to the homesteaders, pioneers, and the rich and diverse Native American cultures that contribute so much to our national identity, rural America has for most of our history been deemed too remote, too sparsely populated, or too inaccessible to be fully connected with our nation's infrastructures.

2. As many of their fellow citizens in more densely populated parts of the country go online for work, education, entertainment, healthcare, civic participation, and much more, too many rural Americans are being left behind. Rural governments and businesses are missing opportunities to function more efficiently and effectively. Even in rural areas where broadband is available, infrastructure deployment has not kept pace with the growing need for faster and more reliable connectivity. At a time when access to affordable, robust broadband services is a fundamental part of efforts to restore America's economic well-being in both rural and urban areas, we must ensure that this capability is available to open the doors of opportunity for everyone.

3. In the 2008 Farm Bill, Congress recognized that the need for broadband in rural areas was becoming ever more critical. That law requires the Chairman of the Federal Communications Commission (Commission), in coordination with the Secretary of the U.S. Department of Agriculture (USDA), to submit "a report describing a comprehensive rural broadband strategy" to Congress. Under the statute, this strategy must include, among other matters, recommendations for improving interagency coordination of broadband policies and initiatives; for assessing broadband needs in rural areas; and for how specific federal agency

programs and resources can overcome the obstacles that currently impede rural broadband deployment. In this Report, I discharge this Congressional mandate.

4. Upon becoming Acting Chairman in January 2009, I directed the Commission staff to begin developing the data and analysis to the extent possible to complete a Report in the limited time remaining. In some respects, events overtook this effort as Congress provided new direction and support for federal broadband policies and initiatives, guidance which frankly has reshaped our approach to the development of this Report. In particular, in the recently passed American Recovery and Reinvestment Act of 2009, also known as the stimulus package, Congress appropriated $7.2 billion for broadband grants, loans, and loan guarantees to be administered by the USDA's Rural Utilities Service (RUS) and the Department of Commerce's National Telecommunications and Information Administration (NTIA). Congress recognized that this funding initiative, though substantial, was still just a down payment on the broadband needs of the country, and that even after this money has been invested, many Americans, including those residing in rural areas, will continue to lack access to critical broadband services. Therefore, the Recovery Act charges this Commission with developing a national broadband plan by next February to ensure that every American has access to broadband capability and establishing clear milestones for reaching this goal.

5. Shortly after President Obama took office, his administration began to play an important leadership role in the effort to expand broadband penetration throughout the nation. Through an interagency working group under the auspices of the National Economic Council, the Administration has brought together representatives from a broad cross-section of federal agencies, including the Commission, to coordinate implementation of the nation's broadband agenda. This working group provides an unprecedented forum for expert staff from different federal agencies to meet, discuss, and harmonize various federal broadband programs and policies, and identify areas for early action. I view these efforts, and other efforts on the part of the administration and Congress, as essential and vital components for the development of a rural broadband strategy.

6. Working together across the federal government and building on the interagency working group's efforts, we view the separate mandates from Congress as an unparalleled opportunity to ensure that the promise of universal access to broadband services is realized for all Americans. Making the most of this opportunity will require coordination of the various responsibilities with which this Commission is charged, as well as coordination among the federal agencies and other stakeholders with roles in achieving the nation's broadband goals. We must ensure that this Report and the national broadband plan, to be delivered to Congress by February 17, 2010, are not isolated outputs. Rather, we must view these initiatives, along with the Commission's other broadband responsibilities, as vital and integral steps in the process of achieving the end result: affordable, value-laden broadband services for all Americans.

7. Consistent with our coordinated approach to broadband policy, and working in particular with the USDA, we build here on the foundation established in the Commission's first step toward the national broadband plan, its recent *National Broadband Plan NOI.* In that notice, the Commission recognized that bold action will be necessary for the nation to achieve its goal of ubiquitous and affordable broadband for all, regardless of location, socioeconomic status, ethnic background, or any other factor. The *National Broadband Plan NOI* recognized "the incredible value of ubiquitous broadband, and the difficulties that lie ahead in ensuring its availability." The nation will need to overcome many obstacles in

ensuring that every American citizen, American business, Tribal and local government, and public safety entity has full access to broadband services. Success in this endeavor will require the input and cooperation of many different entities—individual consumers, businesses and organizations, as well as federal agencies and Tribal, state, and local governments. We must marry the dynamic innovations and flexibility of the private sector with the policy vision of the public sector to create a model of how government and industry can partner to ensure ubiquitous broadband access.

8. I view this Report as a prelude to, and a building block for, the national broadband plan, which will address in greater detail and on a vastly more complete record, the input of all stakeholders and the steps the nation must take to achieve its broadband goals. I also view this Report, which reflects input from a nevertheless substantial array of stakeholders and presents significant information about rural America, as a compendium of information that will be useful in the development of the national broadband plan. The recommendations made today for a rural broadband strategy are a down payment on the development of the national broadband plan, as indeed many of the steps necessary for ensuring access to broadband capability throughout rural America also will be necessary to ensure such access in urban and suburban parts of the nation. Although the national broadband plan will be broader in scope, this Report provides another, critical step in the Commission's efforts to develop an effective, efficient, and achievable national broadband plan. No national broadband strategy can be undertaken without due consideration to the rural broadband infrastructure and the people it must serve. The likely success of rural initiatives is intimately linked to a sound national broadband policy that reflects the complex interdependencies of regulatory policies, economic issues, and technological innovations.

9. In the end, my goal is that all rural Americans, like their counterparts in more densely populated areas of the country, have the opportunity to reap the full benefits of broadband services. Certainly the challenges of deploying broadband across vast rural lands are daunting. But as long as a grade-school child living on a farm cannot research a science project, or a high school student living on a remote Indian reservation cannot submit a college application, or an entrepreneur in a rural hamlet cannot order spare parts, or a local law enforcement officer cannot download pictures of a missing child without traveling to a city or town that has broadband Internet access, we cannot turn back from these challenges. Rural America has shaped this nation's values and institutions, given us many of our greatest leaders, and been an engine of our economic progress, but it will be unable to compete in the digital global marketplace without access to broadband services. Now is the time to start developing solutions for building the critical broadband infrastructure that *all* Americans will need.

10. In developing these solutions, we are not starting from scratch. Many of the technologies and resources needed for rural broadband already exist and, with American ingenuity, will become faster and more powerful in the years to come. High-capacity fiber networks—once found only in dense urban cores—have been redesigned for residential use, and their performance continues to increase. Cable networks are being upgraded to a platform that will support data rates of up to 160 megabits per second (Mbps). While issues remain, broadband over power lines (BPL) continues to emerge as a viable technology option. Wireless technologies are extending broadband into areas unreachable by cables and wires, and enabling consumers to be connected while on the move. Many wireless Internet service providers (ISPs) have used the IEEE 802.11 wireless local area network technologies

(commonly known as Wi-Fi) to offer fixed wireless broadband services in areas not reached by wireline technologies. Wireless providers have been launching new broadband technologies that allow subscribers to access the Internet, while mobile, at speeds that are beginning to rival those on landline networks. We expect to see further advancements on the wireless broadband front as wireless service providers begin to build out networks using advanced technologies—such as Long Term Evolution (LTE) or Worldwide Interoperability for Microwave Access (WiMAX)—that support data rates that may exceed 100 Mbps. Finally, satellite broadband, with its near ubiquitous coverage and downstream data rates between 512 kbps and 5 Mbps, can provide a much-needed connection in rural areas, especially where other broadband solutions are not viable for technical or other reasons.

11. The solutions for rural broadband should reflect consideration of the full range of technological options available, and should not elevate the need for short-term progress over longer-term objectives. Rural broadband likely will include a variety of different technologies that together can support the state-of-the-art, secure, and resilient broadband service that should be our goal for rural America, just as it is for the non-rural parts of the nation. However the rural networks are configured, they should be designed on principles of durability, reliability, openness, scalability, and interoperability so that they can evolve over time to keep pace with the growing array of transformational applications and services that are increasingly available to consumers and businesses in other parts of the country.

12. Although we are at an early stage in the national effort, the Report makes a number of recommendations that I hope will facilitate the rapid and widespread buildout of state-of-the-art broadband access facilities to every street corner and winding road, and every home and business in America. Such a transformation will rival the building of the roads, canals, and ports that made commerce possible in pre-Civil War America; the transcontinental railroads that made us a continental power in the late nineteenth century; the national highway system that opened the way for rapid transportation and demographic migration in the last century; and the immense efforts to extend telephone and electrical service to the far corners of America. Technology, and the ability to be connected, can be an important part of addressing many of the problems rural America faces by, for example, enabling a student at a rural high school to participate in a seminar offered at a distant college, letting a patient in a rural clinic be examined by a specialist located in an urban hospital, or allowing a farm family to use a smart grid to reduce its energy consumption. Thus, delivering ubiquitous broadband to rural areas will help to restore economic growth and opportunity for Americans residing and working in those areas.

Chapter 2 - Both Congress and the President have indicated that access to broadband for all Americans is critically important. Broadband is seen as a critical economic engine, a vehicle for enhanced learning and medicine, and a central component of 21st century news and entertainment. As part of our response to a mandate included in the Internet Tax Nondiscrimination Act of 2004, this report examines the factors that affect the deployment and the adoption of broadband services. In particular, this report provides information on (1) the current status of broadband deployment and adoption; (2) the factors that influence the deployment of broadband networks; (3) the factors that influence the adoption, or purchase, of broadband service by households; and (4) the options that have been suggested to spur greater broadband deployment and adoption.

Chapter 3 - As broadband—or high-speed—Internet use has spread, Internet applications requiring high transmission speeds have become an integral part of the "Information

Economy," raising concerns about those who lack broadband access. This report analyzes (1) rural broadband use by consumers, the community-at-large, and businesses; (2) rural broadband availability; and (3) broadband's social and economic effects on rural areas. It also summarizes results from an ERS-sponsored workshop on rural broadband use, and other ERS-commissioned studies. In general, rural communities have less broadband Internet use than metro communities, with differing degrees of broadband availability across communities. Rural communities that had greater broadband Internet access had greater economic growth, which conforms to supplemental research on the benefits that rural businesses, consumers, and communities ascribe to broadband Internet use.

Chapter 4 - Given the large potential impact broadband access may have on the economic development of rural America, concern has been raised over a "digital divide" between rural and urban or suburban areas with respect to broadband deployment. While there are many examples of rural communities with state of the art telecommunications facilities, recent surveys and studies have indicated that, in general, rural areas tend to lag behind urban and suburban areas in broadband deployment.

Citing the lagging deployment of broadband in many rural areas, Congress and the Administration acted in 2001 and 2002 to initiate pilot broadband loan and grant programs within the Rural Utilities Service (RUS) at the U.S. Department of Agriculture (USDA). Subsequently, Section 6103 of the Farm Security and Rural Investment Act of 2002 (P.L. 107-171) amended the Rural Electrification Act of 1936 to authorize a loan and loan guarantee program to provide funds for the costs of the construction, improvement, and acquisition of facilities and equipment for broadband service in eligible rural communities. The RUS/USDA houses two assistance programs exclusively dedicated to financing broadband deployment: the Rural Broadband Access Loan and Loan Guarantee Program and the Community Connect Grant Program.

The 110th Congress considered reauthorization and modification of the loan and loan guarantee program as part of the farm bill. The Food, Conservation, and Energy Act of 2008 became law on June 18, 2008 (P.L. 110-246). Title VI (Rural Development) contains authorizing language for the broadband loan program.

Meanwhile, on May 11, 2007, RUS released a Proposed Rule seeking to revise the broadband loan program rules and regulations. Some key issues pertinent to a consideration of the RUS broadband programs include restrictions on applicant eligibility, how "rural" is defined with respect to eligible rural communities, how to address assistance to areas with preexisting broadband service, technological neutrality, funding levels and mechanisms, and the appropriateness of federal assistance. The final rule will reflect language in the enacted farm bill statute (P.L. 110-246). Ultimately, modification of rules, regulations, or criteria associated with the RUS broadband program will likely result in "winners and losers" in terms of which companies, communities, regions of the country, and technologies are eligible or more likely to receive broadband loans and grants.

On February 17, 2009, President Obama signed P.L. 111-5, the American Recovery and Reinvestment Act (ARRA). Broadband provisions of the ARRA provide a total of $7.2 billion, primarily for broadband grants. The total consists of $2.5 billion for RUS broadband loans, grants, and loan/grant combinations, and $4.7 billion to the National Telecommunications and Information Administration (NTIA) at the Department of Commerce (DOC) for a newly established Broadband Technology Opportunities Program. On July 1, 2009, RUS and NTIA jointly released the first Notice of Funds Availability

(NOFA) detailing requirements, rules, and procedures for applying for ARRA grants, loans, and loan grant combinations

This report will be updated as events warrant.

Chapter 5 - This is a testimony of Jim Andrew to the House Committee on Agriculture, before the Subcommittee on Specialty Crops, Rural Development and Foreign Agriculture.

Chapter 6- This is a testimony of Kevin Felty, General Manager, Plains Cooperative Telephone Association, before the Subcommittee on Specialty Crops, Rural Development and Foreign Agriculture.

Chapter 7 - This is a testimony of Jack M. Geller, President, Center for Rural Policy and Development, before the Subcommittee on Specialty Crops, Rural Development and Foreign Agriculture.

Chapter 8 - This is a testimony of Denny Law, Eastern Regional Manager, Golden West Telecommunications, before the Subcommittee on Specialty Crops, Rural Development and Foreign Agriculture.

Chapter 9 - This is a testimony of Walter B. McCormick, Jr., President and CEO, U.S. Telecom Association to the House Committee on Agriculture, before the Subcommittee on Specialty Crops, Rural Development and Foreign Agriculture.

Chapter 10 - This is a testimony of Tom Simmons, Senior Vice President of Public Policy, Midcontinent Communications, before the Subcommittee on Specialty Crops, Rural Development and Foreign Agriculture on May 1, 2007.

In: Broadband and Rural America
Editor: Bethany J. Lenz
ISBN: 978-1-61668-637-6

Chapter 1

BRINGING BROADBAND TO RURAL AMERICA

Michael J. Copps

I. INTRODUCTION

1. For many Americans, a world without broadband is unimaginable. For them, broadband Internet access has transformed the way they live their lives. But we have not succeeded in bringing broadband to everyone. For years, large parts of rural America have languished on the sidelines of the digital revolution. Home to the homesteaders, pioneers, and the rich and diverse Native American cultures that contribute so much to our national identity, rural America has for most of our history been deemed too remote, too sparsely populated, or too inaccessible to be fully connected with our nation's infrastructures.

2. As many of their fellow citizens in more densely populated parts of the country go online for work, education, entertainment, healthcare, civic participation, and much more, too many rural Americans are being left behind. Rural governments and businesses are missing opportunities to function more efficiently and effectively. Even in rural areas where broadband is available, infrastructure deployment has not kept pace with the growing need for faster and more reliable connectivity. At a time when access to affordable, robust broadband services is a fundamental part of efforts to restore America's economic well-being in both rural and urban areas, we must ensure that this capability is available to open the doors of opportunity for everyone.

3. In the 2008 Farm Bill, Congress recognized that the need for broadband in rural areas was becoming ever more critical.[1] That law requires the Chairman of the Federal Communications Commission (Commission), in coordination with the Secretary of the U.S. Department of Agriculture (USDA), to submit "a report describing a comprehensive rural broadband strategy" to Congress. Under the statute, this strategy must include, among other matters, recommendations for improving interagency coordination of broadband policies and initiatives; for assessing broadband needs in rural areas; and for how specific federal agency

programs and resources can overcome the obstacles that currently impede rural broadband deployment.[2] In this Report, I discharge this Congressional mandate.

4. Upon becoming Acting Chairman in January 2009, I directed the Commission staff to begin developing the data and analysis to the extent possible to complete a Report in the limited time remaining.[3] In some respects, events overtook this effort as Congress provided new direction and support for federal broadband policies and initiatives, guidance which frankly has reshaped our approach to the development of this Report. In particular, in the recently passed American Recovery and Reinvestment Act of 2009,[4] also known as the stimulus package, Congress appropriated $7.2 billion for broadband grants, loans, and loan guarantees to be administered by the USDA's Rural Utilities Service (RUS) and the Department of Commerce's National Telecommunications and Information Administration (NTIA).[5] Congress recognized that this funding initiative, though substantial, was still just a down payment on the broadband needs of the country, and that even after this money has been invested, many Americans, including those residing in rural areas, will continue to lack access to critical broadband services. Therefore, the Recovery Act charges this Commission with developing a national broadband plan by next February to ensure that every American has access to broadband capability and establishing clear milestones for reaching this goal.[6]

5. Shortly after President Obama took office, his administration began to play an important leadership role in the effort to expand broadband penetration throughout the nation. Through an interagency working group under the auspices of the National Economic Council, the Administration has brought together representatives from a broad cross-section of federal agencies, including the Commission, to coordinate implementation of the nation's broadband agenda. This working group provides an unprecedented forum for expert staff from different federal agencies to meet, discuss, and harmonize various federal broadband programs and policies, and identify areas for early action. I view these efforts, and other efforts on the part of the administration and Congress, as essential and vital components for the development of a rural broadband strategy.

6. Working together across the federal government and building on the interagency working group's efforts, we view the separate mandates from Congress as an unparalleled opportunity to ensure that the promise of universal access to broadband services is realized for all Americans. Making the most of this opportunity will require coordination of the various responsibilities with which this Commission is charged, as well as coordination among the federal agencies and other stakeholders with roles in achieving the nation's broadband goals. We must ensure that this Report and the national broadband plan, to be delivered to Congress by February 17, 2010, are not isolated outputs. Rather, we must view these initiatives, along with the Commission's other broadband responsibilities, as vital and integral steps in the process of achieving the end result: affordable, value-laden broadband services for all Americans.

7. Consistent with our coordinated approach to broadband policy, and working in particular with the USDA, we build here on the foundation established in the Commission's first step toward the national broadband plan, its recent *National Broadband Plan NOI.*[7] In that notice, the Commission recognized that bold action will be necessary for the nation to

achieve its goal of ubiquitous and affordable broadband for all, regardless of location, socioeconomic status, ethnic background, or any other factor. The *National Broadband Plan NOI* recognized "the incredible value of ubiquitous broadband, and the difficulties that lie ahead in ensuring its availability."[8] The nation will need to overcome many obstacles in ensuring that every American citizen, American business, Tribal and local government, and public safety entity has full access to broadband services. Success in this endeavor will require the input and cooperation of many different entities—individual consumers, businesses and organizations, as well as federal agencies and Tribal, state, and local governments. We must marry the dynamic innovations and flexibility of the private sector with the policy vision of the public sector to create a model of how government and industry can partner to ensure ubiquitous broadband access.

8. I view this Report as a prelude to, and a building block for, the national broadband plan, which will address in greater detail and on a vastly more complete record, the input of all stakeholders and the steps the nation must take to achieve its broadband goals. I also view this Report, which reflects input from a nevertheless substantial array of stakeholders and presents significant information about rural America, as a compendium of information that will be useful in the development of the national broadband plan. The recommendations made today for a rural broadband strategy are a down payment on the development of the national broadband plan, as indeed many of the steps necessary for ensuring access to broadband capability throughout rural America also will be necessary to ensure such access in urban and suburban parts of the nation. Although the national broadband plan will be broader in scope, this Report provides another, critical step in the Commission's efforts to develop an effective, efficient, and achievable national broadband plan. No national broadband strategy can be undertaken without due consideration to the rural broadband infrastructure and the people it must serve. The likely success of rural initiatives is intimately linked to a sound national broadband policy that reflects the complex interdependencies of regulatory policies, economic issues, and technological innovations.

9. In the end, my goal is that all rural Americans, like their counterparts in more densely populated areas of the country, have the opportunity to reap the full benefits of broadband services. Certainly the challenges of deploying broadband across vast rural lands are daunting. But as long as a grade-school child living on a farm cannot research a science project, or a high school student living on a remote Indian reservation cannot submit a college application, or an entrepreneur in a rural hamlet cannot order spare parts, or a local law enforcement officer cannot download pictures of a missing child without traveling to a city or town that has broadband Internet access, we cannot turn back from these challenges. Rural America has shaped this nation's values and institutions, given us many of our greatest leaders, and been an engine of our economic progress, but it will be unable to compete in the digital global marketplace without access to broadband services.[9] Now is the time to start developing solutions for building the critical broadband infrastructure that *all* Americans will need.

10. In developing these solutions, we are not starting from scratch. Many of the technologies and resources needed for rural broadband already exist and, with American ingenuity, will become faster and more powerful in the years to come. High-capacity fiber

networks—once found only in dense urban cores—have been redesigned for residential use, and their performance continues to increase. Cable networks are being upgraded to a platform that will support data rates of up to 160 megabits per second (Mbps).[10] While issues remain, broadband over power lines (BPL) continues to emerge as a viable technology option.[11] Wireless technologies are extending broadband into areas unreachable by cables and wires, and enabling consumers to be connected while on the move. Many wireless Internet service providers (ISPs) have used the IEEE 802.11 wireless local area network technologies (commonly known as Wi-Fi) to offer fixed wireless broadband services in areas not reached by wireline technologies.[12] Wireless providers have been launching new broadband technologies that allow subscribers to access the Internet, while mobile, at speeds that are beginning to rival those on landline networks.[13] We expect to see further advancements on the wireless broadband front as wireless service providers begin to build out networks using advanced technologies—such as Long Term Evolution (LTE) or Worldwide Interoperability for Microwave Access (WiMAX)—that support data rates that may exceed 100 Mbps.[14] Finally, satellite broadband, with its near ubiquitous coverage and downstream data rates between 512 kbps and 5 Mbps, can provide a much-needed connection in rural areas, especially where other broadband solutions are not viable for technical or other reasons.[15]

11. The solutions for rural broadband should reflect consideration of the full range of technological options available, and should not elevate the need for short-term progress over longer-term objectives. Rural broadband likely will include a variety of different technologies that together can support the state-of-the-art, secure, and resilient broadband service that should be our goal for rural America, just as it is for the non-rural parts of the nation.[16] However the rural networks are configured, they should be designed on principles of durability, reliability, openness,[17] scalability, and interoperability so that they can evolve over time to keep pace with the growing array of transformational applications and services that are increasingly available to consumers and businesses in other parts of the country.

12. Although we are at an early stage in the national effort, the Report makes a number of recommendations that I hope will facilitate the rapid and widespread buildout of state-of-the-art broadband access facilities to every street corner and winding road, and every home and business in America. Such a transformation will rival the building of the roads, canals, and ports that made commerce possible in pre-Civil War America; the transcontinental railroads that made us a continental power in the late nineteenth century; the national highway system that opened the way for rapid transportation and demographic migration in the last century; and the immense efforts to extend telephone and electrical service to the far corners of America. Technology, and the ability to be connected, can be an important part of addressing many of the problems rural America faces by, for example, enabling a student at a rural high school to participate in a seminar offered at a distant college, letting a patient in a rural clinic be examined by a specialist located in an urban hospital, or allowing a farm family to use a smart grid to reduce its energy consumption.[18] Thus, delivering ubiquitous broadband to rural areas will help to restore economic growth and opportunity for Americans residing and working in those areas.

II. Overview of Recommendations

13. Congress directed the Chairman of the Commission to submit this Report to Congress, and I am pleased to do so. I believe the Commission should do more such reports for Congress—not establishing policies that require Commission approval—but putting forth recommendations, ideas, and options to advise Congress, government, and the public as they consider these important issues. I look forward to continuing to work on these issues with my colleagues and Congress and trust that this Report will help inform the discussion and provide a building block as the Commission develops its national broadband plan. In light of all this, the Report makes the following recommendations:

Coordination of Rural Broadband Efforts. Increasing coordination—among federal agencies; Tribal, state, and local governments; and community groups and individuals—is a critical preliminary step towards ensuring that the various government programs accomplish their broadband goals and objectives in an efficient and effective way.

- **Improving Federal Agency Coordination.**
    - The federal interagency working group, formed by the Obama administration to coordinate the administration's broadband agenda, should continue its ongoing efforts to enhance interagency coordination of rural broadband initiatives; and
    - The Commission and other federal agencies should consider developing their own "rural broadband agendas" consistent with the national broadband plan.

- **Other Coordination Efforts.**

    - **Tribal Coordination.** To ensure a truly comprehensive strategy for addressing rural broadband deployment and adoption, it is important to maintain a continuing dialogue to address the unique issues presented in Tribal areas. As an integral part of their rural broadband initiatives:
        (1) Federal agencies should consider how to maximize existing programs to improve coordination with Tribal governments; and
        (2) The Commission should consult with Tribal governments pursuant to its *Tribal Policy Statement* in developing its national broadband plan and, in particular, in developing the aspects of that plan that affect broadband deployment and subscribership specifically on Tribal lands.

    - **State Coordination.** The Commission and its state counterparts should take advantage of existing coordination mechanisms, such as the Federal-State Joint Conference on Advanced Services (Joint Conference). The Joint Conference should:
        (1) Provide the Commission with its own recommendations for improving federal-state coordination regarding rural broadband;

(2) Include in its recommendations proposals for federal-state coordination to address and ameliorate the unique challenges presented to rural minority communities and persons with disabilities residing in rural areas; and
(3) Compile an inventory of successful state and local projects and "best practices."

- **Coordination with Communities.** In order to be successful in coordinating existing federal programs concerning rural broadband or rural initiatives, it is also critical that the federal government collaborate and coordinate with community and advocacy organizations in rural areas. The minority, disability, and low-income communities in rural areas face particular challenges. Federal agencies should work closely with organizations:
  (1) To help ensure that all members of minority groups residing in rural areas have access to robust and affordable broadband services and that minority-owned businesses participate fully in the buildout of broadband infrastructure in those areas.
  (2) Representing persons with disabilities to help ensure that they have affordable access to broadband services capable of supporting the full array of applications responsive to their needs.
  (3) That serve low-income residents to ensure the opportunities that affordable broadband offers this community do not go unrealized.

- **Streamlining and Improving Existing Federal Programs.** All relevant federal agencies should review their programs to identify what internal barriers, if any, may be making rural broadband deployment more difficult.

- **Promoting Efficient Use of Government Funds and Resources**. Federal agencies should review their non-broadband-related programs that involve rural issues to assess whether those programs provide opportunities to promote rural broadband deployment.

- **Coordinating Program Criteria.** So that dissimilar definitions and criteria across related or complementary programs do not unnecessarily hinder interagency coordination, federal agencies involved in rural broadband should coordinate key terminology (*e.g.*, rural) across programs, consistent with their legislative mandates.

- **Government Websites.** One barrier to rural broadband deployment and adoption is a lack of easily-accessible and coordinated information about government resources available for promoting broadband. To help address this problem, we recommend that the Commission expand its website to include a comprehensive set of links to all federal government programs related to rural broadband. We also suggest expanding the Commission's and USDA's existing "Broadband Opportunities for Rural America" website to include a comprehensive list of all federal government programs related to rural broadband.

- *Assessing Broadband Needs*. Congress directed that this Report make recommendations "to address both short- and long-term needs assessments" for rural broadband. The Report does this by addressing the challenges of rural broadband today and the needs of rural broadband going forward.

  - **Technological Considerations.** Every rural area presents its own special challenges, and a particular technological solution may be well-suited to one situation and poorly-suited to another. Decision makers therefore should proceed on a technology-neutral basis—by considering the attributes of all potential technologies—in selecting the technology or technologies to be deployed in a particular rural area.

  - **Information on Broadband Availability.** One significant challenge to ubiquitous broadband deployment in rural areas is obtaining accurate information on broadband service and infrastructure availability and the demand for broadband services. Pursuant to the Broadband Data Improvement Act (BDIA) and the Recovery Act, the Commission should work to collect this information to better inform decision making, in coordination with the administration, and Tribal and state governments.[19]

  - **Broadband Mapping.** In the rural context, broadband mapping is a necessary tool for identifying and tracking broadband service availability and infrastructure deployment, yet it is only as accurate and reliable as its underlying data. Pursuant to the Recovery Act and the BDIA, the Commission and the administration should continue their efforts to coordinate federal, Tribal, state, local, and private mapping efforts.

  - **Stimulating and Sustaining Demand for Broadband.** Various factors may affect demand for broadband services in rural areas, including a lack of knowledge regarding the benefits of Internet access, lack of training on how to use a computer, socioeconomic and demographic factors, and affordability. To help stimulate and sustain demand for broadband services in rural areas, both public and private entities should consider developing consumer education and training initiatives, broadband affordability programs, and other incentives to achieve sustainable penetration rates.

  - **Addressing Network Costs.** Relying on market forces alone will not bring robust and affordable broadband services to all parts of rural America. Therefore, all levels of government should explore ways to help overcome the high costs of rural broadband deployment.

- ***Overcoming Challenges to Rural Broadband Deployment.*** Because the national broadband plan is not due until February 2010, it is prudent for the Commission to identify any pending and proposed Commission proceedings affecting rural broadband. These pending proceedings include universal service reform, network openness, spectrum access, middle mile/special access reform, intercarrier compensation, access to poles and rights of way, tower siting, and video programming proceedings. The Commission should consider all these proceedings as it develops the national broadband plan, balancing the desire to resolve these matters with the need to address rural broadband in the context of a much broader and forward-looking national broadband plan. Of particular note, I continue to support comprehensive reform of the universal service program. It is of great interest to Congress, consumers, industry, and the Commission. I have long held the view that it is time for universal service to meet the communications challenge of the 21st century—broadband deployment—just as it did the communications challenge of the 20th century—telephone service.

- ***Status Report.*** To help inform Congress of any needed changes to the recommendations in this Report in light of additional efforts to address rural broadband issues, including the completion of the national broadband plan, the next Commission Chairman should consider completing a status report on rural broadband approximately one year from now.

III. BACKGROUND

A. Rural Broadband: Why It Matters

14. Broadband Internet access, often on small, easy-to-use devices, is rapidly transforming the way we communicate, work, learn, and play.[20] Touch-screen mobile devices, e-books, and Voice over Internet Protocol (VoIP) have become standard features in many people's daily lives. As broadband speeds have increased, we use the Internet to complete our education, receive medical care, communicate with our government, including on matters of public health and safety, run web-based businesses, and do our work. In addition, we not only surf the web, send e-mail, form social networks, and download music and video, but also become ourselves creators of web-based content as we blog about local or world events, and become publishers in our own right.

15. In rural areas, however, many Americans have no access to these applications and services, and by extension, to the global community. Rural communities have long been unserved or underserved by broadband technology, but the full implication of this divide has only emerged as the Internet has become less and less a novelty, and more and more a necessity. The Obama administration and Congress recognize that broadband access will benefit all Americans in much the same way that the nation as a whole benefited from our past successes in overcoming infrastructure challenges. Bringing ubiquitous and affordable broadband services to rural America will improve the quality of education, healthcare, and

public safety in rural America, among other benefits. On a larger scale, ensuring that all Americans, including those in rural areas, have access to such services will help to improve America's economy, its ability to compete internationally, and its unity as a nation.

16. Deploying broadband throughout rural America will fundamentally benefit the nation's economy.[21] One study estimates that communities having access to mass-market broadband grew disproportionately in employment, the number of information technology-oriented businesses, and the number of businesses overall.[22] Economic development in rural areas, in turn, will foster demand for broadband subscriptions, broadband-delivered content, and broadband-capable devices. As with the basic telephone network, the more people that connect to the broadband network, the more value the network has for everyone on it, including initial users.[23] And just as rural electrification created a new group of home appliance consumers, so will a broadband-connected rural America want Internet Protocol (IP)- enabled phones, smart meters, telehealth, distance learning, video relay services, online music, streaming movies, interactive gaming, and a host of other broadband-related products and services. Simply put, broadband buildout to rural Americans promotes and encourages sustained economic development, to the benefit of us all.

17. The benefits of broadband extend particularly to small businesses in rural areas. With broadband, running a small web-based business in a rural location becomes a reality.[24] Craftspeople, particularly those with a national or international following, can deal directly with their customers— quickly and without expensive middlemen.[25] Local tourism authorities can more effectively market their attractions to potential visitors, shopkeepers can better manage their inventories, and travel to commercial centers can be dictated more by efficiency than necessity.

18. It is clear that access to fixed and mobile broadband services also has the potential to enhance the efficiency and productivity of a number of agricultural activities in rural areas. Farmers, particularly those with smaller operations or in more remote locations, can materially benefit from real- time access to weather and crop reports and to databases of local and national agricultural extension services.[26] For example, farmers can be warned if there is a heightened risk of a plant blight, a livestock disease, or an insect infestation; examine the problem; and know immediately what they need to do to address it.[27] They can call upon Internet-driven tools and applications to consult with experts and precisely calculate the additional inputs they might need to enable their fields to flourish and their animals to thrive—with less waste or risk to the environment.[28] Detailed online market information can help farmers time the sales of their products more profitably. Web-based interaction can help consumer- focused farmers develop a personal relationship with their clientele to assist in bringing the right products to markets at the right time—saving time, fuel, and frustration.

19. Broadband buildout to rural America also can enhance educational opportunities and the likelihood of academic achievement. Students without access to broadband cannot do the same type of homework as their counterparts who enjoy access to broadband, and students in certain rural areas are often many miles away from advanced educational institutions, such as colleges and universities. Broadband can significantly improve the quality of education by providing students in rural America with the ability to do online research, interact with their

teachers and schools from home, and obtain college credit and college degrees, even though they are not physically on campus.[29]

20. Bringing broadband to rural residents can also help improve healthcare. Often, the full range of health care services is not available in rural areas, because vast distances and low population densities make access to care, particularly specialty care, uneconomic. With sufficiently robust broadband services, clinics in rural areas can have access to facilities and specialists in more-densely populated areas. At a time when rising health care costs are a national challenge, this ability to leverage our health care resources would benefit all Americans.

21. Broadband also is essential to meeting our public safety and homeland security needs. Experience has made us acutely aware of the importance of communications in the event of an accident, natural disaster, or terrorist attack. We also know that the Internet can continue to function during these types of disasters, providing more reliable communications for first responders and the public.[30] First responders, public health officials, and government workers increasingly use broadband to share critical, time-sensitive public safety and security information with each other and with their communities.[31] All Americans, including those in rural areas, must have access to broadband to be put on alert and kept in contact in the event of a local, state, or national emergency.

22. Broadband access to the Internet throughout rural America would be particularly valuable in assisting people with disabilities to realize their full potential.[32] For example, broadband can allow individuals with hearing disabilities to use their computers and video equipment to communicate in real time in a visual and unfiltered manner, rather than on a delayed basis through a specialty device such as a TTY. Broadband also can let distant sign-language interpreters assist these individuals with essential tasks, such as communicating with doctors or participating in business meetings.[33] With broadband connections, people who are blind or visually impaired can use adaptive equipment that directs them to the information on a web page that they are likely to find most useful.[34] Broadband connections also can enable individuals who have difficulty walking to visit the library or shopping mall, to name two uses among many.

23. Furthermore, broadband offers potentially enormous environmental benefits. Broadband network technology distributes information efficiently. Its ubiquitous deployment and use would have immense potential to help create a greener planet through the miles not driven or flown. Telecommuting, teleconferencing, and videoconferencing all could reduce drastically both commuting and general purpose travel, and the associated carbon emissions.[35] Increased use of electronic documents lessens the need for paper, leaving more trees growing in forests.[36] Broadband enabled smart-grid technologies help utilities balance their network loads more accurately, making energy distribution more efficient.[37] E-commerce complements and simplifies physical distribution networks and can make both business and consumer transactions both more efficient and environmentally friendly.[38]

24. Finally, broadband deployment in those rural areas where poverty is historical and structural is particularly important.[39] Properly implemented, connection via broadband to the

wider world offers a boost to people caught in a cycle of poverty. The need for distance learning and telemedicine is most acute in such areas. We must see that broadband infrastructure and the means to use it reach rural communities that have been redlined, neglected, or segregated from better-served areas.

25. The many benefits to the public demonstrate the necessity of ensuring that robust and affordable broadband is available to all Americans. Further, America's economy depends on ensuring that all Americans, including those in rural areas, have access to broadband and are able to compete in this connected, global economy. According to Dr. Vinton Cerf, in the future, billions of devices will be accessible through the Internet at ever-increasing speeds,[40] using applications that, for the most part, have yet to be conceived.[41] Broadband is the interstate highway of the 21st century for small towns and rural communities, the vital connection to the broader nation and, increasingly, the global economy. Our nation as a whole will prosper and benefit from a concerted effort to bring broadband to rural America.

B. The State of Broadband in Rural Areas

26. Our efforts to bring robust and affordable broadband to rural America begin with a simple question: what is the current state of broadband in rural America? We would like to answer this question definitively, and detail where broadband facilities are deployed, their speeds, and the number of broadband subscribers throughout rural America. Regrettably, we cannot. The Commission and other federal agencies simply have not collected the comprehensive and reliable data needed to answer this question. As the Commission has indicated, more needs to be done to obtain an accurate picture of broadband deployment and usage in America, including its rural areas.[42] As discussed below, Congress recently enacted legislation that will help us obtain more complete data on the deployment and adoption of broadband service: the BDIA and the Recovery Act.[43]

27. While these two pieces of commendable legislation together will improve the federal government's ability to collect broadband data, the information we have today demonstrates that broadband service in rural America is generally inadequate.[44] Although inexact, currently available data and studies suggest that, in comparison to non-rural areas, broadband services are less extensively adopted in rural areas generally, and that this stems in part from less extensive deployment of broadband capability in rural areas. The *2008 Pew Broadband Adoption Study* found that a much larger proportion of urban and suburban residents have broadband at home (5 7-60 percent) than rural residents do (38 percent).[45] And according to NTIA, while 54 percent of urban households had broadband in the home in 2007, only 39 percent of rural households did.[46] Mobile broadband networks also cover more urban than rural areas. We estimate that, although mobile broadband networks cover 95.6 percent of the total U.S. population, they cover only 82.8 percent of the U.S. rural population compared with 99.0 percent of the non-rural population.[47]

28. We also are unable to provide the state of broadband deployment in rural areas among, for example, minorities, those with disabilities, and those residing on Tribal lands.

Broadband deployment to these groups is of particular concern because they may benefit the most from broadband deployment and adoption.[48] With respect to minorities, the data we have show that minorities typically have lower adoption rates than non-minorities.[49] Low subscription rates may be explained partly by a lack of computer ownership.[50] However, the most accurate marker for low broadband adoption is most likely low income.[51]

29. Similarly, we know even less about broadband access in rural areas for people with disabilities, in part because we lack properly disaggregated information. Some research data show that fewer than 30 percent of people with disabilities over the age of 15 use the Internet; by comparison, more than 60 percent of those with no disability use the Internet.[52] Moreover, data also show that people with disabilities, both inside and outside metropolitan areas, use the Internet less than their geographic counterparts with no disability, and people with disabilities outside metropolitan areas have a very low rate of Internet use (26.7 percent).[53]

30. Many Tribal lands also lack broadband service, but we lack data describing availability or subscribership.[54] Many Native Americans are still struggling to receive basic telephone service and, based on the Commission's data, have the lowest percentage of basic telephone service subscribers by group.[55] Indeed, 2007 data from the Census Bureau's Current Population Survey reveal that nationwide, American Indian/Alaskan Native households have a broadband subscription rate of only 30 percent, by far the lowest subscription rate among any ethnic group identified.[56] The situation is likely to be even worse in rural areas.[57]

31. Although we do not have accurate data on broadband deployment in rural America, MMTC presents one community, Weirwood, Virginia, and through Weirwood gives "rural broadband deployment" a human face.[58] As MMTC describes it, Weirwood is an isolated rural community on Virginia's Eastern Shore, on the site of a former cotton plantation.[59] Weirwood is only a mile and a half from U.S. Route 13, along which lies a broadband Internet backbone.[60] The residents of Weirwood, however—mostly African-American descendents of former slaves—lack access to broadband. MMTC states that Weirwood has "absolutely no ability to raise internally" the funds needed to build a broadband node to the community from the existing backbone line.[61] Pending acquisition of thorough, reliable, and disaggregated data, we glimpse through Weirwood the state of broadband deployment in impoverished rural areas.

32. Even without detailed maps of broadband service availability, we know that Weirwood is not unique. Whether we are discussing a historically African-American community like Weirwood, Tribal lands that even now lack access to voice telephone service, or individuals with disabilities whose access to broadband is essential, overall, there needs to be an active federal governmental role if all Americans are to have access to robust and affordable broadband services. The challenge we face is determining ways to adjust our efforts to ensure that the residents in places like Weirwood, or anywhere in rural America, are able to take advantage of the opportunities that come with broadband. As we have done in the past, America will overcome our broadband infrastructure challenges and reach our goal of providing broadband to all Americans, including those in rural areas.

C. America Overcomes Its Infrastructure Challenges

33. From the country's earliest days, building the nation's infrastructure has required federal resources and leadership, and this federal role continues. In generations past, the United States has faced and solved the challenges of connecting a vast and growing nation. In its earliest days, the young republic subsidized the delivery of newspapers by having the Postal Service deliver them for a fraction of the postal rate for letters. In the 1 800s, the federal government partnered with private industry to lay the first transcontinental railroad. Beginning in the 1930s, the Rural Electrification Administration (REA) and its successor agency, the RUS financed hundreds of locally-owned utility cooperatives, enabling them to bring electric power and telephone service to rural America. The next generation saw the federal government work with the states to build the interstate highway system. Federal agencies also cooperated with scientists and universities, first to seed the Internet and later to grow its first backbone. At their inceptions, some of these projects were controversial. Many considered them too expensive; others doubted their efficacy. Today, few would question their value, but each of these undertakings depended on a strong and coordinated national vision.

1. The Postal system and newspapers

34. America's founders believed that a postal system—the communications network of the eighteenth century—was vital to a democracy where people would govern themselves. The federal government passed laws and advanced policies to bring postal service even to remote areas of the country. From the United States government's early actions, we know the importance it attached to the postal system and the newspaper industry. As Benjamin Rush wrote, "[f]or the purpose of diffusing knowledge, as well as extending the living principle of government to every part of the united states— every state—city—county—village—and township in the union, should be tied together by means of the post-office. . . . It should be a constant injunction to the postmasters, to convey newspapers free of all charge for postage. They are not only the vehicles of knowledge and intelligence, but the centinels of the liberties of our country."[62]

35. To that end, President Washington signed into law the Post Office Act of 1792, which not only established the U.S. Postal Service, but also heavily subsidized newspaper delivery and exchange.[63] Abandoning the principle that each post office must be self-sustaining, Congress used the more- established post offices to subsidize postal roads and offices in remote areas of the expanding nation.[64] In less than fifty years, this prescient national vision had created the greatest postal system and news network in the world.[65] Newspapers, easily obtainable because of discounted postage rates, enjoyed rapidly expanding subscribership, leading to the creation of still more newspapers.[66] The young government's policy quickly built a delivery system that rapidly expanded the newspaper industry, did not run a deficit, and reached "towns and villages deep in the interior" of the country.[67]

2. The transcontinental railroad

36. A visionary public-private partnership also produced the world's first transcontinental railroad.[68] In 1862, in a nation divided by civil war, President Lincoln signed the Pacific

Railway Act to connect gold-rich California to the Union.[69] Combining government bonds and land grants, the Pacific Railway Act commissioned two start-up rail companies to build the line: the Central Pacific starting from the west, and the Union Pacific, starting from the east.[70] In 1869, six years after the project began, a transcontinental rail line connected America across its harsh backland and challenging terrain, allowing travelers to go coast-to-coast in seven days. The next transcontinental railway lines would not be completed until 12 years later.[71]

3. Rural electrification

37. Today we take the pervasiveness of electricity largely for granted, but in 1936, electricity reached only 11.6 percent of U.S. farms.[72] Private investment had dried up as a result of the Depression, and private utility companies believed that rural, low-density areas were insufficiently profitable.[73] In response, President Franklin Roosevelt signed the Rural Electrification Act, envisioning a nationwide electric power generation, transmission, and distribution network. The Act created the REA, which financed rural electrification through federal low interest rate loans and loan guarantees.[74] Hundreds of locally-owned utility cooperatives formed across rural America and became the chief mechanism to compensate for the lack of private investment in low-profit rural areas.[75] The project was wildly successful; nearly 100 percent of U.S. farms and rural households now have electrical services.[76] The early successes of the rural electrification program led Congress in 1949 to expand the REA's mission to include the financing of rural telephone networks. People in rural areas would not have adequate, dependable, or urban-quality telephone service without federal government initiative and support.[77]

38. The REA public-private partnership is a distinguished chapter in the economic development of the United States. It transformed rural life by boosting the productivity of rural farms and businesses, and allowing rural families to stay connected to the rest of the nation through telephone calls and radio broadcasts. Virtually all industries, businesses, and homes benefited. A shortsighted policy that brought the convenience, innovation, and new opportunities of electricity and telephone service only to urban and a smattering of rural areas could have created two Americas of utility haves and have-nots. We cannot make this mistake today.

4. The Interstate highway system

39. The interstate highway system provides another example of the success public-private partnerships have had in stimulating infrastructure development. The Federal-Aid Highway Act of 1956 initially funded construction of the nation's interstate highways.[78] Funding continued until 1996, when the originally-planned network was near completion.[79] State governments received the bulk of the funds, owned the interstate highways, and awarded contracts to the private sector to construct the segments that lay within their boundaries. President Eisenhower, recognizing the capacity of a highway system to unify the country, said of the project, "[t]ogether, the uniting forces of our communication and transportation

systems are dynamic elements in the name we bear—United States. Without them, we would be a mere alliance of many separate parts."[80]

40. A concrete goal and steadfast leadership helped overcome the challenges that faced the interstate highway project in its early years. Many viewed it as an untried and risky project, immense in scope, impact, and cost.[81] Advocates emphasized its necessity for economic growth and national unity.[82] As it became obvious that the private sector and state governments could not keep pace with the needs of the country, debate grew over how to build and fund the project.[83] Different urban and rural needs and the existing patchwork of state-by-state approaches led to disagreements.[84] Stakeholders bickered over the mix of federal and state subsidies, fuel and other excise taxes, tolls, and financing through loans and bonds. Leaders were constantly challenged to adequately explain the project's nuances and address public concerns in an informative and diplomatic manner.[85] The project moved forward despite these disagreements, and eventually became the interconnected network of roadways that Americans use extensively every day.

5. The internet and internet backbone

41. The story of the Internet also starts with federal vision and funding. Responding to the 1957 launch of Sputnik, President Eisenhower started the Advanced Research Projects Agency (ARPA, renamed the Defense Advanced Research Projects Agency (DARPA) in 1972).[86] About a decade later, DARPA created ARPANet to link university computers, which created a network for communication among research centers.[87] Although decentralized, the ARPANet used the existing AT&T phone system, with DARPA leasing dedicated lines to avoid cumbersome dial-up connections.[88] The network infrastructure required that each participating research institution have dedicated computers to route the traffic—the interface message processor (IMP)—to save the universities' main computers from having to handle the burden.[89] In other words, DARPA was in charge of the backbone: it designed and implemented the network, leased the dedicated data lines, and financed the construction and implementation of the IMPs.[90] Each research institution was responsible for hooking up its computer to its respective IMP.[91] By 1973, the number of entities using the network had grown from four to 40.[92]

42. The federal government also played a role in the first Internet backbone when the National Science Foundation's NSFNet became the *defacto* Internet backbone in 1986.[93] The networks created by and in response to DARPA's efforts, and the "internetworking" capability offered by TCP/IP, made NSFNet the first network available to every U.S. campus.[94] Usage exploded.[95] Because NSFNet could not be used for profit,[96] the need for commercial Internet backbones arose. As a result, several commercial, interconnecting, national Internet backbone networks comprise the Internet backbone today.[97] Federal funding and visionary federal agencies created a noncommercial enterprise that would ultimately generate enormous sums of money, transform nearly every aspect of our daily lives, and change how the entire world communicates. This is the heritage we bring to the current challenge of comprehensive broadband deployment in rural areas.

D. Recent Legislative Developments

43. Recognizing the benefits of universal broadband access to the nation as a whole and the concomitant lack of robust broadband deployment in many parts of the country, Congress passed several important pieces of legislation in 2008 and 2009 to address the issue. Below, we discuss the 2008 Farm Bill, the BDIA, and the Recovery Act. Collectively, these pieces of legislation require the development of a rural broadband strategy, improvements in broadband data collection, and a national broadband plan.

1. 2008 Farm Bill—Rural Broadband Strategy

44. On June 19, 2008, Congress enacted the 2008 Farm Bill, which includes a provision requiring the Chairman of this Commission, in coordination with the Secretary of the USDA, to submit "a comprehensive rural broadband strategy" to Congress by May 22, 2009.[98] This strategy must include the following:

(1) recommendations—
 (A) to promote interagency coordination of Federal agencies in regards to policies, procedures, and targeted resources, and to streamline or otherwise improve and streamline the policies, programs, and services;
 (B) to coordinate existing Federal rural broadband or rural initiatives;
 (C) to address both short- and long-term needs assessments and solutions for a rapid build-out of rural broadband solutions and application of the recommendations for federal, state, regional, and local government policymakers; and
 (D) to identify how specific Federal agency programs and resources can best respond to rural broadband requirements and overcome obstacles that currently impede rural broadband deployment; and
(2) a description of goals and timeframes to achieve the purposes of the report.[99]

The Chairman of this Commission must also, in coordination with the Secretary of the USDA, update and evaluate the rural broadband report during the third year after enactment of the 2008 Farm Bill.[100]

2. BDIA—Improving the Collection of Broadband Data

45. On October 10, 2008, Congress passed the BDIA, which provides for improved federal data on the deployment and adoption of broadband services.[101] The BDIA mandates improvements in the Commission's Section 706 Reports.[102] Previously, that section directed the Commission to conduct a regular inquiry as to whether advanced telecommunications capability is being deployed to all Americans in a reasonable and timely fashion.[103] The BDIA now requires the Commission to issue its Section 706 Reports "annually" instead of "regularly" and adds several types of data regarding broadband services that the Commission must produce and evaluate.[104] On March 31, 2009, the Commission released a Public Notice seeking comment on how it should implement certain of the requirements in the BDIA.[105]

3. Recovery Act—National Broadband Plan

46. On February 17, 2009, Congress passed the Recovery Act.[106] The Recovery Act charges the Commission with developing a national broadband plan that "shall seek to ensure that all people of the United States have access to broadband capability and shall establish benchmarks for meeting that goal."[107] By February 17, 2010, the Commission must deliver to Congress a national broadband plan that seeks to ensure that every American has access to robust broadband capability and establishes clear benchmarks for meeting that goal. The Commission's plan must include an analysis of several specific elements of broadband deployment. First, the Commission must analyze "the most effective and efficient mechanisms for ensuring broadband access by all people of the United States."[108] Second, the Commission must include "a detailed strategy for achieving affordability of such service and maximum utilization of broadband infrastructure and service by the public."[109] Third, the Commission must include "an evaluation of the status of deployment of broadband service, including progress of projects supported by the grants made pursuant to this section."[110] Finally, the Commission must include "a plan for use of broadband infrastructure and services in advancing" a broad array of public interest goals.[111]

47. The Recovery Act also provides significant support for programs that will accelerate the deployment of and subscription to broadband services throughout the nation. The Recovery Act charges RUS and the NTIA with making grants and loans to expand broadband deployment and for other important broadband projects. As noted above, by February 2011, the Recovery Act also directed NTIA to create a "comprehensive nationwide inventory map of existing broadband service capability and availability" that shows the geographic extent to which that capability is deployed and available for each state.[112] NTIA must make this inventory map accessible by the public on an NTIA website in a form that is interactive and searchable. These programs are discussed in detail below.[113]

IV. COORDINATION OF RURAL BROADBAND EFFORTS

48. As part of its mandate to develop a rural broadband strategy, Congress tasked us with developing recommendations to promote interagency coordination and to streamline and improve federal agencies' policies, programs, and services.[114] Congress also tasked the Commission with developing recommendations for coordinating existing federal rural broadband initiatives.[115] These legislative requirements were established before the 2008 election and before the enactment of the Recovery Act. As there has been considerable interagency coordination in the weeks following the passage of the Recovery Act, the recommendations here are to support and promote the continued interagency coordination that now exists with the expectation that this will result in improvements to existing and new federal programs.

49. Much like extending the reach of railroads across the country or bringing electricity and telephones to rural areas, ensuring that broadband service is available to all Americans is a massive undertaking in which there is a significant role for government. In fact, a number of

federal agencies have already developed and implemented programs related to the buildout of broadband in rural areas.

50. *The Role of the USDA's RUS.* The USDA's RUS plays a particularly important role in administering programs that aim to expand utilities and new technologies to rural communities. In particular, RUS administers three programs aimed specifically at improving broadband access: the Rural Broadband Access Loan and Loan Guarantee Program;[116] the Community Connect Grant Program;[117] and the Distance Learning and Telemedicine Loan and Grant Program.[118] The funding available under these programs to expand broadband coverage helps offset prohibitively high deployment costs that plague many rural areas. As part of the Recovery Act, Congress authorized an additional $2.5 billion in funding for these programs.[119]

51. Numerous RUS programs have already started to increase rural broadband deployment. For example, as early as 2004, RUS began to work with International Broadband Electric Communications, Inc. (IBEC) in Huntsville, Alabama, to fund deployment of broadband solutions in a number of rural locations.[120] As a result, residents in the targeted communities will be able, many for the first time, to access the Internet at speeds of up to 5 Mbps. Another company that has successfully worked with the RUS is Rural Telephone Service Co. (Rural Telephone), a Lenora, Kansas-based incumbent local exchange carrier (LEC) that began working with the RUS shortly after its incorporation in 1951. Rural Telephone has used RUS funding to become an operator of 29 exchanges in a rural area that averages two households per square mile.[121] Through its competitive LEC affiliate, Nex-Tech, Rural Telephone also has used RUS funding to deploy fiber to the home infrastructure in neighboring towns. In many cases, Nex-Tech is bringing broadband to customers for the very first time.[122]

52. RUS also has worked with Air Advantage LLC (Air Advantage), a wireless ISP headquartered in Frankenmuth, Michigan. Air Advantage uses a broad portfolio of wireless solutions to deliver reliable and secure broadband connectivity to thousands of rural businesses and residents in rural Eastern Michigan.[123] Over the past six years, the company has secured three RUS Community Connect Broadband Grants and used the Broadband Loan Program to expand the footprint of its wireless network and increase broadband connectivity to local businesses and residents.[124] Related developments include the creation of computer labs in two local community centers that provide the residents with free broadband Internet access[125] and the deployment of distance learning technologies throughout seven partner rural school systems using funding from an RUS Distance Learning/Telemedicine Grant.[126] In addition, the Air Advantage network has linked several rural hospitals to enable data and record sharing, as well as enabling the Saginaw Valley State University to establish distance learning professional development courses at these healthcare facilities.[127] Air Advantage is using its most recent Community Connect Broadband Grant, approved in 2008, to provide wireless broadband access to the small community of Applegate, Michigan, where it is working with local officials to expand the village hall to include a new community center that will host a computer lab with free Internet access.

53. The RUS broadband programs face several challenges. Unlike some of RUS 's other infrastructure programs, the RUS broadband programs only have a limited ability to offer projects combining loan and grant funds. The primary RUS broadband program is the Broadband Loan Program. Under the Broadband Loan Program, it is difficult for RUS to reach small remote places, like Weirwood, Virginia, because the community lacks the needed resources to make a broadband loan work. Places like Weirwood are better candidates for the Community Connect Broadband Grant program, but the funding for that program is severely limited ($13.4 million for the 2009 fiscal year). However, the Recovery Act provides RUS the resources to administer a broadband program that offers assistance in the form of grants and loan/grant combinations designed to assist communities like Weirwood.

54. *NTIA.* The NTIA also has an important role to play in broadband development. The Recovery Act appropriates $4.7 billion to the NTIA to "establish a national broadband service development and expansion program" called the "Broadband Technology Opportunities Program" (BTOP).[128] This program will award grants[129] to states, non-profit organizations, and broadband providers to fulfill the broadband deployment goals of the Recovery Act.[130] The NTIA must award, "to the extent practical," at least one grant in each state,[131] and in doing so must consider a variety of factors, including affordability and speed, as well as improved access for healthcare, education, and children,[132] and "whether the applicant is a socially and economically disadvantaged small business concern."[133] Grantees under this program will also be subject to "non-discrimination and network interconnection" obligations.[134]

55. In addition, the Recovery Act requires NTIA to "develop and maintain a comprehensive nationwide inventory map of existing broadband service capability and availability in the United States that depicts the geographic extent to which broadband service capability is deployed and available from a commercial provider or public provider throughout each State."[135] Such a map is required to be made accessible in an interactive and searchable format on the web by February 17, 2011.[136]

56. *Other Agencies.* There are several other programs run by various federal agencies, including the Commission, that provide or have provided broadband-related funding.[137] Developed essentially in administrative silos and existing in parallel within their respective administrative agencies, these programs are responsible for significant annual spending. We believe that in enacting the 2008 Farm Bill, Congress implicitly recognized that many of these programs do share common broadband purposes and goals; however, coordination between federal agencies has been lacking. As Congress recognized, greater coordination among agencies, policies, programs, and staff is needed to realize the full benefit of federal resources that have already been dedicated to bringing broadband to rural America and to best leverage further investments of both taxpayer monies and private capital. We recognize positive changes that have occurred in recent months with regard to broadband coordination to address these prior shortcomings.

A. Promoting Interagency Coordination

57. Lack of interagency coordination is a significant challenge to the deployment of broadband in rural areas. Consequently, we believe that increasing coordination—among federal departments and agencies; Tribal, state, and local governments; community groups; and individuals—is a critical preliminary step toward ensuring that the various government programs accomplish their broadband goals and objectives in an efficient and effective way. Promoting and fostering a culture that encourages the exchange of information within and between federal agencies and, where appropriate, with non-federal government organizations, Tribal governments, and the public lie at the core of effective coordination. Successfully implemented, the coordination strategy we outline below should enable the federal government and other stakeholders to work together to maximize resources, harness expertise, and avoid duplication of effort in facilitating the deployment of broadband in rural areas.

58. In highlighting areas where better coordination may be needed, we hope to underscore the particular challenges faced by rural America in the deployment and adoption of broadband services that should be considered in conjunction with every agency's broadband policy. To this end, the Commission has invited suggestions from the public on both formal and informal means of coordination among federal departments and agencies; Tribal, state, and local governments; and community groups and individuals to achieve Congress's goal that *all* Americans have access to broadband.[138]

1. Federal interagency coordination

59. Shortly after President Obama took office, his administration undertook an important leadership role in the effort to expand broadband penetration throughout the nation. In early 2009, the Obama administration formed an interagency working group under the auspices of the National Economic Council to bring agencies together to discuss broadband issues of common interest.

60. This interagency working group is focused on coordinating the country's broadband agenda and has sought input from agencies regarding their broadband programs to this end. The formation of this group gives expert staff in different federal agencies an unprecedented opportunity to meet, discuss, and coordinate complementary federal broadband programs and policies, and prioritize agency actions to avoid wasted resources. Thus far, a significant focus of the group has been on ensuring that the Recovery Act stimulus programs related to broadband are implemented in a coordinated manner. It also facilitated discussions among federal agencies and the submission of information in connection with the development of this Report. The administration should be commended for these efforts. We recommend that the interagency working group continue, devoting special attention as appropriate to enhancing interagency coordination on rural broadband initiatives.[139] We believe, at a minimum, that such a focus could serve as a vital step in ensuring that federal agencies do not work at cross purposes and form an important component of a national broadband plan.

61. In addition to the interagency working group, we believe that there are additional steps that can be taken to promote interagency coordination. For example, joint hearings, such as those held recently by the NTIA and USDA regarding the Recovery Act's broadband provisions, can help to promote interagency coordination and provide a forum for the public to provide input toward the shaping and implementation of complementary federal programs. Joint hearings may also facilitate federal, Tribal, state, and local cooperation. To that end, we suggest, for example, that the Commission may want to hold joint public meetings or hearings with USDA and NTIA to gain valuable public input regarding their complementary broadband agendas. We further recommend that the Commission and other federal agencies consider developing their own "rural broadband agendas," consistent with the national broadband plan. This agenda could include the agency's pending (and perhaps planned) proceedings affecting rural broadband. The agencies could share these agendas with each other and, to the extent they do not include confidential information, with the public.

2. Additional coordination

a. Coordination with tribal governments

62. It is critical that federal agencies devote increased attention to improving coordination and collaboration with Tribal governments,[140] consortia, and organizations regarding broadband deployment in rural Tribal areas.[141] Encouraging broadband deployment in these areas presents unique issues.[142] To ensure a truly comprehensive strategy for addressing rural broadband, it is important that the federal government maintain a continuing dialogue with Tribal governments to address these issues. The Commission has taken a series of steps, through regulatory action, the publication of consumer information, and Tribal outreach, to address the lack of communications deployment and subscribership throughout Indian Country. For example, in the *Tribal Policy Statement*, the Commission reaffirmed its recognition of Tribal sovereignty and the trust relationship between the Commission as part of the federal government and Tribal Nations.[143] This policy statement indicates the Commission values a governmentto-government relationship with federally recognized Indian Tribes and Alaska Native communities. Close consultation between the Commission and the Tribes is the Commission's principal means of identifying and working to resolve communication policy issues relevant to Indian Country. The *Tribal Policy Statement* describes the Commission's consultative and trust responsibilities toward the Tribes. It also expresses the Commission's commitment to work cooperatively with Tribal, state, and local governments as well as other federal departments and agencies to address and remedy communication problems in Indian Country, such as low telephone and broadband penetration rates and poor service quality.

63. In addition to the *Tribal Policy Statement*, the Commission created Tribal Land bidding credits to assist those Tribal communities with the greatest need for telecommunications services. The Tribal Land bidding credits provide winning bidders in spectrum auctions that agree to deploy facilities and provide service in certain Tribal areas with a discount on their spectrum.[144] The Commission also established the Indian Telecommunications Initiative (ITI) to help improve telecommunications services on Tribal

lands. The ITI seeks to increase telephone subscribership rates on Tribal lands, upgrade the telecommunications infrastructure on those lands, and inform Tribal consumers about the financial support available through federal programs, such as the Universal Service programs.[145] The ITI also seeks to promote understanding, cooperation, and trust among Tribal Nations, government agencies, and the communications industry in addressing issues facing Tribal lands. As part of the ITI, Commission staff maintains regular contact with Tribal telecommunications professionals and Tribal representatives. Commission senior officials and other staff also attend and participate in a variety of meetings on telecommunications issues with Tribal officials and representatives.

64. Although much has been done to foster cooperation, collaboration, and communication with Tribal governments, much work remains. Not only are residents of Tribal lands lagging behind the country as a whole in broadband access, but many still do not even have access to voice service;[146] Native American communities have the lowest reported levels of telephone subscribership in the country.[147] Thus, as an initial matter, we suggest that federal agencies consider how to maximize already existing programs to improve coordination with Tribal governments. We also recommend that the Commission consult with Tribal governments pursuant to the *Tribal Policy Statement* in developing its national broadband plan and, in particular, in developing the aspects of that plan that affect broadband deployment and subscribership specifically on Tribal lands. Further, before promulgating any regulation, agencies may want to consider conducting formal consultations with Tribal governments as may be required under Executive Order No. 13175.[148]

b. Coordination with state and local authorities

65. As part of the nation's rural broadband strategy, we believe that federal agencies should seek to establish an ongoing dialogue with state and local authorities. State and local governments have knowledge unique to their areas. Many states have broadband initiatives, and in some cases, entities are already working to address the broadband needs of unserved and underserved rural areas.[149] A dialogue among state and local entities and federal agencies enables state and local governments to gain valuable information concerning federal resources and initiatives. At the same time, federal-state coordination allows the federal government access to the states' valuable knowledge and experience,[150] helps inform the federal government in establishing rural broadband policies and initiatives, and prevents wasted time and resources from duplicating states' efforts.

66. When the Commission is fully constituted, we recommend that the states take full advantage of existing mechanisms for coordination, such as the Joint Conference on Advanced Services.[151] We further recommend that the Joint Conference provide the Commission with its own recommendations for improving federal coordination with states regarding rural broadband deployment.[152] We understand that the state members of the Joint Conference have already begun the process of compiling an inventory of "best practices" and successful state and local projects as an aid to industry, consumers, and fellow governmental entities.[153] We recommend that the Joint Conference continue this program. We further suggest that the government at all levels work to develop an inventory of resources, "best practices," and success stories to inspire and motivate others to undertake the difficult but

ultimately rewarding task of bringing broadband to rural communities across this nation. Finally, we also recommend that the Joint Conference include in its recommendations suggested proposals to address and ameliorate the unique challenges presented to rural minority communities, rural low-income communities, and persons with disabilities residing in rural areas.

c. Coordination with communities

67. Our goal of ubiquitous, affordable, and robust broadband for all will be achieved faster and more easily when everyone involved has access to the essential information and resources necessary to make informed decisions. Thus, in order to be successful in coordinating existing federal programs concerning rural broadband or rural initiatives, it is critical that the federal government collaborate and coordinate with community and advocacy organizations in rural areas. The federal government should work closely with these organizations to help ensure that all minority group members residing in rural areas have access to robust and affordable broadband services and that minority-owned businesses participate fully in the buildout of broadband infrastructure in those areas. The federal government also should work closely with organizations representing persons with disabilities to help ensure that they have affordable access to broadband services capable of supporting the full array of applications responsive to their needs. Finally, the federal government should work closely with organizations that serve low-income residents to ensure the opportunities that affordable broadband offers this community do not go unrealized.

68. We know that community and local advocacy groups are an essential component to the success of deploying broadband in rural areas.[154] Further, public-private partnerships can play a critical role in bringing broadband to rural areas.[155] Community and advocacy groups and public-private partnerships can function as valuable information sources for local communities, businesses, and consumers in rural areas, and various groups have developed guidance on how to deploy broadband in those areas. For example, the Commonwealth of Virginia has produced an online "Community Broadband Tool-Kit" that provides step-by-step guidance on how a community can deploy broadband services.[156] This tool-kit has information on broadband applications and case-studies from Virginia localities that have successfully deployed broadband facilities. Another group, called Connecting Rural Communities, publishes a guidebook that explains in detail how to bring broadband services to rural communities.[157] The Michigan Department of Information Technology has released its own "Action Plan for Deploying Broadband Internet to Michigan Local Governments," which similarly details how developing goals is essential for building a broadband network.[158]

69. The federal government should collaborate with these organizations and ones like them to fully understand the challenges in deploying broadband in rural areas and develop solutions that overcome those challenges. We suggest that the federal government continue to take a leadership role alongside individuals, groups, businesses and other governmental organizations seeking to fit together all the pieces needed to bring state-of-the-art broadband services to rural areas.

B. Streamlining, Improving, and Coordinating Existing Federal Programs

70. Given the numerous federal programs currently addressing rural broadband, it is critical that those programs function efficiently and effectively to maximize consumer benefits. With so many programs addressing mostly complementary, but occasionally overlapping, aspects of rural broadband deployment there is the risk of duplication of effort and inefficient use of government resources. Given this, all relevant federal agencies should review their programs to identify what internal barriers, if any, may be making rural broadband deployment more difficult. Further, all federal agencies with responsibility for rural issues may wish to consider the need and opportunity for rural broadband deployment in designing their programs, as well as how their programs support the ability of minority-owned and small or disadvantaged businesses to employ broadband services. We note that all federal agencies have an opportunity to learn from the grantees and recipients of RUS and NTIA broadband funds and should identify the business models most successful in rural areas.[159]

71. In addition to these overarching strategic elements to streamline and improve existing federal programs, we believe that there are additional specific considerations that should be taken into account. We discuss some of these considerations below.

1. Efficient use of government funds and resources

72. We recommend that federal agencies review their non-broadband programs regarding rural issues to see if those programs provide opportunities to promote rural broadband deployment. For example, given that one of the largest deployment costs of underground fiber networks is the expense associated with digging the trenches to lay the fiber, some commenters suggest that costs of those networks could be reduced if fiber could be installed at a time when roads are already being constructed or repaired. Thus, these commenters suggest that agencies responsible for road construction or repair should consider how these projects can be used to facilitate broadband deployment in rural areas.[160] In addition, relevant agencies could consider offering collocation space in federally-owned buildings to rural broadband providers that wish to interconnect with broadband facilities terminating in those buildings,[161] developing expedited procedures for granting rights of way on public lands for broadband projects,[162] or wiring government-supported housing for broadband.[163] We also suggest federal agencies consider how existing federal programs and resources can be maximized to spur rural broadband deployment.

2. Coordinating program criteria

73. Federal agencies involved in rural broadband initiatives should consider coordinating key terminology, such as the term "rural," across their programs, consistent with their legislative mandates.[164] Dissimilar definitions and criteria across complementary programs can complicate the flow of funds to rural broadband users and hinder federal interagency coordination. Coordinating key terminology across related programs will help to ensure that federal programs work consistently and in concert with one another.

3. Government websites

74. One of the challenges to rural broadband deployment and adoption is a lack of easy public access to comprehensive information about all the government resources available to help communities and individuals obtain access to broadband services. Although the federal agencies involved have attempted, under great time pressures, to publicize and educate the public about the programs implementing the Recovery Act, other broadband-related programs exist that would benefit from the same type of information dissemination. Further, the public would benefit enormously from the availability of one access point that serves as a central repository for information about all federal programs addressing rural broadband deployment. Consequently, to improve the efficiency and effectiveness of federal broadband initiatives and programs, and enable the public to take full advantage of available federal resources, we recommend the Commission, in coordination with other federal agencies, consider the development of a comprehensive website that will provide a centralized access portal for information concerning all federal programs addressing broadband.[165]

75. In the meantime, as an aid in these efforts, we recommend that the Commission expand its website to include a comprehensive set of links to all federal government rural broadband-related programs.[166] The Commission and USDA have already taken some preliminary steps toward this end. In 2005, the Commission's Wireless Telecommunications Bureau and RUS created the Joint Federal Rural Wireless Outreach Initiative to coordinate activities and essential information on programs and financial and other assistance regarding telecommunications opportunities for rural communities. This initiative seeks to encourage greater access and deployment of wireless services to enhance economic development throughout rural America. Since February 2008, the Commission and USDA have launched the "Broadband Opportunities for Rural America" website and held four regional educational workshops on rural broadband.[167] The website is designed to provide those in rural America looking to bring the benefits of broadband services to their communities with the expertise and resources of the Commission and USDA in a single, easily-accessible location and user-friendly format.[168] We propose that the existing "Broadband Opportunities for Rural America" website be expanded to include a comprehensive list of all federal government programs related to rural broadband. This list could later be incorporated in, or linked to, a central website.

4. Delay caused by other federal requirements

76. The 2008 Farm Bill directs that this Report recommend ways to "address short- and longterm needs assessments and solutions for a rapid build-out of rural broadband solutions"[169] and "identify how specific Federal agency programs can best... overcome obstacles that currently impede rural broadband deployment."[170] Meeting the goal of rapid deployment of broadband to rural areas will require federal agencies not only to resolve numerous issues, but also to implement their decisions quickly. We recommend that federal agencies having responsibility for addressing rural broadband matters consider reviewing their rules, regulations, or other requirements to identify those that might impede quick

implementation of rural broadband. Agencies also might consider whether any of their routine processes or functions could be streamlined when rural broadband deployment is implicated.

V. Assessing Rural Broadband Needs

77. While the Commission and USDA, as well as other federal, Tribal, state, and local government agencies have made efforts in recent years to address the lack of broadband in rural areas— through funding programs, regulatory actions, and outreach and data gathering initiatives—the need for rural broadband remains. The 2008 Farm Bill directs that this Report include recommendations "to address both short- and long-term needs assessments and solutions for a rapid buildout of rural broadband solutions and application of the recommendations for Federal, State, regional, and local government policymakers."[171] We set forth below our analysis of the most common problems that have been brought to our attention. We discuss activity to date with respect to each issue and offer recommendations.

A. Technological Considerations

78. The number and range of technological solutions available to speed the deployment of affordable broadband in rural areas is increasing. As in other parts of the country, a number of technologies are available to provide the various components of broadband access in rural areas.[172] Each technology has specific cost and performance attributes that, coupled with compatibility and appropriateness of existing infrastructures and demand expectations, have an impact on its suitability for deployment in a particular rural area.[173] In particular, the introduction of new technologies and broadband deployment approaches increasingly enables providers using wireless, wireline, or satellite-based networks, or a combination thereof, to realize significant savings. This is particularly true with respect to providers building new networks in previously unserved rural areas.[174] Nevertheless, every technology has inherent capabilities and limitations. Those characteristics vary greatly among technologies. Similarly, every rural area presents its own special challenges, and a particular technological solution may be well-suited to one situation and poorly-suited to another. Therefore, decision makers should proceed on a technology-neutral basis—by considering the attributes of all potential technologies—in selecting the technology or technologies to be deployed in a particular rural area.

79. Rural broadband networks are fundamentally similar to broadband networks in other areas in that, in order to have broadband access to the Internet, they must include local access, or last- mile, broadband access to the end user and backhaul, or middle-mile, capabilities to an available Internet peering point. The last-mile network connects residential and business end users to a local ISP. In this configuration, the middle-mile or backhaul component connects the local ISP to an Internet peering point or node.[175] In rural settings, either or both of these components may not support robust broadband connectivity.[176] The choice of any local access or "middle-mile" technology in a rural setting must take into account factors

including desired capacity, cost, reach, and the need for additional resources like radiofrequency spectrum, electronic equipment, access to poles and rights of way, and power.

80. The best choice for any particular area in all likelihood will reflect, in addition to the population density and terrain of the area, the capabilities and limitations of the technology or technologies under consideration. The technology choice should take into account the ability of each feasible solution to provide cost-effective broadband connectivity in a given area based on consistent, high-quality performance that ideally will be capable of evolving over time to meet the growing requirements of Internet access and may well combine a variety of wireline and wireless elements. Some of the technology issues to consider are discussed below.

81. *Latency*. Latency is the time delay from when an end user sends a signal to the moment that the signal reaches its intended destination and vice versa. It is relevant in all parts of the network. Network technologies that create significant time delay can arguably degrade the performance of many interactive Internet applications.[177] The extent of the degradation increases with the extent of the delay. Latency is particularly important for voice applications such as VoIP[178] and Video Relay Service (VRS)[179] where a high degree of latency can degrade voice communication to an unintelligible level. Other non-voice, interactive Internet applications may also be less tolerant of the effects of latency. These include some educational applications,[180] some telework applications,[181] telepresence,[182] many telemedicine applications,[183] and interactive online gaming.[184] We note that these industries currently represent only a small portion of total Internet traffic.[185] However, all are burgeoning and have the potential to employ thousands of Americans and generate tremendous economic activity.[186] Latency therefore should be important in considering the best options for rural America.

82. *Scalability*. Some technologies are easier to upgrade than others. An amply sized conduit can be filled with more cables to increase capacity; a conduit that is too small has to be supplemented with a new conduit to increase capacity. Fiber networks can easily be upgraded by swapping out the optoelectronic equipment. Copper broadband loops may be capable of economically feasible upgrades with new network and consumer premises equipment. Certain wireless networks can be upgraded with the latest technology through the relatively simple replacement of the network's software cards rather the entire cell sites. In addition, an existing, multi-use radio tower may be able to support additional wireless infrastructure that helps provide new services, improve coverage, and increase the bit rate and efficiency of spectrum usage. Remote software upgrades are easier to accomplish than corresponding hardware upgrades, particularly in remote areas. Given the high fixed costs of constructing broadband networks, once built, they are not likely to be replaced, especially in rural areas that are unserved today. As a consequence, we believe that networks deployed in rural areas should not merely be adequate for current bandwidth demands. Instead, they also should be readily upgradeable to meet bandwidth demands of the future. An international comparison suggests significant additional capacity may be necessary. For example, while the average download speed for residential broadband subscribers in the United States is currently 2.3 Mbps, residential subscribers in Japan now average 63 Mbps.[187] Moreover, service providers in Hong Kong, Japan, South Korea, and Singapore either offer 1 Gbps

residential service now or are planning to have comprehensive 1 Gbps residential service in the near future,[188] and South Korea is complementing its fiber rollout with 10 Mbps wireless 4G services for mobility.[189] Bandwidth-intensive applications could very quickly become the norm in the U.S.—even in rural areas. Technologies that cannot be upgraded easily could make Internet applications less than five years from now look like the dial-up downloads of today.

83. *Weather and Environmental Conditions.* Unfavorable weather and environmental conditions can affect transmission technologies. Rain, snow, extreme temperatures, salt, pollution, and wind can degrade some technologies' broadband performance or even render a technology unusable until the conditions change. However, these weather conditions are considered in the design phase of each system, and service outages are typically very limited in duration, on the order of a few minutes or less per month.[190] Technologies that are adversely affected by typical weather conditions are less useful than alternatives that are not affected by weather.[191]

84. *Survivability, Redundancy, and Security.* As noted, investment in rural technologies will be for the long term. Ideally, decision makers should consider the survivability of the technology under adverse conditions. Extreme weather conditions can damage broadband infrastructure and could be especially devastating to the extent that a rural area might not have the resources to repair broadband facilities disrupted over a wide region. Disasters such as Hurricane Katrina have shown that infrastructure that is vulnerable to adverse weather events can leave large numbers of people without critical communications capability for long periods of time.[192] Overall economics should be balanced with planning to ensure that critical facilities supporting large numbers of subscribers are adequately protected from foreseeable weather events. Middle mile and backhaul facilities specifically should be capable of surviving harsh environments and foreseeable disaster events. However, this may be difficult in rural environments. For example, there may be only a single right of way, making implementation of diverse strategies difficult. In such instances, hardening of critical facilities, *e.g.*, trenching fiber to a depth of 3 feet or more, may be an alternative in building critical infrastructure. Moreover, all technologies present security concerns. America's future broadband network will be used for everything from our electrical grid to our emergency systems and will require infrastructure that is secure at every step.[193]

85. *Distance and Topography.* How well a technology performs over extended distances and in the context of the local topography could be a critical factor in many rural areas. For instance, because some wireless technologies need uninterrupted lines of sight between nodes, terrain can block the proper transmission of signals: a technology that works well on an open plain may not be appropriate if mountains or dense forests occupy a significant portion of the transmission route. In addition, wireless signals using frequency bands below 1 GHz generally penetrate environmental obstructions better than signals using higher bands, making the lower-band technologies particularly well-suited for rural deployment.[194] Furthermore, wireless technologies that rely on additional infrastructure, such as a series of repeaters or amplifiers, to regenerate a signal over long distances may not be appropriate in some settings. Wireline technologies typically require poles or trenches for deployment, which may be more costly in areas of permafrost and rocky terrain.

86. *Maintenance and Repair.* Some technologies are more expensive to maintain and repair than others. All structures eventually deteriorate over time, but some materials and technologies are naturally more durable than others. Because sustainability is a serious consideration for many rural areas, the lower the maintenance and repair costs, the more likely the technology will prove to be cost-effective over time. For example, optical technology, because of its lack of outside plant electronics, relative immunity to moisture, and sophisticated diagnostic capabilities, offers significant maintenance advantages.

87. *Resource Contention and "Micro-Congestion."* Some technologies operate on a shared last-mile platform and some offer dedicated last-mile capacity. Shared technologies often provide greater peak performance than dedicated technologies. However, when the access network is congested, the performance an individual user experiences on a shared-technology network can be significantly worse than that over a dedicated last-mile network. When comparing the speed of broadband access networks, it is important to consider peak performance, typical performance, and minimum performance. Resource contention may also exist in the backhaul or middle-mile portions of the network. Moreover, some technologies are more susceptible to "micro-congestion" or "jitter," which occurs "whenever a large number of packets come from a faster network link to a slower network link or where several networks links merge to a single link."[195] High jitter can make use of interactive applications difficult.[196]

B. Assessment of Broadband Deployment

1. Broadband data collection

88. As recognized above,[197] we do not have comprehensive and reliable data on the extent of broadband availability and subscribership in rural areas. Nor do we have sufficient information on rural broadband demand, transfer speeds, and prices, or on the infrastructure available to help provide broadband services to unserved and underserved rural areas. This lack of information constitutes a significant challenge to ubiquitous and robust broadband deployment in rural areas.[198] Policymakers, entrepreneurs, community groups, and consumers all need accurate information in these information categories in order to make informed decisions. We recommend that the Commission work to collect this information, in coordination with the administration and Tribal and state governments. We provide a brief overview of each of these information categories below.

89. *Availability and Subscribership.* Determining the extent to which broadband service is available in rural areas, and the extent to which consumers and businesses subscribe in areas where it is offered, are key elements of assessing broadband needs and deployment. Data collected for smaller geographic areas and for discrete population subgroups, such as Tribal Nations, minority groups, and persons with disabilities, are generally more informative than data collected for larger areas or groups. In addition, provider-specific and technology-specific data on broadband availability and subscribership would help policymakers evaluate issues such as the level of broadband competition in rural areas and the extent to which

certain technology platforms are better suited to serve certain types of rural markets. It may also be important to identify the availability and subscribership of broadband at typical "anchor institutions," such as schools, libraries and health institutions, which play a unique role in both introducing the benefits of broadband to a community as well as stimulating further subscriber growth. Furthermore, overlaying or correlating granular availability and subscribership data with data for other metrics, such as population density, income, and terrain, will allow policymakers to understand the steps needed to increase broadband availability and subscribership in particular rural areas.

90. *Speeds.* Information on the data transfer speeds available to and experienced by rural broadband users is an important component in assessing rural broadband deployment. The Commission has recognized that the broadband connection speeds that customers experience are neither constant nor identical to the advertised speeds or the theoretical maximums of a given network or particular service configuration.[199] Ideally, data collected on broadband speeds would acknowledge such differences and attempt to accurately reflect the average or typical data transfer rates that broadband users experience.

91. *Prices.* Because the price of broadband service affects the consumer's decision whether or not to subscribe, detailed pricing information could be helpful in analyzing the lack of broadband subscribership in rural areas. We recognize that collecting and analyzing pricing information raises various complexities, such as how to determine the price of broadband service offered as part of a service bundle, how to account for introductory offers and promotions, and how to keep pace with frequent price fluctuations.[200] Nevertheless, we believe information on broadband prices, including how prices vary among rural communities and service tiers, could be helpful because of the key role that prices play in broadband demand and adoption.

92. *Demand.* Policymakers also should consider obtaining detailed information on the demand for broadband services. As discussed in detail below,[201] promoting broadband deployment will require steps to increase broadband demand as well as supply. Addressing broadband demand effectively likely will require specific survey data on the interest or lack of interest in broadband service among non- subscribers, and the reasons for that interest or lack of interest. Such data ideally would be available for various demographics such as age, location, and income; for businesses; and for certain classes of institutional users, such as schools, libraries, public safety agencies, and hospitals.[202] A special focus also must be given to Tribal members and minorities, where the lack of broadband subscribership appears particularly acute.

93. *Infrastructure.* The metrics described above focus on evaluating the different components of the rural broadband marketplace. In addition to examining data for these metrics, a needs assessment also could examine information on the existing infrastructure in rural areas that is or could be used to provide broadband service. For example, policymakers could consider gathering data on the Internet backbone and middle-mile Internet access points,[203] including fiber routes and fiber-lit locations; locations of base stations, towers, switches, and collocation facilities; and locations of non- communications infrastructure, such as water towers, railroads, and highways, that could support broadband network facilities.

94. *Federal Efforts to Assess Broadband Availability*. The Commission has been tracking broadband subscribership and deployment since 2000 through its Form 477 local competition and broadband deployment reporting program. Providers of high-speed Internet access service—formerly defined as data speeds exceeding 200 kbps in at least one direction—are required to submit to the Commission semi-annually data regarding several metrics, including their number of broadband subscribers, data rates, and technology platforms. In its Section 706 Reports, the Commission has used these data to assess whether "advanced telecommunications capability" is being deployed to all Americans in a reasonable and timely manner.[204]

95. Although past Section 706 Reports included an incomplete analysis of the broadband market and relied on data that lacked sufficient granularity,[205] the changes to the Commission's broadband data collection rules adopted in the *2008 Data Gathering Order* and the new requirements for Section 706 Reports included in the BDIA should improve those reports going forward. In the *2008 Data Gathering Order*, the Commission required most broadband providers to file subscribership information, including their number of subscribers—broken down by technology, speed tier, and business/residential—on a Census Tract level.[206] That order increased the number of speed tiers (both upload and download) for broadband reporting purposes in order to provide a more granular depiction of the data rates at which broadband consumers access the Internet.[207] Providers of mobile wireless broadband services must submit their broadband subscriber totals on a state-by-state basis as well as a list of the Census Tracts covered by their mobile broadband networks.[208] In addition to reporting their number of broadband-capable mobile devices in use, mobile broadband providers now must specify the percentage of those users that have devices and subscription packages that permit them to access the lawful Internet content of their choice.[209] Broadband providers subject to these reporting requirements must file Form 477 twice each year. The Commission staff is in the process of analyzing the first round of Form 477 filings under these new rules, which were due March 16, 2009. The next Form 477 data filings are due on September 1, 2009.

96. In addition to collecting the Form 477 broadband data, the Commission tracks the deployment of mobile wireless broadband networks in its *Annual CMRS Competition Reports*.[210] With this information, the Commission is able to estimate, at the Census Block level, the percentage of the U.S. population covered by various mobile broadband network technologies.[211] The Commission annually collects data on cable system broadband capability, including subscriber numbers and capacity, using FCC Form 325.[212] The Commission also collects data on the satellite industry for its *Annual Satellite Competition Reports*, which examine the reach of satellite-based, two-way broadband to the home.[213]

97. In October 2008, Congress passed the BDIA, which provides for improved federal data on the deployment and adoption of broadband services.[214] That Act requires the Commission to issue its Section 706 Reports "annually" instead of "regularly" and adds several types of data regarding broadband services that the Commission must produce and evaluate. Specifically, the BDIA requires that the Commission's Section 706 Report discretely identify "unserved" areas, as well as the population, population density, and average per capita income of each of these areas.[215] Further, the Commission must conduct a

consumer survey at least annually that includes questions regarding broadband technology choices, prices, speeds, applications, consumer decisions and options.[216] The BDIA also requires the Commission to conduct and evaluate an international comparison of broadband speeds and prices.[217] On March 31, 2009, the Commission released a Public Notice seeking comment on how it should implement the BDIA's consumer survey and international comparison requirements.[218]

98. *Broadband-Related Efforts in the States.* Many efforts have been made at the state level to review and improve broadband deployment. For example, in California, Governor Schwarzenegger commissioned a Broadband Task Force to "remove barriers to broadband access, identify opportunities for increased broadband adoption, and enable the creation and deployment of new advanced communications technologies."[219] This Task Force's final report, issued in January 2008, includes broadband availability and speed maps as well as recommendations for improving broadband deployment throughout California.[220] Similarly, in the Commonwealth of Virginia, Governor Kaine established a Broadband Roundtable to advance the goal of broadband access for all Virginia businesses by 201 0.[221] The Roundtable has met with local and regional leaders and collected information concerning broadband deployment, community needs, and barriers to deployment throughout Virginia. As discussed above, in response to those meetings, the Roundtable created an online resource to guide participants in community-led broadband initiatives,[222] and in September 2008 issued a final report on broadband issues facing Virginia.[223]

99. Other efforts at the state level have focused on using public-private partnerships to expand access to broadband, while also assessing the demand for broadband services.[224] These efforts include ConnectKentucky's efforts to bring broadband services to previously unserved areas within Kentucky.[225] Other states, including Minnesota, Ohio, South Carolina, Tennessee, and West Virginia, have adopted public-private partnerships following the ConnectKentucky model.[226]

100. Some states have provided tax incentives to encourage investment in broadband infrastructure and other related equipment and expenses;[227] created dedicated funding that leverages private sector funds to increase investment in broadband infrastructure;[228] and engaged local communities to identify and increase demand for broadband deployment in unserved areas.[229] Finally, there have been a number of regional efforts to increase rural broadband deployment. For example, in Southeastern Wisconsin, an advisory committee was established to implement a regional broadband plan covering seven counties, of which about 64 percent of the land area is rural.[230] The Appalachian Regional Commission and the Delta Regional Authority also have made efforts to encourage broadband deployment in their respective regions.[231]

2. Broadband mapping

101. Mapping provides a powerful and expressive way to convey information, and may be uniquely suited to explaining the status and progress of the evolution of broadband deployment and to targeting deployment obstacles. In the rural context, broadband mapping is a necessary tool for identifying and tracking broadband service availability and infrastructure

deployment. However, as with any visualization or other presentation of information, the utility of a map is only as good as the underlying data, and the accuracy and reliability of such data must be verifiable. Similarly, broadband mapping efforts must reconcile the sometimes competing concerns of technical limitations, confidentiality, and infrastructure security.

102. The Recovery Act directs NTIA to create a "comprehensive nationwide inventory map of existing broadband service capability and availability" that shows the geographic extent to which that capability is deployed and available for each state.[232] By February 2011, NTIA should to the extent practical make this inventory map accessible by the public on an NTIA website in a form that is interactive and searchable.[233]

103. A number of different organizations have begun to map broadband availability in several states.[234] The entities vary, although most of them are public-private partnerships,[235] or task forces established by a governor and similarly comprised of community, government, and broadband industry representatives.[236] In addition, a handful of state public utility commissions also map broadband availability,[237] as do certain state-sponsored initiatives.[238] Apart from mapping, these organizations generally seek to promote broadband deployment throughout their states, particularly in rural areas. These organizations typically rely on voluntary submissions of data on residential broadband availability, and the entities collocating the data often sign non-disclosure agreements and make other commitments (such as agreeing to depict only aggregated data) in response to providers' requests for confidentiality.[239] Many of these mapping efforts have resulted in useful maps of network broadband availability, although we are aware of no state mandate that all broadband providers must submit broadband availability data. We note also that certain commenters in the Commission's availability mapping proceeding question the independence and effectiveness of many of the non-governmental broadband mapping organizations.[240]

104. Several federal agencies have mapping programs for various initiatives. Federal mapping efforts can and should play a pivotal role in overcoming the challenges that currently impede rural broadband deployment. Elsewhere in this Report, we discuss in detail the importance of coordination among governmental organizations and private parties.[241] Such coordination will be critical to making the best use of the data being gathered through the efforts discussed above.[242] We recognize the importance of including all rural areas, particularly Tribal lands, in federal mapping efforts. Pursuant to the Recovery Act and the BDIA, the Commission and the Administration should continue their efforts to coordinate federal, Tribal, state, local, and private mapping efforts.

C. Stimulating and Sustaining Demand for Broadband

105. A critical component in addressing solutions to promote rural broadband deployment is ensuring the sustainability of rural broadband networks. That is, once built, the networks must generate enough revenue to cover their costs. Several surveys show that a substantial percentage of consumers in the United States have access to broadband services, but choose not to subscribe.[243] Given that sustained deployment of broadband services is

unlikely without sufficient consumer demand for broadband services, a strategy designed to promote rural broadband adoption must examine and address the discrepancy between broadband availability and broadband adoption. Rural areas have diverse populations and varied terrains.[244] They also have variable access to resources, and are served, if at all, by various types of providers (*e.g.*, a large incumbent LEC, a small local cooperative, or a WISP), each of which may have differing levels of technical expertise, business experience, and access to capital markets. We discuss below the various factors that may affect demand for, and sustainability of, broadband services in rural areas.

106. Several factors may contribute to low consumer demand for broadband services.[245] First, a lack of training and knowledge regarding the benefits of Internet access may hinder broadband adoption.[246] Many individuals in non-broadband households view broadband as either unimportant or difficult to use.[247] For example, TIA claims that approximately 42 percent of rural residents without broadband at home fail to subscribe because of a "perceived lack of need."[248] The Pew American Home Life Project survey found similar perceptions among those that did not have Internet access, finding that an estimated 40 percent of non-users do not use the Internet because "they are not interested" or view it as "a waste of time."[249]

107. Another demand and sustainability factor is the affordability of broadband services to consumers, which may include continuing subscription costs, computer equipment costs,[250] and the costs of other customer premises equipment necessary to access broadband services.[251] Some studies show that many dial-up users believe they cannot afford broadband services.[252] According to CFA/CU, only 15 percent of rural households with annual incomes less than $25,000 have broadband subscriptions whereas 45 percent of rural households with annual incomes greater than $25,000 have broadband access in the home.[253] These studies demonstrate that the inability of consumers to afford either broadband service or the computer equipment necessary to access broadband service at current market prices is a likely barrier to broadband adoption and sustainability in certain rural markets.

108. Demand-side programs can effectively promote the adoption and use of broadband among underserved and rural populations.[254] Congress already has taken important steps in this regard. For example, the BTOP program, established by the Recovery Act, provides funding for, among other measures, grants to provide broadband education, awareness, training, access, equipment, and support to educational institutions, libraries, healthcare providers, and other community support organizations to facilitate greater use of broadband. The Recovery Act also directs grants to support organizations and agencies that provide outreach, access, equipment, and support services to facilitate greater use of broadband service by low-income, unemployed, aged, and otherwise vulnerable populations.[255] The Recovery Act provides, in addition, at least $250 million in funding for innovative programs to encourage sustainable adoption of broadband service.[256] In singling out these purposes, Congress recognized the importance of consumer affordability and education in ensuring the adoption and sustainability of rural broadband networks.

109. In its *National Broadband Plan* proceeding, the Commission is exploring long-term solutions to address the lack of demand for broadband services.[257] Specifically, the *National*

Broadband Plan NOI sought comment on improving digital literacy and media literacy skills, increasing broadband access device ownership, and the effect of content and copyright protections on broadband network deployment and usage.[258] The *National Broadband Plan NOI* also sought comment on the extent to which a centralized clearinghouse for outreach and computer and broadband training initiatives should be a component of the national broadband plan.[259]

110. Some training and education programs are already underway to increase adoption of broadband in local communities.[260] Rural libraries, which currently serve as an access point for broadband Internet service, can provide another avenue for Internet training and education, as librarians are well-positioned to educate and train individuals on the benefits of Internet access.[261] Further, libraries can stimulate demand for broadband services by hosting community fora and providing training on accessing specific information.[262] Rural libraries can also function as public computing centers, providing broadband Internet access to patrons, which in turn can help stimulate further demand for consumer broadband services.[263] We suggest that Internet education focus on general digital literacy as well as the many Internet-based resources that can benefit users, such as access to healthcare information, education of children and adults, financial planning, online banking, online shopping, and electronic mail. Further, it is imperative that education and training initiatives be forward-looking to ensure continued and sustained subscription to broadband services in rural areas.

111. Both public and private entities, including broadband providers, should consider programs designed to stimulate broadband adoption and sustainability among targeted groups. Pennsylvania has established a special program that aggregates demand in communities looking to achieve an agreed-upon deployment threshold.[264] This program provides state grants for outreach and demand aggregation activities, including programs on the benefits and use of broadband services.[265] Communities may also want to consider ways to aggregate or consolidate demand as part of developing a strategy for a sustainable broadband network.[266] Participants in this effort could include individual consumers, businesses, educational institutions, health care facilities, and government agencies. Entities that can function as anchor tenants with adequate demand to both spur broadband infrastructure investment and ensure sustainability can function as an integral part of a rural broadband strategy.

112. Furthermore, public and private entities also should consider programs, such as making computers or laptops available at a discount to qualifying households or discounting monthly service, designed to make broadband affordable to those with low incomes.[267] The Commission has also sought comment on establishing a Broadband Lifeline/Link Up pilot program to examine how the low-income universal service support program can be used to enhance access to broadband Internet access services for low-income Americans.[268]

D. Addressing Network Costs

113. As a general matter, the costs involved in deploying a broadband network are significant, requiring providers to purchase electronic equipment,[269] obtain access to rights of

way,[270] interconnect with other networks, and construct the actual network.[271] Yet, rural networks can often be even more expensive to deploy and potentially more expensive to maintain than networks in non-rural areas for a variety of reasons, which can serve as a formidable barrier to rural broadband deployment.[272] Rural broadband networks typically serve far fewer customers per square mile than urban and suburban networks,[273] and often cover larger land areas that may include challenging topographies and climate conditions,[274] making it extremely costly to provide broadband service to remote areas.[275] For example, the topography in such areas may limit the reach of wireless transmission facilities on individual towers, and, consequently, more towers may need to be built, thereby increasing the cost of wireless broadband deployment. Similarly, because radio signals using spectrum below 1 GHz generally penetrate environmental obstructions better than signals using higher bands,[276] licensees with spectrum only in the higher frequency bands may need to deploy more infrastructure, including towers, to cover the same land area, which can also lead to higher deployment costs in rural areas. In addition, the cost of extending telecommunications lines to remote areas contributes to higher deployment costs in rural areas. It is notable that costs may vary among, for example, fiber,[277] digital subscriber line (DSL),[278] and cable deployment.[279]

114. Access to adequate and affordable "middle-mile" broadband facilities—the facilities that are commonly used to connect the "last mile" ISP with an Internet backbone service provider—is a necessary precursor to a provider's being able to deploy broadband services to its customers. Although rural broadband networks are fundamentally similar to broadband networks in non-rural areas in that they involve both a local access or distribution network[280] and a backhaul component,[281] rural broadband networks are also typically built in locations that are geographically more removed from Internet backbone nodes.[282] In many cases, because of this more distant location, the rural broadband provider will need to obtain backhaul transport, or "middle mile" facilities, from more than one provider, often over facilities that were designed for voice telephone or cable television services.[283] Some of these "middle mile" facilities may have insufficient capacity, causing the transmission speed on otherwise adequate last-mile broadband facilities to come to a crawl or stall before the data reach the Internet backbone.[284] Overcoming this may require the construction of a dedicated facility, which drives up costs and can deter last-mile broadband investments.[285] Moreover, even when the last-mile provider acquires access to adequate middle-mile facilities, that access may be prohibitively expensive.[286] Consequently, backhaul transport costs in rural areas can be significantly higher than for networks in other areas.[287]

115. There is no one solution to addressing middle mile transport costs in rural areas. Commenters suggest explicitly encouraging middle-mile buildout,[288] revising universal service funding to help cover costs of the middle-mile,[289] and using current or potential infrastructure more effectively by coordinating with other infrastructure projects to shrink deployment costs, and reforming interconnection obligations.[290] Others suggest that these facilities be offered at cost-based or favorable rates on nondiscriminatory terms and conditions.[291]

116. The regulatory process associated with obtaining access to rights of way on "Indian lands" may add to the costs of bringing broadband services to Tribal members. These high costs of deploying broadband to rural areas can be compounded by potential providers'

experiencing difficulties gaining access to capital.[292] Gaining access to capital can be a problem for rural providers that have fewer customers over whom they can spread the costs of providing broadband service. This entry barrier increases with the more capital that is needed to cover minimum costs,[293] which can be particularly pronounced in rural areas.

117. Although the free market has many benefits, such as driving down the costs of services for consumers and improving service quality, it also can leave behind geographic areas with high costs and lower profit potential.[294] Such is the case with many rural areas. Market forces often demand returns commensurate with investment risk. In many parts of rural America, the relatively high deployment costs per potential customer make relying on market forces alone an inadequate strategy for promoting the deployment of broadband services. The repercussions of relying on market forces alone ultimately would be felt throughout America because many of rural broadband's benefits go beyond the direct benefits to the rural subscriber and the direct profits to the broadband deployer.[295] Ultimately, as the Commission recognized regarding the need for universal telephone service,[296] the more individuals that are on the network, the greater value of the network itself.[297] Moreover, the more extensive a network, the greater the benefits to be derived from the network.[298] The same is true of an extensive broadband network.[299] Therefore, we believe that government action is needed to encourage deployment of broadband to rural areas.

118. To that end, various government entities have implemented programs to encourage the buildout of broadband infrastructure to rural areas. Recent legislation has created several opportunities for organizations seeking to build out broadband infrastructure and provide services to unserved and underserved areas to receive grants, loans, or loan guarantees to help defray deployment costs. In addition, a number of states have enacted tax incentives designed to help offset deployment costs in order to encourage the provision of broadband services to consumers.[300] Many rural cooperatives are deploying broadband to rural areas through collaborative efforts and by obtaining federal or state funding support.[301]

119. An alternative solution that has been adopted in some areas is government sponsorship or ownership of broadband networks.[302] Commenters suggest that many successful government-sponsored investments in the United States have involved the deployment of fiber networks in cities;[303] however, some rural areas also have benefited from government-sponsored broadband deployment.[304] Although many have expressed concerns regarding the provision of government-sponsored or government-owned broadband services, raising questions about the appropriate role of government as a broadband service provider,[305] the potential for market distortion,[306] and the consequences of unfair competition,[307] others assert that government sponsorship is the best solution in rural areas where existing service providers and private entities have not invested in building broadband infrastructure.[308] Internationally, we note that several countries have undertaken government-sponsored efforts to provide broadband services nationwide. One example is the Australian government's recent announcement that it plans to upgrade its broadband infrastructure to deliver up to 100 Mbps to 90 percent of homes and offices in Australia.[309] Remote and sparsely populated regions of that country will receive wireless broadband service with speeds up to 12 Mbps.[310] Similarly, France's "dead zone programme," created in 2003, allows operators to use public funds to provide mobile telephony coverage ultimately to 3,000 towns in France.[311] Sweden

has adopted the “stadsnätt” urban area network model under which a city builds and administers fiber infrastructure, which it then rents at cost to service providers that deploy their own transmission equipment.[312]

120. A complementary government role in broadband deployment can yield advantages that a free market solution cannot achieve alone. For example, government involvement in rural broadband may increase the efficiency and reliability of local services, such as law enforcement and emergency services;[313] promote job growth and economic development by attracting and retaining businesses and increasing use of technology in a community;[314] provide educational benefits, both for local schools and those seeking education online;[315] and generally generate the indirect benefits to America that private deployers may not consider in their cost-benefit profit assessments.[316] Finding creative solutions to overcome the hurdles presented by high deployment costs will require the development of a multi-faceted approach, including, among other things, collaborative efforts among federal, Tribal, state, and local governments, community organizations, businesses, and individuals; federal and state funding; government ownership or sponsorship; and continuation of the important contributions made by the rural cooperatives spread across this nation.

VI. Overcoming Challenges to Rural Broadband Deployment

121. The 2008 Farm Bill directs that this Report include recommendations “to identify how specific Federal agency programs and resources can best respond to rural broadband requirements and overcome obstacles that currently impede rural broadband deployment.”[317] Throughout this Report, we have identified how Federal agency programs can respond better and overcome obstacles that impede rural broadband deployment.[318] In this part, we provide recommendations concerning the Commission’s existing programs and resources.

122. During recent years, the Commission has not had a comprehensive strategy regarding how its programs and resources can best respond to rural broadband requirements or promote rural broadband deployment. The Commission’s approach generally has been to prevent or remove “economic” regulation of broadband services. For example, the Commission established a deregulatory environment for the provision of broadband Internet access service by classifying that service as an information service regardless of the nature of the platform over which it is provided,[319] and by eliminating legacy regulation that had applied to facilities-based wireline providers (but not to other providers) of that service.[320] Despite this generally deregulatory approach, the Commission has continued to impose regulatory obligations on broadband telecommunications services providers (although not on broadband Internet access service providers) to the extent it was needed to further important public policies.[321]

123. The theory behind this generally deregulatory approach was that reliance on market forces, rather than regulation, was the best way to increase investment in broadband networks and make affordable broadband services available to consumers. In the *National Broadband*

Plan NOI, the Commission has sought comment on competition as a mechanism to achieve the goals of the Recovery Act.[322] The results of this inquiry may require changes in specific Commission programs affecting rural broadband.

124. Because the national broadband plan is currently under consideration and is not due until February 2010, we have included here a compilation of pending Commission proceedings affecting rural broadband, including proceedings related to universal service, open network policies, spectrum access, intercarrier compensation, special access, pole attachments, tower siting, and video programming. The discussion of these proceedings here should not be construed as prejudging outcomes, which in any event must be based on the record before the Commission in those proceedings.

125. The 2008 Farm Bill requires that the Chairman of the Commission, in coordination with the Secretary of Agriculture, provide Congress with a full, updated rural broadband report two years from now.[323] We recommend that the next Commission Chairman consider completing a status report on rural broadband approximately one year from now. This status report will help inform Congress of the Commission's progress on these proceedings and of any needed changes to the recommendations in this Report in light of additional efforts to address rural broadband issues, including the completion of the national broadband plan.

A. Universal Service Programs and Reform

126. The 1996 Act codified the historical commitment of the Commission and state regulators to promote universal service by ensuring that consumers in all regions of the nation have access to affordable, quality telecommunications services.[324] The 1996 Act added section 254 to the Communications Act, which directs the Commission, after consultation with the Federal-State Joint Board on Universal Service (Joint Board), to establish specific, predictable, and sufficient support mechanisms to preserve and advance universal service.[325] In addition, in section 254(b), Congress provided a list of principles upon which the Commission must base policies for the preservation and advancement of universal service.[326] Among other things, section 254(b) directs that quality services should be available at just, reasonable, and affordable rates; access to advanced telecommunications and information services should be provided in all regions of the nation; and consumers in all regions of the nation, including those in rural areas, should have access to telecommunications and information services that are reasonably comparable to those services provided in urban areas.[327]

127. To implement the universal service goals outlined in the 1996 Act, the Commission established the High-Cost Program, the Low-Income Program, the E-rate Program, and the Rural Health Care Program.[328] The universal service programs are funded by contributions from telecommunications carriers providing interstate telecommunications services and from certain other providers of interstate telecommunications.[329] While the universal service programs have primarily been focused on ensuring the availability of telecommunications services, the Commission has made efforts to expand broadband availability through

universal service policies and is considering how to reform the programs to further expand broadband availability.[330] The four universal service programs currently treat the support of broadband differently. The Rural Health Care Program and the E-rate Program explicitly support the provision of broadband. The High-Cost Program indirectly supports the provision of broadband, and the Low-Income Program does not currently support broadband.

1. Rural Health Care Program

128. The Rural Health Care Program provides reduced rates for eligible rural health care providers for telecommunications and Internet services necessary for the provision of health care.[331] When the program was established, the Commission adopted an annual cap of $400 million for universal service support for rural health care providers.[332] Despite modifications the Commission has made to the Rural Health Care Program, the program continues to be greatly underutilized and is not fully realizing the benefits intended by the Telecommunications Act of 1996 and the Commission's rules. For example, from 1997 to 2007, the program generally disbursed less than 10 percent of the authorized funds, and in 2008, the program disbursed slightly more than 10 percent of authorized funds.[333]

129. In 2007, in response to the underutilization of the program, the Commission established the Rural Health Care Pilot Program (Pilot Program) to provide, among other things, funding for the construction of state or regional broadband networks and for the advanced telecommunications and information services provided over those networks for health care providers.[334] The goal of the Pilot Program is to stimulate the deployment of the broadband infrastructure necessary to support innovative telemedicine services to rural America. In coordination with the U.S. Department of Health and Human Services (HHS), the Pilot Program also will support the advancement of HHS 's health information technology (health IT) initiatives for electronic health records and create vital broadband links for disaster preparedness and emergency response to any large-scale emergency or public health crisis.[335] Under the Pilot Program, 67 projects are eligible to receive funding for telehealth networks serving 6,000 health care facilities in 42 states and three U.S. territories, using broadband technology to bring state-of-the-art medical practices to isolated rural communities.[336] Most recently, in April 2009, the Commission announced the approval of funding for: 1) the buildout of five networks that will link hundreds of hospitals regionally in Iowa, Minnesota, Montana, Nebraska, North Dakota, South Dakota, South Carolina, Wisconsin, and Wyoming; and 2) the design of a telehealth project in Alaska.[337]

2. E-rate program

130. The E-rate Program provides discounts for telecommunications, Internet access, and internal connections for qualifying schools and libraries throughout the nation, including insular areas and isolated regions such as Tribal lands.[338] Schools and libraries can obtain E-rate funding for broadband under either the telecommunications or the Internet access category.[339] E-rate participants can also receive discounts on e-mail service; web hosting services; cabling, connectors, and related components used for eligible voice, video, and data transmission; and components used to distribute information from Internet access facilities to individual classrooms or public areas of a library.[340]

131. Since the inception of the E-rate Program, schools and libraries have received more than $23.7 billion in funding commitments.[341] E-rate funding has provided millions of school children, teachers, and library patrons access to modern telecommunications and information services. In a study released in 2006, the National Center for Education Statistics (NCES) found that nearly 100 percent of public schools in the United States had Internet access, and 97 percent of these schools used broadband connections to access the Internet.[342] Today, Internet access in public schools is nearly ubiquitous; before the E-rate Program was established in 1997, only 78 percent of public schools had any Internet access at all. In 1997, only 27 percent of public school instructional classrooms had Internet access; that figure increased to 94 percent by 2005.[343] A 2006 study found that 99 percent of public library branches are connected to the Internet, 98 percent of which offer public Internet access.[344]

132. As these figures demonstrate, the E-rate Program has been successful in connecting schools and libraries to the Internet using broadband. As those connections become obsolete or insufficient, however, the E-rate Program should continue to ensure that American schools and libraries remain competitive and on the forefront of technology. Schools will need E-rate funding for both the initial implementation of broadband access, as well as for the on-going costs to maintain and continuously improve their networks.[345] Broadband is essential to handle the applications available today, including online distance learning and videoconferencing. One of the next steps will be to determine what level of broadband is adequate for each school's needs.

133. In some schools, broadband is an integral part of the everyday curriculum and is used by students, teachers, and administrators to access online and distance learning; online assessment, data, and other administration resources; special education; and professional development. For example, the Bering Strait School District (BSSD) in Alaska uses SchoolAccess, a satellite network that provides basic connectivity and videoconferencing.[346] The BSSD is located in an area of coastal northwest Alaska that is about the size of Great Britain. It administers 15 schools that collectively have approximately 1,800 students.[347] Today, BSSD has satellite connections of at least three Mbps from each school to the district office in Unalakleet that are used to create a coherent sense of community throughout the district, provide educational opportunities to students, and allow teachers and administrators to meet without having to fly between communities.[348] Broadband has been critical to providing BSSD's students with the same high- quality educational opportunities available in non-rural communities.

3. High-Cost Program

134. The High-Cost Program is designed to ensure that consumers in rural, insular, and high- cost areas have access to telecommunications services at rates that are affordable and reasonably comparable to rates charged for similar services in urban areas.[349] The High-Cost Program does not explicitly support the provision of broadband; however, a carrier providing broadband services indirectly receives the benefits of high-cost universal service support when its network provides both the supported voice services and broadband services.[350]

4. Low-Income Program

135. The Low-Income Program provides discounts on telephone installation and monthly telephone service to low-income consumers through two programs.[351] The Commission's Link Up Program provides federal support to reduce eligible consumers' initial connection charges by up to one half and the Lifeline Program reduces qualifying consumers' monthly charges.[352] Qualified subscribers living on Tribal lands can receive discounts of up to $35 per month on basic monthly telephone service,[353] and depending on current rates, may receive basic local phone service for as little as $1 a month.[354] In addition, qualified subscribers on Tribal lands may receive a one-time discount on initial installation or activation of a wireline or wireless telephone for their primary residence.[355]

136. Although the Low-Income Programs do not currently provide support for broadband services, the Commission sought comment in 2008 on establishing a Broadband Lifeline/Link Up Pilot Program to examine how the universal service Low-Income Program can be used to enhance access to broadband Internet access services for low-income Americans.[356] Specifically, the Commission sought comment on making available $300 million each year for three years to enable eligible telecommunications carriers to provide discounts on broadband Internet access service and the necessary access devices to low-income consumers.[357]

5. Reform efforts

137. The Communications Act specifies that "[u]niversal service is an evolving level of telecommunications service" that should be revisited periodically, and the Commission has been considering comprehensive universal service reform.[358] In 2007, the Joint Board recommended including broadband service as a supported service under the High-Cost Program and proposed permitting states to use various methods to allocate available universal service funds for broadband projects to reach unserved areas, including funding broadband projects through a competitive bidding system designed to select the most efficient provider of such service.[359] The Commission declined to adopt the recommendations of the Joint Board.[360] In 2008, the Commission released a Further Notice seeking comment on ways to comprehensively reform the federal universal service fund.[361] For example, the Commission sought comment on requiring recipients of high-cost support to offer broadband Internet access service throughout their service areas.[362] In the *National Broadband Plan NOI*, the Commission also sought comment on universal service reform.[363]

138. Many commenters suggest that universal service should be used as a mechanism to ensure that broadband services are available in rural areas but these commenters disagree about the changes that should be made to the universal service programs.[364] While this Report is not intended to address the merits of each of these issues, I continue to support the need for comprehensive reform of the universal service fund. It is of great interest to Congress, consumers, industry, and the Commission, and it is time for universal service to meet the challenges of the 21st century—broadband deployment—just as it did the challenge of the 20th century: telephone service. And while there are a variety of ways to comprehensively reform the system, adding broadband to both the contribution and distribution sides of the ledger, eliminating the identical support rule, and conducting effective auditing and oversight,

along with a Congressional change to include intrastate as well as interstate revenue as part of the fund, would accomplish a great deal in addressing the sustainability and integrity of the fund for the long term and promote broadband in the areas served by the fund.

B. Network Openness

139. Legacy and next generation applications are converging on the communications platform we know as the Internet, and how this critical infrastructure is managed now and in the future will determine whether it remains an open platform for innovation. The positive externalities and network effects of ubiquitous broadband will not be realized if consumers are all constrained by careful bundling, packaging, and discriminatory practices that whittle away the end-to-end structure of the public Internet. "Openness" is not just another bromide, but a principle we must tenaciously preserve.

140. The value of open networks is not a novel concept, but the Commission must act to ensure that the genius of the open Internet is not lost. Over the course of the Commission's history, powerful network operators have argued that harm will result from any reduction in their absolute control over the network. For example, the Commission had to intervene in cases like *Carterphone* to allow even basic attachments to the telephone network that incumbents argued would compromise the entire network.[365] As broadband networks developed, the few companies that controlled the on-ramps to the Internet could interfere with and distort the development of technology, opportunities for entrepreneurship, and the choices available to consumers. As the Commission re-categorized telecommunications services as information services, this only amplified the potential for interference. Fortunately, the Commission adopted the *Internet Policy Statement* containing the basic rights of Internet end-users to access lawful content, run applications and services, connect devices to the network, and enjoy the benefits of competition.[366] While these protections have been critically important as a first step forward, another step is needed.

141. I have long advocated adopting a fifth principle of nondiscrimination. The principle would allow for reasonable network management but make crystal clear that the transformative power of the Internet is not to be limited by such techniques. The Commission also should establish a systematic, expeditious, case-by-case process for adjudicating claims of discrimination. Such a principle is particularly important in a rural context where a citizen may have only one option for broadband Internet access. We need to guarantee the openness of the Internet, and the Commission appropriately raised the question of how to do this in its *National Broadband Plan NOI*.[367]

C. Spectrum Access

142. Wireless service will play a critical role in ensuring that broadband reaches rural areas. Because wireless infrastructure costs are frequently less significant than comparable wired broadband deployments, wireless broadband can be an efficient means of delivering

both backhaul and "last-mile" access services in rural areas.[368] It can also enable mobility or portability. Consequently, wireless broadband service can offer cost-effective connectivity where no broadband exists, as well as complementary or competitive service where it does.[369]

143. For a number of years, the Commission's spectrum policies have attempted to promote wireless broadband deployment in rural areas. These policies have included increasing spectrum availability in rural areas, and adopting licensing, technical, auction eligibility, bidding credit, and spectrum leasing rules that have sought to encourage wireless service in rural and Tribal areas.

144. *Spectrum Availability.* In recent years, the Commission has made available significant amounts of licensed spectrum that can be used to provide wireless broadband services.[370] The Commission's actions have included, among other things, an examination of public safety needs to develop a nationwide, interoperable wireless broadband public safety network in the 700 MHz Band.[371] In addition, unlicensed spectrum use is currently permitted in several spectrum bands.[372] Most recently, in November 2008, the Commission adopted rules allowing unlicensed radio transmitters to operate in the unused portions of the broadcast television spectrum, the so-called "TV white spaces."[373] The Commission has also adopted an innovative, non-exclusive licensing scheme in the 3650-3700 MHz band that allows multiple entrants to use the band simultaneously.[374]

145. *Technical Rules.* In addition to increasing the amount of spectrum available for wireless broadband services, the Commission has also revised its technical rules for certain spectrum bands to improve wireless coverage in rural areas.[375] For instance, the Commission permits Cellular, Broadband Personal Communications Service (PCS), AWS, and certain 700 MHz licensees to employ twice the power in rural areas than is permitted in non-rural areas.[376] In addition, the Commission has allowed licensees in several bands to use a power spectral density model to measure power levels,[377] and has adopted "average" rather than "peak" measurements of power limits, which enables better services in rural areas through an ability to use increased power in most instances.[378]

146. *Service and Licensing Rules.* The Commission, in recent years, has adopted rules governing auctioned spectrum licenses that are meant to encourage broadband deployment in rural areas. These policies have included adopting smaller license sizes when creating band plans and more stringent buildout requirements for licensees.[379] Furthermore, the Commission has provided wireless licensees with the flexibility to deploy the technologies and services that best fit their business plan and meet the needs of consumers.[380] The Commission's rules permit licensees to transfer their licenses, or partition or disaggregate their licenses, in the secondary market with Commission approval.[381] The Commission's secondary markets rules also provide flexibility to a wide array of wireless licensees, including broadband providers, to enter into spectrum leasing arrangements with other providers that seek access to spectrum in rural areas.[382]

147. *Bidding Credits.* To provide greater incentives for certain entities—including small rural telephone companies and small businesses that may have an interest in serving rural areas—to access spectrum, the Commission has offered bidding credits to such companies at

spectrum auctions.[383] In doing so, the Commission has recognized that bidding preferences provide these smaller companies with an opportunity to compete successfully against large, well-financed entities at auction, and in some cases to subsequently deploy wireless service in rural areas.[384]

148. *Tribal Land Bidding Credits.* In 2000, the Commission created Tribal Land bidding credits to assist Tribal communities with the greatest need for access to telecommunications service.[385] The bidding credits are available to winning bidders in wireless spectrum auctions that agree to deploy facilities and provide service in certain Tribal areas,[386] namely federally-recognized Tribal areas with telephone penetration rates equal to or less than 85 percent.[387] In order to ensure that applicants awarded bidding credits actually deploy facilities and provide service to Tribal lands, the Commission requires that a licensee construct and operate its system to cover 75 percent of the population of the qualifying Tribal Land within three years of the grant of the license or face repayment penalties and, potentially, license termination.[388]

149. Despite the various efforts of the Commission to promote spectrum access in rural areas, as described above, commenters have noted that these policies have not been consistently successful in promoting rural broadband deployments. In particular, parties have noted that there remains a lack of available, affordable, and suitable spectrum for rural wireless broadband;[389] that our secondary market rules do not always promote spectrum trading and re-use;[390] and that backhaul costs between wireless points of presence are prohibitively high.[391] Commenters have also noted that further modifications to rules for power limits in rural areas would promote wireless broadband deployment in those areas.[392]

150. We recommend the Commission conduct a thorough inventory of the spectrum it has already licensed, examining how, why, and where it is used, and identifying distinct geographic areas where service has not been deployed or where the spectrum is being used inefficiently. The Commission could then consider various ways to redeploy this spectrum for more efficient use, including possible modifications to secondary markets rules. For example, the Commission could examine whether requiring licensees to make "good faith" efforts to negotiate with potential spectrum lessees, particularly in rural areas, would promote the development of broadband service in rural markets.[393]

D. Middle Mile/Special Access

151. As discussed above, an ISP providing service to subscribers in a rural area must obtain connections to a node of an Internet backbone service provider.[394] The facilities making this connection are among those commonly referred to as "middle-mile" facilities. Many ISPs typically obtain access to these facilities by purchasing special access services from one or more incumbent LECs, wireless services providers, or other carriers. These services may be subject to tariffed rates or may be obtained pursuant to contract, depending upon the regulatory status of the provider.[395]

152. Certain commenters argue that the prices charged for access to middle-mile facilities are excessive.[396] In 2005, the Commission released the *Special Access Order and NPRM* to broadly reexamine the special access regime for price cap carriers.[397] The Commission asked parties to refresh the record in this *Special Access Reform* proceeding in 2007.[398] Numerous price cap carriers have received relief from the price cap regime, in the form of pricing flexibility or forbearance relief, for certain special access services that could be used to provide middle-mile connectivity.[399] The issues raised in the *Special Access Reform* proceeding and the forbearance relief discussed above directly affect the rates that price cap carriers may charge for access to middle-mile and other dedicated facilities for various types of broadband providers. We recommend that the Commission consider the impact special access prices have on rural broadband deployment and affordability as a part of its overall review of special access rates in the *Special Access Reform* proceeding.

153. Some commenters suggest that wireless backhaul could provide a middle-mile alternative to wireline special access services in rural areas.[400] In its *Second Report and Order* in the *White Spaces* proceeding, the Commission determined that fixed unlicensed TV band devices (TVBDs) will be allowed to provide wireless broadband services (*e.g.*, wireless Internet access) and other services using multiple vacant TV channels and will be permitted to operate on a fixed, point-to-point, or point-to-multipoint basis.[401] The Commission decided to limit fixed unlicensed TVBDs to a peak transmitter output power of one watt with a maximum antenna gain of 6 decibels isotropic (dBi) and require that transmitter power be reduced by the same amount in decibels (dB) that the maximum antenna gain exceeds 6 dBi, allowing fixed unlicensed TVBDs to operate with the equivalent of 4 watts Equivalent Isotropic Radiated Power (EIRP).[402] The Commission determined that these power levels would allow some improved coverage for wireless broadband service providers. The Commission also found that higher power levels would increase the risk of interference in congested areas and that it was prudent to set power limits at levels that would minimize the risk of interference to authorized TV band users. It was unclear that the current rules would provide for middle-mile coverage. However, recognizing that there are advantages, such as reduced infrastructure costs and increased service range, to operation of unlicensed TVBDs at even higher power levels, the Commission noted that it would further explore through a separate Notice of Inquiry whether higher powered unlicensed operation might be accommodated in the TV white spaces in rural areas.[403]

154. We recommend that the Commission consider additional actions to address middle-mile connectivity as it analyzes the records being developed in open proceedings at the Commission.

E. Intercarrier Compensation

155. Interconnection arrangements between carriers are currently governed by a complex system of intercarrier compensation regimes. Critics of the existing regimes argue that they create regulatory arbitrage opportunities that distort the marketplace and impede the deployment of broadband facilities.[404] The Commission's efforts to develop a more unified

intercarrier compensation regime began in 2001, with the release of the *Intercarrier Compensation NPRM*.[405] These efforts have included a *Further NPRM*,[406] additional rounds of comments,[407] and, most recently, efforts to adopt a comprehensive approach to universal service and intercarrier compensation reform.[408] The record assembled on comprehensive reform is extensive, the issues are difficult, and agreement on solutions has proven elusive. We believe a critical factor in evaluating any specific set of proposed reforms should be the effect they will have on the deployment of broadband services in rural areas.

F. Access to Poles and Rights of Way

156. Congress first directed the Commission to ensure that the rates, terms, and conditions for pole attachments by cable television systems are just and reasonable in 1978 when it added section 224 to the Communications Act.[409] The 1996 Act expanded the reach of section 224 to grant attachers an affirmative right to access utility poles.[410] Congress also added "telecommunications carrier" as a category of attacher.[411] The 1996 Act specified separate rate formulas for cable operators and telecommunications carriers,[412] and permitted electric utilities to deny access, on a nondiscriminatory basis, where there is insufficient capacity and for reasons of safety, reliability, and generally applicable engineering purposes.[413] When the Commission implemented section 224's new access provision,[414] it adopted a few broad policy guidelines and rules of general applicability, but concluded that the reasonableness of particular conditions of access imposed by a utility should be resolved on a case- specific basis.[415]

157. In 2007 the Commission issued the *Pole Attachments NPRM* to comprehensively review these issues.[416] In the *Pole Attachments NPRM*, the Commission tentatively concluded that all attachments used to provide broadband Internet access services should be subject to a single rate formula, regardless of the kind of company (*e.g.*, cable operator, telecommunications carrier) providing those services.[417] In the pending *Pole Attachments NPRM* proceeding, the Commission also sought comment on the interplay between section 224 pole attachment regulation and Congress's intention in section 706 to promote broadband infrastructure deployment.[418] Timely and reasonably priced access to poles and rights of way is critical to the buildout of broadband infrastructure in rural areas. We recommend that the Commission consider this factor in analyzing the record in the *Pole Attachments NPRM* proceeding.

G. Tower Siting

158. Wireless broadband development in rural areas will depend in part on the ability of providers to access towers and other structures for the deployment of their network facilities, either through new tower construction or collocation on existing towers or other structures. For instance, one study concludes that, in order to achieve ubiquitous mobile broadband coverage, approximately 16,000 new towers will need to be constructed, disproportionately in

rural areas.[419] We note that there are several open Commission proceedings that may affect the pace or cost of tower construction.[420]

H. Video Programming

159. Access to video programming is often an important element in a customer's decision to purchase broadband service. Accordingly, access to video programming could become an issue that has an impact on the potential competitiveness of the service offerings of rural broadband providers and thus on rural broadband deployment.

160. The Commission has several ongoing proceedings examining the video programming marketplace. For instance, on January 16, 2009, the Commission requested data on the status of competition in the market for the delivery video programming for 2007; similar data for years 2008 and 2009 were requested on April 9, 2009.[421] In addition, the Commission has ongoing proceedings on program access and program carriage requirements, both of which can affect the ability of rural multichannel video programming distributors (MVPDs) to provide video programming.[422] We recommend that when addressing these issues, the Commission consider the impact of any actions it may take on rural broadband deployment.

VII. CONCLUSION

161. Since January, the Obama administration has taken a proactive, leadership role in broadband-related matters, and the federal government has begun to develop a coordinated approach. The Recovery Act allocated money for broadband development and also directed the Commission to prepare a national broadband plan. We hope that this Report also makes a useful contribution to the national broadband plan. Infrastructure deployment is something Americans do well; it plays to our national strengths. We have built out canals, bridges, electricity, telephone service, roads, and highways. Now, with much history to learn from and with an array of technological resources at our disposal, we can and will do it again.

APPENDIX A. LIST OF COMMENTERS GN DOCKET NO. 09-29

Commenter	Abbreviation
Access Humboldt	Access Humboldt
Alaska Communications Systems	ACS
Allan J. Hastings	Allan J. Hastings
AlphaStar International, Inc.	Alphastar
American Cable Association	ACA
American Farm Bureau Federation	AFBF
American Homeowners Grassroots Alliance	AHGA
American Library Association	ALA

American Petroleum Institute	API
Amateur Radio Relay League, Inc.	ARRL
Art Menius	Art Menius
AT&T Inc.	AT&T
Benton Foundation	Benton Foundation
Big Think Strategies	Big Think Strategies
Brian Wanless	Brian Wanless
Bhawana Sharma	Bhawana Sharma
California Emerging Technology Fund	CETF
Cisco Systems, Inc.	Cisco
Common Sense Media, Center for Safe and Responsible Internet Use, Children Now, The Children's Partnership, Enough is Enough, ETPRO/C3 Institute, Family Online Safety Institute, Internet Keep Safe Coalition, LOOKBOTHWAYS Inc., National Institute on Media and the Family, and Web Wise Kids	Common Sense Media *et al.*
City of Shafter, CA	City of Shafter
Commonwealth of Massachusetts Department of Telecommunications and Cable	Massachusetts Commission
Commonwealth of Pennsylvania	Pennsylvania
Connected Nation, Inc.	Connected Nation
Consumer Federation of America and Consumers Union	CFA/CU
CTIA—The Wireless Association®	CTIA
David Mason (on behalf of Skylight Associates)	David Mason
Dero Forslund	Dero Forslund
DigitalBridge Communications Corp.	DigitalBridge
Edyael Casaperalta	Edyael Casaperalta
Embarq	Embarq
Expand Networks, Inc.	Expand Networks
Fergus County, Montana	Fergus County, MT
FiberTower Corporation, Rural Telecommunications Group, Inc., COMPTEL, and Sprint Nextel Corporation	FiberTower *et al.*
Fibertech Networks, LLC and Kentucky Data Link, Inc.	Fibertech/KDL
Frank Odasz (on behalf of Lone Eagle Consulting)	Frank Odasz
General Communication, Inc.	GCI
Halo Wireless	Halo
Harris Corporation	Harris
HierComm, Inc.	HierComm
iClick2Media, Inc.	iClick2Media
Independent Telephone & Telecommunications Alliance	ITTA
Iowa Telecommunications Services, Inc.	Iowa Telecom
Jay April	Jay April
Jim Miller	Jim Miller
Joseph Schroeder	Joseph Schroeder
Joyce Hospodar	Joyce Hospodar
Kodiak-Kenai Cable Company, LLC	KKCC
Link Shadley (on behalf of representatives and broadband advocates in North Central Oregon)	Link Shadley
Liz McGeachy	Liz McGeachy
M2Z Networks, Inc.	M2Z
Main Street Broadband LLC	Main Street Broadband
Mark Bayliss (on behalf of Visual Link Internet)	Mark Bayliss
Mary Wildfire	Mary Wildfire
Michigan Public Service Commission	Michigan Commission

(Continued)

Commenter	Abbreviation
Microsoft Corporation	Microsoft
Mimi Pickering	Mimi Pickering
Miriam Lederer	Miriam Lederer
Minority Media and Telecommunications Council	MMTC
MSS/ATC Coalition	MSS/ATC Coalition
National Association of State Utility Consumer Advocates	NASUCA
National Association of Telecommunications Officers and Advisors and the National Association of Counties	NATOA/NACo
National Association of Tower Erectors	NATE
National Cable & Telecommunications Association	NCTA
National Exchange Carrier Association, Inc.	NECA
National Rural Electric Cooperative Association	NRECA
National Rural Telecommunications Cooperative	NRTC
National Telecommunications Cooperative Association	NTCA
Nebraska Public Service Commission	Nebraska Commission
New America Foundation	New America Foundation
New Mexico Media Literacy Project	NMMLP
New York State Office for Technology and the New York Public Service Commission	New York Commission
Nicholas John Slater	Nicholas Slater
Nickolaus E. Leggett	Nickolaus E. Leggett
Office of Advocacy, U. S. Small Business Administration	SBA
Open Range Communications Inc.	Open Range
Organization for the Promotion and Advancement of Small Telecommunications Companies	OPASTCO
PCIA—The Wireless Infrastructure Association and The DAS Forum	PCIA/DAS
Personal Broadband Industry Association	PBIA
PocketiNet Communications	PocketiNet
QUALCOMM Incorporated	Qualcomm
Regulatory Commission of Alaska	Alaska Commission
Rob Glidden	Rob Glidden
Rural Broadband Policy Group (Center for Rural Strategies, Appalshop, Access Humboldt, the Benton Foundation, the California Center for Rural Policy, and the Main Street Project)	Rural Broadband Policy Group
Rural Fiber Alliance	RFA
Rural Iowa Independent Telephone Association	RIITA
Rural Telecommunications Alliance	RTA
Rural Telecommunications Group, Inc.	RTG
Sacred Wind Communications, Inc.	Sacred Wind
Sasha Costanza Chock	Sasha Costanza-Chock
Satellite Industry Association	SIA
Shel Anderson	Shel Anderson
Source Communications of America, LLC	Source Communications
Sprint Nextel Corporation	Sprint Nextel
Standing Rock Sioux Tribe	Standing Rock Sioux Tribe
State Members of the Federal-State Joint Board on Universal Service	Joint Board
State Members of the Federal-State Joint Conference on Advanced Services	Joint Conference
Stayton Cooperative Telephone Company	SCTC
Stephouse Networks	Stephouse Networks
Telecommunications Industry Association	TIA

Texas Statewide Telephone Cooperative, Inc.	TSTCI
Tim Marema	Tim Marema
Todd M. Jagger	Todd Jagger
Todd Starling	Todd Starling
Tom Jacobson	Tom Jacobson
TracFone Wireless, Inc.	TracFone
United States Postal Service	USPS
United States Telecom Association	USTA
Universal Service for America Coalition	USA Coalition
Valerie Fast Horse (on behalf of Coeur d'Alene Tribe)	Valerie Fast Horse
Verizon Communications Inc. and Verizon Wireless	Verizon
Vickie Edwards (on behalf of InLine)	Vickie Edwards
WildBlue Communications, Inc. and Intelsat Corporation	WildBlue/Intelsat
Windstream Communications, Inc.	Windstream
Wireless Internet Service Providers Association	WISPA
XO Communications, LLC and Nextlink Wireless, Inc.	XO/Nextlink
Yaana Technologies, LLC	Yaana

APPENDIX B. SELECTED FEDERAL PROGRAMS ADDRESSING BROADBAND ACCESS

Federal Agency	Description of Program
Federal Communications Commission	Manages several universal service fund programs that assist in broadband deployment: 1. *Rural Health Care Program*—funds broadband access for rural health care centers; 2. *E-rate Program*—provides grants for broadband to schools and libraries; 3. *High Cost Program*—does not explicitly fund broadband 4. infrastructure, but subsidies can be used to upgrade existing telephone networks.
Department of Agriculture (USDA) Rural Utilities Service (RUS)	Administers several programs addressing broadband including: 1. *Rural Telephone Loan and Loan Guarantee Program*—provides loans and loan guarantees for telephone voice service, but requires funding recipients to provide DSL broadband at a rate of at least 1 Mbps; 2. *Distance Learning and Telemedicine Loans and Grants*—provides seed money for loans and grants to rural community facilities (e.g., schools, libraries, hospitals) for advanced telecommunications systems that can provide
	health care and educational benefits to rural areas; 3. *Rural Broadband Access Loan and Loan Guarantee Program*—provides loans and loan guarantees for facilities and equipment providing broadband service in rural communities; 4. *Community Connect Broadband Grants*—provides grants to applicants proposing to provide broadband service on a "community-oriented connectivity" basis to rural communities of fewer than 20,000 inhabitants.
Department of Commerce National Telecommunications Information Administration (NTIA)	Manages several programs and projects addressing broadband including: 1. *Broadband Technology Opportunities Program*—provides grants to service providers to increase access and adoption of broadband infrastructure and services; 2. *Public Telecommunications Facilities Program*—provides grants for

(Continued)

Federal Agency	Description of Program
	planning, acquiring, installing and modernizing public telecommunications facilities through its; *3.* *Public Safety Interoperable Communications Grant Program*—provides funding to states and territories to enable and enhance public safety agencies' interoperable communications capabilities; *4.* Coordinating federal rights of way for telecommunications infrastructure projects with other agencies.
Economic Development Administration	*Investments for Public Works and Economic Development Facilities*—provides grants to economically distressed areas for construction of public facilities and infrastructure, including broadband deployment and other types of telecommunications enabling projects.
Department of Education Office of Elementary and Secondary Education	*Education Technology State Grants*—provides grants to state education agencies for development of information technology to improve teaching and learning in schools.
Office of Assistant Secretary for Educational Research and Improvement	*Ready to Teach Program*—provides grants to carry out a national telecommunication-based program to improve the teaching in core curriculum areas.
Office of Special Education and Rehabilitative Services	Technology and Media Services for Individuals with Disabilities—supports the development and application of technology and education media activities for disabled children and adults.
Department of Energy	Required to spend $4.5 billion to modernize the electric grid, which will likely include some broadband for Smart Grid. See Recovery Act, Pub. L. No. 111- 5,123 Stat. 115, 133 (2009)
Environmental Protection Agency	Exchange Network Grant Program—provides funding to states, territories, and federally recognized Indian Tribes to support the development of an Environmental Information Exchange Network, including broadband infrastructure.
Department of Health and Human Services Health Resources and Services Administration	Administers several programs addressing broadband including: 1. *Telehealth Network Grant Program*—provides grants to develop sustainable telehealth programs and networks in rural and frontier areas, and in medically unserved areas and populations. Grantees are required to apply for the Commission's universal service funds; 2. Telehealth Resource Center Grant Program—provides grant funds for regional centers to assist health care providers in the development of their Telehealth services, including decisions regarding the purchase of advanced telecommunications services; 3. Licensure Portability Grant Program—provides grant support for state professional licensing boards to carry out programs under which the boards of various states cooperate to develop and implement State policies to reduce the statutory and regulatory barriers to telemedicine.
National Library of Medicine, National Institute of Health	*Medical Library Assistance Program*—provides funds to strengthen library and information services, facilitate access to and delivery of health science information, and plan and develop advanced information networks.
Department of Housing and Urban Development	Administers several programs addressing broadband including: 1. *Community Development Block Grants*— provides loans to deploy broadband networks and make them affordable for low to moderate income residents; 2. *Indian Community Development Grants*—funds broadband technology infrastructure projects.
Department of the Treasury	Distributes grants for Smart Grid investments, which includes a broadband component. See Recovery Act, Pub. L. No. 111-5, § 405, 123 Stat. 115, 144 (2009)
Department of Transportation Research and Innovative Technology Administration	*Next Generation (NG9-1-1) Initiative*—a Research and Development project that will define a system architecture for emergency communication that enables the transmission of voice, data or video from

	different types of communication devices to Public Safety Answering Points (PSAPs) and onto emergency responder networks. The NG9-1-1 Initiative will also develop a transition plan that considers responsibilities, costs, schedule and benefits for deploying this architecture.
National Foundation on the Arts and the Humanities Institute of Museum and Library Services, Office of Library Services	Administers several programs addressing broadband including: 1. *State Library Program*– provides grants to state library administrative agencies for promotion of library services that provide all users access to information through State, regional, and international electronic networks; 2. *Native American and Native Hawaiian Library Services*—supports library services that include linking libraries to electronic networks.
Appalachian Regional Commission	Provides project grants for Appalachian communities to support the physical infrastructure necessary for economic development and improved quality of life
Delta Regional Authority	Provides project grants to support self-sustaining economic development of eight states in the Mississippi Delta Region
Denali Commission	Provides grants through a federal and state partnership designed to provide critical infrastructure and utilities throughout Alaska, particularly in distressed communities.

* This non-comprehensive list was prepared using information provided in Congressional Research Service, *Broadband Internet Access and the Digital Divide: Federal Assistance Programs*, RL307 19, Tables 1 and 2 (Mar. 19, 2009), *available at* http://assets

** For additional information on federal agency broadband programs, *see* Interagency Questionnaire Responses, attached to Letter from Kathy D. Smith, Chief Counsel, NTIA, to Marlene H. Dortch, Secretary, FCC, GN Docket No. 09-29 (filed May 18, 2009).

End Notes

[1] Food, Conservation, and Energy Act of 2008, Pub. L. No. 110-246, § 6112, 122 Stat. 923, 1966 (2008) (2008 Farm Bill).

[2] 2008 Farm Bill § 6112; *see infra* Part III.D (Recent Legislative Developments).

[3] For the purposes of the preparation of this Report, staff opened a special docket (GN Docket No. 09-29) and solicited comments from the public. *See Comment Date Established for Report on Rural Broadband Strategy*, GN Docket No. 09-29, Public Notice, 24 FCC Rcd 2987 (2009). See Appendix A for a list of commenters. Though the Report uses "I" and "we" interchangeably, the comments, suggestions, and recommendations contained in this Report reflect the views of Acting Chairman Michael J. Copps, and not of the Federal Communications Commission.

[4] American Recovery and Reinvestment Act of 2009, Pub. L. No. 111-5, 123 Stat. 115 (2009) (Recovery Act). The Recovery Act was signed into law on February 17, 2009.

[5] Recovery Act, Division A, Title II, National Telecommunications and Information Administration (NTIA Appropriations); Recovery Act, Division A, Title I, Rural Utilities Service (RUS Appropriations).

[6] Recovery Act § 600 1(k).

[7] *A National Broadband Plan for Our Future*, GN Docket No. 09-51, Notice of Inquiry, 24 FCC Rcd 4342 (2009) (*National Broadband Plan NOI*).

[8] *Id.* at para. 123.

[9] *See* H.R. REP. NO. 110-256, pt. 1, at 231 (2007) (stating that businesses "from banks, to automotive repair shops, to new age technology industries need broadband service to compete in the digital global marketplace").

[10] DOCSIS (Data Over Cable Service Interface Specification) 3.0, which is capable of supporting downstream rates up to 160 Mbps or higher and upstream data rates of 120 Mbps, is being rolled out in cable systems across the country. *See* Press Release, CableLabs, CableLabs Issues DOCSIS 3.0 Specification Enabling 160 Mbps (Aug. 7, 2006), *available at* http://www.cablelabs.com/news/pr/2006/06_pr_docsis30_080706.html.

[11] *See Amendment of Part 15 Regarding New Requirements and Measurement Guidelines for Access Broadband over Power Line Systems; Carrier Current Systems, Including Broadband Over Power Line Systems*, ET Docket Nos. 04-37, 03-104, Report and Order, 19 FCC Rcd 21265, 21266, paras. 1–2 (2005). *But see* ARRL Comments at 3-6 (noting that there are obstacles to deployment of BPL systems that will need to be addressed, including interference with amateur radio and other licensed services).

[12] *See* WIRELESS BROADBAND ACCESS TASK FORCE, FCC, CONNECTED & ON THE GO 31–32 (2005), *available at* http://hraunfoss.fcc.gov/edocs_public/attachmatch/DOC-257247A1.pdf (WIRELESS BROADBAND ACCESS TASK FORCE REPORT).

[13] For example, some Code Division Multiple Access (CDMA) carriers have deployed 1xEV-DO (EV-DO) Revision A (Rev A) technology across their networks, which has average download speeds between 600 kilobits per second (kbps) and 1.4 Mbps, and average upload speeds between 350 and 800 kbps. In addition, the data rates offered by one mobile WiMAX service providers average between 2 and 4 Mbps downstream and between 500 kbps and 1.5 Mbps upstream. *See Implementation of Section 6002(b) of the Omnibus Budget Reconciliation Act of 1993; Annual Report and Analysis of Competitive Market Conditions with Respect to Commercial Mobile Services*, WT Docket No. 08-27, Thirteenth Report, DA 09-54, paras. 134–52 (WTB rel. Jan. 16, 2009) (*Thirteenth CMRS Competition Report*); Sascha Segan, *Sprint Xohm (Mobile WiMax)*, PC MAGAZINE, Oct. 10, 2008, http://www.pcmag.com/article2/0,2817,2331483,00.asp.

[14] LTE can support up to 58 Mbps for upper link transmission and 173 Mbps for downlink transmission with 20 megahertz of spectrum and a 2x2 Multiple Input Multiple Output (MIMO) antenna structure. *See* RYSAVY RESEARCH AND 3G AMERICAS, EDGE, HSPA AND LTE—THE MOBILE BROADBAND ADVANTAGE 16 (Sept. 2007), *available at* http://www.3gamericas.com/pdfs/2007_Rysavy_091007.pdf. Initial versions of the mobile WiMAX technology can support peak downlink data rates up to 63 Mbps and peak upper link data rates up to 28 Mbps in a 10 MHz channel. *See Mobile WiMAX – Part I: A Technical Overview and Performance Evaluation, Mobile WiMAX Forum, August 2006*, at 10, *available at* http://www.wimaxforum.org/technology/downloads/Mobile_WiMAX_Part1_Overview_and_Performance.pdf (last visited May 19, 2009).

[15] There are currently three licensed satellite operators providing satellite-based broadband Internet access services. *See* HughesNet, High Speed Internet Service Plans and Pricing, http://go.gethughesnet.com/plans.cfm (last visited April 13, 2009) (showing list of HughesNet service packages ranging from 1-5 Mbps downstream and 128-300 kbps upstream); WildBlue Communications, Detailed Package Comparison Chart, http://www.wildblue.com/getWildblue/details.jsp (last visited May 19, 2009) (showing list of WildBlue service packages ranging from 512 kbps-1 .5 Mbps downstream and 128-256 kbps upstream); Starband Inc., New Starband Nova Series, http://starband.com/services/ (last visited April 13, 2009).

[16] *See, e.g.*, CFA/CU Comments at 2.

[17] *See infra* Part VI.B (Network Openness).

[18] *See* Benton Foundation Comments at 16 (stating that in a smart grid, information flows within a customer's house to thermostats, appliances, and other devices, and from the house back to the utility and that a smart grid includes smart meters, smart appliances, renewable energy resources, and energy efficiency resources).

[19] Broadband Data Improvement Act of 2008, Pub. L. No. 110-385, 122 Stat. 4097 (2008) (codified at 47 U.S.C. §§ 1301–04) (BDIA).

[20] Broadband connections have grown at a remarkable rate. *See* INDUST. ANALYSIS & TECH. DIV., FCC, HIGHSPEED SERVICES FOR INTERNET ACCESS: STATUS AS OF DECEMBER 31, 2007, at tbl. 2 (rel. Jan. 2009) (2009 HIGH SPEED REPORT) (showing that Internet connections with speeds exceeding 200 kbps in both directions grew from 4 million in 2000 to 80 million in 2007).

[21] *See, e.g.*, CONNECTED NATION, THE ECONOMIC IMPACT OF STIMULATING BROADBAND NATIONALLY (2008), *available at* http://www.connectednation.com/_documents/Connected_Nation_EIS_Study_Full_Report_02212008.pdf (finding that a 7% increase in broadband penetration to underserved areas could have a direct economic impact of $134 billion annually).

[22] WILLIAM H. LEHR *et al.*, MEASURING BROADBAND'S ECONOMIC IMPACT 16 (2006), http://www.andrew.cmu.edu/user/sirbu/pubs/MeasuringBB_EconImpact.pdf. There are other examples of rural communities that have benefited from broadband by attracting large businesses to their area. *See, e.g.*, Cecilia Kang, *Rural Riddle: Do Jobs Follow Broadband Access?*, WASH. POST, Apr. 23, 2009, at A1 (noting that Lebanon, Virginia was able to attract two large companies, creating 700 well-paying jobs for residents, after broadband was installed throughout the town, yet a neighboring town that also installed broadband failed to attract large businesses and only created a "handful" of jobs, though some local businesses benefited).

[23] This is called a "network effect" and is discussed at note 297, *infra*.

[24] *See, e.g.*, Black Dinah Chocolatiers, Welcome, http://www.blackdinahchocolatiers.com (last visited May 18, 2009) (linking to a gourmet chocolate business run from an island off the coast of Maine).

[25] For example, craftspeople can build and design their own Internet store on eBay. The eBay website allows merchants to track sales, provides discounting strategies, and provides a way to promote their stores on eBay or other Internet sites. *See* eBay, Stores, http://pages.ebay.com/storefronts/shop.html (last visited May 14, 2009). Similarly, craftspeople can sell their goods on Amazon.com through the Amazon.com Marketplace. *See* Amazon, Create a Product Detail Page, http://www.amazon.com/gp/help/customer/display.html/ref=mm_sys_faq_create?nodeId=10683361 (last visited May 19, 2009). Merchants can also create their own websites through several Internet service companies that provide a website template, an Internet address, and custom e-mail addresses for a monthly fee. *See, e.g.*, Homestead, http://www.homestead.com/ (last visited May 14, 2009).

[26] For example, the National Agricultural Statistics Service (NASS) provides weekly growing reports listing planting, fruiting, and harvesting progress and the overall condition of selected crops in major agricultural producing states. This information is available via e-mail subscription to farmers. *See* USDA, NASS, http://usda.mannlib.cornell.edu/MannUsda/viewDocumentInfo.do?documentID=1048 (last visited May 14, 2009).

[27] Weekly weather reports for 44 states are available from the National Agricultural Statistics Service. The reports provide farmers with a summary of relevant information ranging from the number of days suitable for fieldwork, yearly comparisons to past harvests, and whether farmers were using alternative methods to maintain their crops. *See* USDA, NASS, Weekly Weather National Summary, http://usda.mannlib.cornell.edu/MannUsda/viewDocumentInfo.do;jsessionid=2889DCE2E86B04AD9EB296E C533 07AB1?documentID=1573 (last visited May 19, 2009). *See also* ANIMAL AND PLANT HEALTH INSPECTION SERVICE, USDA, FOOT-AND-MOUTH DISEASE (Feb. 2007), http://www.aphis.usda.gov/ publications/animal_health (last visited May 15, 2009) (providing a fact sheet on a particular disease).

[28] For example, USDA provides reports and databases on dairy, poultry, and livestock, including chemical usage, losses due to animal death, and historical tracking records. *See* USDA, National Agricultural Statistics Service, Historical Track Record—Livestock, http://usda.mannlib.cornell.edu/MannUsda/ viewDocumentInfo.do? documentID=1 595 (last visited May 19, 2009).

[29] One non-profit organization has claimed dramatic results with its broadband efforts in one rural area. The One Economy Corporation provides laptops and broadband connections throughout rural Greene County, North Carolina. Among other noted improvements in student achievement, the program notes that the percentage of high school students applying to college was greatly improved from when the program began, and the high school dropout rate has been reduced. *See* One Economy Corporation, Bring IT Home, http://www.one-economy.com/sites/all/files/a_replicable_model (last visited May 14, 2009).

[30] Benton Foundation Comments, Attach. at 28 (stating that on September 11, 2001, while 95% of cell phone calls made at 11:00 a.m. failed to get through, only 2% of Internet addresses remained off-line for an extended period, illustrating the Internet's overall resilience).

[31] For example, today, some public safety answering points (PSAPs) can send a map showing a distressed caller's location to the nearest fire department or police station or can alert Internet-connected devices to a hazardous spill or weather condition. *See* DEP'T OF TRANSP., NEXT GENERATION 9-1-1 SYSTEM PRELIMINARY CONCEPT OF OPERATIONS 8–10, 17–19 (2005), http://www.its.dot.gov/NG911/pdf/ConOps.pdf; *see also* Department of Transportation, Research and Innovative Technology Administration, Intelligent Transportation Systems, Examples and Scenarios, http://www.its.dot.gov/NG911/ng911_ex.htm (last visited Apr. 13, 2009).

[32] *See* Deborah Kaplan, former executive director of the World Institute on Disability, *Broadband: A Vital Link for People with Disabilities*, *available at* http://www.wid.org/publications/broadband- (last visited April 14, 2009). Both Congress and the Commission have recognized the tremendous value that broadband networks bring in improving communications with and among people with disabilities and have undertaken efforts to bring more opportunities to them. *See, e.g.*, The Assistive Technologies Act of 2004, Pub. L. No. 108-364, 118 Stat. 1707 (2004) (codified at 29 U.S.C. §§ 3001-07) (supporting state efforts to improve provision of assistive technology to individuals with disabilities); The Americans with Disabilities Act of 1990, Pub. L. No. 101-336, 104 Stat. 327, Title IV (1990) (codified at 47 U.S.C. § 225) (requiring common carriers to provide telecommunications relay services for individuals with hearing and speech disabilities); *Amendment of the Commission's Rules Governing Hearing Aid-Compatible Mobile Handsets; Petition of American National Standards Institute Accredited Standards Committee C63 (EMC) ANSI ASC C63®*, WT Docket No. 07-250, First Report and Order, 23 FCC Rcd 3406 (2008) (adopting hearing aid compatibility requirements for mobile wireless devices); 47 C.F.R. § 64.601–06 (providing Commission's telecommunications relay service rules).

[33] The Benton Foundation urges the Commission to initiate and expand programs and policies that extend broadband to persons with disabilities, "those that are generally on the wrong side of the digital divide." Benton Foundation Comments at 32. We also note that, in the recently released *National Broadband Plan NOI*, we sought comment on, among other things, what it means for a person with disabilities to "have access" to broadband capabilities, and whether, and if so, how, to ensure that the technical characteristics of current and future broadband networks align with the needs of disabled citizens. *See National Broadband Plan NOI*, at para. 28.

[34] *See* American Foundation for the Blind, AFB Senior Site, Using Technology, http://www.afb.org/seniorsite.asp? SectionID=67&DocumentID=3336 (last visited May 15, 2009).

[35] JOSEPH P. FUHR JR. & STEPHEN B. POCIASK, AMERICAN CONSUMER INSTITUTE, BROADBAND SERVICES: ECONOMIC AND ENVIRONMENTAL BENEFITS 2 (2007), http://www.theamericanconsumer.org/2007/10/31/ broadbandservices-economic-and-environmental-benefits/.

[36] JOSEPH ROMM *et al.*, THE CENTER FOR ENERGY AND CLIMATE SOLUTIONS, THE INTERNET ECONOMY AND GLOBAL WARMING: A SCENARIO OF THE IMPACT OF E-COMMERCE ON ENERGY AND THE ENVIRONMENT 5–6 (1999), http://www.p2pays.org/ref%5C04%5C03784/0378401.pdf (JOSEPH ROMM *et al.*).

[37] *See, e.g.*, Peter Slevin and Steven Mufson, *Stimulus Dollars Energize Efforts To Smarten Up the Electric Power Grid*, WASH. POST, Mar. 10, 2009, at A1.

[38] JOSEPH ROMM *et al.* at 6–7.

[39] *See generally* MMTC Comments.

[40] *See* Youtube, Dr. Vinton G Cerf, Internet, Infinity and Beyond, http://www.youtube.com/watch?v=-4R9FeBh9Kc (last visited May 5, 2009); *see also* The Connections Show, Connections #52—History of the Internet: Part III (Future), at 4:40–5:00 (posted Mar. 22, 2009), http://connections.thepodcastnetwork.com/2009/03/22/connections052-history-of-the-internet- (last visited May 5, 2009). *See also* Carl Fussman, *What I Have Learned: Vint Cerf.*, ESQUIRE, Apr 24, 2008, http://www.esquire.com/features/what-ive-learned/vint-cerf-0508?click=main_sr; Youtube, Vint Cerf on the Internet's Future, at :38–:58 (posted Sept. 17, 2007), http://www.youtube.com/watch?v=Yv6igXryFSw&feature=related (last visited May 5, 2009). Dr. Vinton G. Cerf is Vice President and Chief Internet Evangelist for Google where he identifies new Internet applications. He is co- creator of TCP/IP (with Robert Kahn) and is widely known as "Father of the Internet." *See* National Institute of Standards and Technology, Visiting Committee on Advanced Technology, Vinton G. Cerf, http://www.nist.gov/director/vcat/cerf.htm (last visited May 19, 2009).

[41] *See* Youtube, Vint Cerf on the Internet's Future, at 2:20–3:15 (posted Sept. 17, 2007), http://www.youtube.com/watch?v=Yv6igXryFSw&feature=related (last visited May 5, 2009).

[42] *See, e.g.*, *Development of Nationwide Broadband Data to Evaluate Reasonable and Timely Deployment of Advanced Services to All Americans, Improvement of Wireless Broadband Subscribership Data, and Development of Data on Interconnected Voice over Internet Protocol (VoIP) Subscribership,* WC Docket No. 07-3 8, Notice of Proposed Rulemaking, 22 FCC Rcd 7760, 7765–66, para. 10 (2007).

[43] *See supra* Parts III.D (Recent Legislative Developments) & V.B (Assessment of Broadband Deployment).

[44] *See, e.g.*, CFA/CU Comments at 2 ("[R]ural Americans are severely disadvantaged in broadband access").

[45] *See* PEW INTERNET & AMERICAN LIFE PROJECT, HOME BROADBAND ADOPTION 2008, at 3 (2008), *available at* http://www.pewinternet.org/~/media (2008 PEW BROADBAND ADOPTION STUDY). The survey was conducted by phone from April 8, 2008 to May 11, 2008 among 2,251 American adults, 1,153 of whom were broadband users. *Id.* at 22.

[46] *Compare* NTIA, HOUSEHOLDS USING THE INTERNET IN AND OUTSIDE THE HOME, BY SELECTED CHARACTERISTICS: TOTAL, URBAN, RURAL, PRINCIPAL CITY, 2007 (2008), tbl. 3, http://www.ntia.doc.gov/reports/2008/table_householdinternet2007.pdf (last visited Apr. 10, 2009) (NTIA, HOUSEHOLDS USING THE INTERNET) (providing broadband statistics for urban areas from the U.S. Census Bureau's (Census Bureau) Current Population Survey) *with id.* tbl. 5 (providing broadband statistics for rural areas). State surveys also suggest that people who live in rural areas lack broadband service more often than urban dwellers do, but results vary widely, and, again, may reflect different definitions and methodologies. *See, e.g.*, CONNECT OHIO TECHNOLOGY ASSESSMENT: EXECUTIVE SUMMARY at 9 (June 27, 2008), http://connectoh.org/_documents/Res_OHExecutiveSummary06252008_FINAL.pdf; CALIFORNIA BROADBAND TASK FORCE, THE STATE OF CONNECTIVITY, BUILDING INNOVATION THROUGH BROADBAND, FINAL REPORT at 18 (Jan. 2008), www.calink.ca.gov/pdf/CBTF_FINAL_Report.pdf (last visited Apr. 1, 2009) (CALIFORNIA BROADBAND REPORT).

[47] These estimates are based on an analysis of network coverage maps submitted to the Commission by American Roamer in February 2009 and on 2000 Census Block information. American Roamer is an independent consulting firm that tracks service provision for mobile voice and mobile data services. The Commission describes how it analyzes mobile network coverage in the *Thirteenth CMRS Competition Report. See Thirteenth CMRS Competition Report*, paras. 37–39, 144–47.

[48] *See, e.g.*, Benton Foundation Comments, Attach. at 32; MMTC Comments at 2.

[49] For example, the *2008 Pew Broadband Adoption Study* found that African-Americans had a lower broadband adoption rate than similarly situated whites (43% versus 5 7%). 2008 PEW BROADBAND ADOPTION STUDY at 3. Connected Nation found a broadband adoption rate of 33% for rural minorities and 40% for rural non-minorities. *See* CONNECTED NATION, THE CALL TO CONNECT MINORITY AMERICANS: A CONNECTED NATION POLICY BRIEF, AT 1 (2009) (CONNECTED NATION POLICY BRIEF), http://www.connectednation.org/_documents/cn_minority. Internet for Everyone reported that "40% of racial and ethnic minority households subscribe to broadband, while 55% of non-Hispanic white households are connected." INTERNETFOREVERYONE.ORG, ONE NATION ONLINE at 1 (2008), http ://www.freepress.net/files/IFE_Brochure.pdf (last visited May 14, 2009).

[50] *See* CONNECTED NATION POLICY BRIEF at 1 (showing computer ownership and subscription rates of 76% and 52% among Whites, 70% and 47% among Hispanics, and 65% and 46% among African-Americans); U.S. CENSUS BUREAU, COMPUTER AND INTERNET USE IN THE UNITED STATES: 2003, at tbl. A (2005) (showing that overall 62% of households have a computer compared to 44% of Hispanic households), http://www.census.gov/prod/2005pubs/p23-208.pdf.

[51] Overall, fewer than 35% of households earning a family income of less than $50,000 subscribe to broadband services, compared to 76% of households earning a family income more than $50,000. NTIA, HOUSEHOLDS USING THE INTERNET at tbl. 1. *See also* Connected Nation Comments at 9; Verizon Comments at 13.

[52] *See* JENIFER SIMPSON, TELECOMMUNICATIONS & TECHNOLOGY POLICY, AMERICAN ASSOCIATION OF PEOPLE WITH DISABILITIES (AAPD), FACTORS PROMOTING BROADBAND USE BY PEOPLE WITH DISABILITIES at 1 (2008), www.aapd.com/TTPI/Broadband_Policies_and__PWDs_by_Jenifer_Simpson.pdf.

[53] *Id.*

[54] For purposes of this Report, the term "rural areas" includes Indian Country, although not all of Indian Country is rural. To the extent that sections of Indian Country are rural in nature, they are likely to face the same—and some additional—difficulties in achieving increased broadband deployment as faced by "rural areas." Generally, we recognize that Indian Country is politically distinct and as a result historical and legal circumstances pose unique barriers to expanded broadband deployment. "Indian Country" means: (a) all land within the limits of any Indian reservation under the jurisdiction of the United States Government, notwithstanding the issuance of any patent, and, including rights-of-way running through the reservation; (b) all dependent Indian communities within the borders of the United States whether within the original or subsequently acquired territory thereof, and whether within or without the limits of a state; and (c) all Indian allotments, the Indian titles to which have not been extinguished, including rights-of-way running through the same. 18 U.S.C. § 1151. For the purpose of this document Indian Country also includes Alaska Native Villages, Native Hawaiian Homeland, and Trust lands. While section 1151 is a criminal statute, its definition of Indian Country applies in the civil context as well. In this Report, we use "Indian Country," "Tribal areas," and "Tribal lands" interchangeably.

[55] *See* INDUS. ANALYSIS & TECH. DIV., FCC, TRENDS IN TELEPHONE at tbl. 16.5 (2008), http://hraunfoss.fcc.gov/edocs_public/attachmatch/DOC-284932A1.pdf (2008 TRENDS IN TELEPHONE); Standing Rock Sioux Tribe Comments at 2 (discussing comparable lack of basic telephone service for Indian households).

[56] NTIA, HOUSEHOLDS USING THE INTERNET at tbl. 1.

[57] For example, whereas 39.6% of American Indian/Alaskan Native urban households using the Internet subscribe to broadband service, only 16.9% of American Indian/Alaskan Native rural households using the Internet subscribe to broadband service. *Id.* at tbls. 3, 5.

[58] MMTC Comments at 2–3; *see also* Letter from Jocelyn Tate, Chair, Rural Broadband Working Group, Minority Media and Telecommunications Council, to Marlene H. Dortch, Secretary, FCC, GN Docket 09-29, at 1 (filed Mar. 30, 2009).

[59] MMTC Comments at 2. [60] *Id.*

[61] *Id.* at 2–3.

[62] *See* BENJAMIN RUSH, ADDRESS TO THE PEOPLE OF THE UNITED STATES (1787), *available at* http://teachingamericanhistory.org/library/index.asp?document=1779; *see also* THE FEDERALIST NO. 84, at 517 (Alexander Hamilton) (Clinton Rossiter ed., 1961) (foreseeing that "public papers will be expeditious messengers of intelligence to the most remote inhabitants of the Union").

[63] "The 1792 postal law, modified slightly in 1794, allowed newspapers—regardless of size, weight, or advertising content—to circulate within one hundred miles, or anywhere in the state of publication, for 1 cent; those mailed outside the state and beyond one hundred miles paid 1.5 cents. Letter postage, in contrast, was divided into nine zones, ranging from a minimum of 6 cents per sheet for delivery up to thirty miles to a maximum of 25 cents per sheet for any distance beyond 450 miles. . . . The privilege accorded the press was striking, especially considering the differences in size." *See* STEVEN L. VAUGHN, ED., ENCYCLOPEDIA OF AMERICAN JOURNALISM 399 (2007).

[64] *Id.*

[65] *See* PAUL STARR, THE CREATION OF THE MEDIA: POLITICAL ORIGINS OF MODERN COMMUNICATIONS 86–90 (2004) (CREATION OF THE MEDIA). The number of U.S. post offices per 100,000 residents skyrocketed from 4.5 in 1775 to 74 in 1828. This 1828 number far exceeded the post offices per 100,000 persons in Great Britain at the time, 17, and France, 4. *Id.* at 87–88. In the mid-1800s, the French authorized a new post office only where it could generate $200 in revenue annually, a principle that would have closed 90% of the post offices in the United States. RICHARD R. JOHN, SPREADING THE NEWS: THE AMERICAN POSTAL SYSTEM FROM FRANKLIN TO MORSE 52 (1996).

[66] *See* CREATION OF THE MEDIA at 87–88. British historian Michael G. Mulhall claimed that by 1840 the total weekly circulation of newspapers for the 17 million residents of the United States surpassed the weekly circulation for all of Europe, which had 233 million inhabitants at the time. *Id.*; MICHAEL G. MULHALL, THE PROGRESS OF THE WORLD 90 (Adamant Media Corp. 1998) (1880).

[67] *See* CREATION OF THE MEDIA at 86.

[68] *See* DEBORAH CADBURY, DREAMS OF IRON AND STEEL 153(2005).

[69] *See* Pacific Railway Act of 1862, ch. 120, 12 Stat. 489 (1862) (Pacific Railway Act), *available at* http://memory (last visited Apr. 28, 2009). The Pacific Railway Act also commissioned a transcontinental telegraph, which was built alongside the transcontinental rail line. *See id.* at 490–92. A telegraph network serving much of the country had emerged in the 1 840s and 1 850s out of commercial need, in part due to the large newspaper industry that the country had created. *See* CREATION OF THE MEDIA at 169–71.

[70] Kathy Weiser, Legends of America, Old West Legends (Feb. 2009), http://www.legendsofamerica.com/rr-railroadcompanies.html (last visited Apr. 24, 2009) (Weiser, Old West Legends). In 1885, the Central Pacific Railroad was leased by the Southern Pacific Company, though it technically remained a corporate entity until 1959 when it was formally merged into Southern Pacific, which in turn formally became a part of Union Pacific in 1996. *See* STUART DAGGETT, CHAPTERS ON THE HISTORY OF THE SOUTHERN PACIFIC 153 (1922); Weiser, Old West Legends.

[71] The Atchison, Topeka, and Santa Fe; Southern Pacific; and Northern Pacific lines were completed between 1881 and 1883. Weiser, Old West Legends.

[72] RUS, USDA, A BRIEF HISTORY OF THE RURAL ELECTRIC AND TELEPHONE PROGRAMS at A-1 (1982), http://www.rurdev.usda.gov/rd/70th/rea-history.pdf (last visited Apr. 20, 2009) (A BRIEF HISTORY OF THE RURAL ELECTRIC AND TELEPHONE PROGRAMS).

[73] *Id.*

[74] REA was RUS's predecessor.

[75] USDA, USDA MARKS 70TH ANNIVERSARY OF LANDMARK RURAL LEGISLATION (2005), http://www.rurdev.usda.gov/rbs/pub/jul05/70th.htm (last visited Apr. 20, 2009). In the words of Wally Beyer, former administrator for the Utilities Program, during 1999 Congressional testimony, "Our agency exists in large part because the commercial marketplace was unable or unwilling to serve many rural Americans at a price that was affordable and conducive to economic development." As of 2005, about 37 million people in 47 states still received their electricity through cooperatives. *Id.* Of the rural electric systems established and supported by the approximately $57 billion dollars in USDA debt financing between 1935 and 2000, about 96% were nonprofit cooperatives owned and operated by the consumers they served. RUS, USDA, CONNECTING RURAL AMERICA (2000), http://www.rurdev.usda.gov/rd/pubs/pa1684.pdf.

[76] A BRIEF HISTORY OF THE RURAL ELECTRIC AND TELEPHONE PROGRAMS at A-1.

[77] *Id.* The 2002 Farm Bill established the USDA Rural Development Broadband program under a similar debt financing system. USDA, USDA RURAL DEVELOPMENT: BRINGING BROADBAND TO RURAL AMERICA 4(2007), http://www.rurdev.usda.gov/rd/pubs/RDBroadbandRpt.pdf.

[78] Facts on the construction, funding, and ownership of the interstate highways are available at the Department of Transportation website. U.S. Department of Transportation, Federal Highway Administration, History of the Interstate Highway System, http://www.fhwa.dot.gov/interstate/history.htm (last visited Apr. 20, 2009).

[79] *Id.*

[80] Press Release, White House, *Address to Congress on the Interstate Highway System* (Feb. 22, 1955), *available at* http://www.eisenhower.utexas.edu/Research/Digital_Documents/InterstateHighways/New%20PDFs/1955_02_22_Message_to_Congress.pdf.

[81] W.L. Mertz and Joyce Ritter report that when President Eisenhower signed a 1954 highway act that preceded the main 1956 act he said, "That gets us started, but we must do more." *See* W.L. MERTZ & JOYCE RITTER, U.S. DEP'T OF TRANSP., BUILDING THE INTERSTATE 3–4 (2006), *available at* http://www.fhwa.dot.gov/infrastructure (last visited Apr. 24, 2009) (BUILDING THE INTERSTATE).

[82] *See* Remarks of Senator Edward Martin, *as reported by* BUILDING THE INTERSTATE at 4–5.

[83] *See* Remarks of Senator Francis Case at the 1957-1958 meeting of the American Association of State Highway Officials, *as reported by* BUILDING THE INTERSTATE at 20–21.

[84] *See id.*

[85] *See* Remarks of Rep. George Fallon at the 1957-1958 meeting of the American Association of State Highway Officials, *as reported by* BUILDING THE INTERSTATE at 22–23. "An enlightened and understanding public is not only essential to the success of your local highway programs but of immense help to members of Congress in their efforts to shape acceptable legislation. We need the support of the folks back home." *See id.* at 23.

[86] *See* DARPA, DARPA History, http://www.darpa.mil/history.html (last visited Apr. 20, 2009); *see also* Mitch Waldrop, *DARPA and the Internet Revolution*, *in* DARPA: 50 YEARS OF BRIDGING THE GAP 83 (2008), *available at* http://www.darpa.mil/Docs/Internet_Development_2008071 80909255.pdf (last visited Apr. 28, 2009) (Waldrop, *DARPA and the Internet Revolution*).

[87] The four-node network was completed December 5, 1969, and connected the University of California, Los Angeles (UCLA), the Stanford Research Institute, the University of California–Santa Barbara, and the University of Utah. UCLA sent the first transmission to the Stanford Research Institute on October 29, 1969 at 22:30 PST. The first network e-mail using the "username@hostname" format was sent in 1971. *See* Waldrop, *DARPA and the Internet Revolution* at 83.

[88] "[N]o one master computer [was] responsible for sorting the packets and routing them to their destination." *Id.*.

[89] *See id.*

[90] *See id.*

[91] *See id.*

[92] ARPANet was the first of several packet-switched DARPA networks that would be created in the following 15 years. In order to allow these networks to talk to each other, traffic from each discrete network would go through a gateway computer—a "router"—that translated digitized packets from the originating network to the receiving network, so long as both networks met the interface standards. This was achieved January 1, 1983,

when ARPANet began using version 4 of DARPA members Robert Kahn and Vinton Cerf's Transmission Control Protocol/Internet Protocol (TCP/IP), allowing "inter-networking," and arguably marking the birth of the Internet. Another government funded project—NSFNet—made ARPANet obsolete and ARPANet was formally discontinued in 1989. *See id.* at 83–8 5; *see generally* VINTON G. CERF, PACKET SATELLITE TECHNOLOGY REFERENCE SOURCES, REQUEST FOR COMMENTS: 829 (1982), *available at* http://community

[93] *See* Waldrop, *DARPA and the Internet Revolution* at 85. The purpose of this TCP/IP-based network was to link the existing regional networks to a new high-speed network of supercomputer centers. *Id.*

[94] *See id.*

[95] *See id.* The logarithmic increase in traffic caused the lines among the supercomputer centers to require two upgrades in four years. The capacity of the lines was upgraded from 56 kbps in 1987, to 1.5 Mbps in 1988, to 45 Mbps in 1991. *See* NSF, The Launch of NSFNet, http://www.nsf.gov/about/history/nsf0050/internet (last visited Apr. 20, 2009).

[96] *See* THE NSFNET BACKBONE SERVICES ACCEPTABLE USE POLICY § 10 (June 1992), *available at* http://w2.eff.org/Net_culture (last visited Apr. 23, 2009).

[97] The interconnection of the commercial Internet backbones first occurred at a "Commercial Internet eXchange," or CIX, router in 1991. Although this particular router has been retired, there currently are over 150 similar types of nodes worldwide, referred to as Internet eXchange Points or IXPs. *See generally* Farooq Hussain, *Historic Role of The Commercial Internet eXchange Router And Its Impact On the Development of Internet eXchange Points [IXs]* (Oct. 3, 2003), http://www.farooqhussain.org/projects/Historic%20Role%20of%20the%20CIX%20Router%2010- 03 -03 .pdf (last visited Apr. 22, 2009); *see also* EP.NET, llc Information, Exchange Point Information, http://www.ep.net/ep-main.html (last visited May 15, 2009) (providing a worldwide repository of IXPs); Packet Clearing House, Resources, Internet Exchange Directory, https://prefix.pch.net/applications/ixpdir/ (last visited May 19, 2009).

[98] We note that the 2008 Farm Bill was initially enacted on May 22, 2008. *See* Pub. L. No. 110-234, 122 Stat. 923 (May 22, 2008) (May 22, 2008 Bill). The May 22, 2008 Bill, as enacted, however, did not include one title (*i.e.*, Title III—Trade) that Congress had intended to include. The June 18, 2008 Bill corrected this omission by repealing the May 22, 2008 Bill and enacting a statute that includes Title III but otherwise is identical to the May 22, 2008 Bill. *See* Pub. L. No. 110-246, 122 Stat. 1651 (June 18, 2008) (June 18, 2008 Bill or 2008 Farm Bill). The June 18, 2008 Bill specified that it would take effect on the earlier of the enactment date of that Bill or the enactment date of the May 22, 2008 Bill. *See* Pub. L. No. 110-246, § 4(b), 122 Stat. at 1664. We treat May 22, 2009, as the deadline for the rural broadband strategy report required by the Act.

[99] 2008 Farm Bill § 6112(a).

[100] *Id.* § 61 12(b).

[101] BDIA § 101. *See infra* Part V.B (providing additional information on the BDIA).

[102] BDIA § 103(a).

[103] 47 U.S.C. § 157 nt. Since 1996, the Commission has initiated and completed five Section 706 Reports to Congress. *See, e.g., Inquiry Concerning the Deployment of Advanced Telecommunications Capability to All Americans in a Reasonable and Timely Fashion, and Possible Steps to Accelerate Such Deployment Pursuant to Section 706 of the Telecommunications Act of 1996*, GN Docket No. 07-45, Fifth Report, 23 FCC Rcd 9615 (2008) (*Section 706 Fifth Report*), *pet. for recon. pending*. In all five Section 706 Reports, the Commission has found that advanced services are being deployed in a reasonable and timely fashion. *Id.* at 9616, para. 1.

[104] BDIA § 103(c). *See infra* para. 97.

[105] *Comment Sought on International Comparison and Consumer Survey Requirements in the Broadband Data Improvement Act*, GN Docket No. 09-47, Public Notice, 24 FCC Rcd 3908 (2009) (*BDIA Public Notice*).

[106] *See supra* para. 4 (describing passage of the Recovery Act).

[107] Recovery Act § 6001(k)(2).

[108] *Id.* § 600 1(k)(2)(A).

[109] *Id.* § 600 1(k)(2)(B).

[110] *Id.* § 6001(k)(2)(C).

[111] *Id.* § 600 1(k)(2)(D).

[112] *Id.* § 6001(1).

[113] *See infra* Parts IV, V.C (discussing the Recovery Act).

[114] 2008 Farm Bill § 61 12(a)(1)(A).

[115] *Id.* § 6112(a)(1)(B).

[116] The Rural Broadband Access Loan and Loan Guarantee Program provides loans and loan guarantees to fund the "cost of construction, improvement, or acquisition of facilities and equipment for broadband service," with priority given to areas where broadband service is not available or is inadequate. RUS is in the process of promulgating regulations to implement program requirements mandated by the 2008 Farm Bill. *See* 7 C.F.R. § 1738.10–11; USDA Telecommunications Program: Rural Development Broadband Loan and Loan Guarantee Program, http://www.usda.gov/rus/telecom/broadband.htm (last visited May 19, 2009). We note that on April 13, 2009, USDA's Inspector General released an audit report regarding RUS's broadband loan program, finding that the agency had not implemented eight of fourteen recommendations from a 2005 audit report and

expressing concerns about the future of the program. *See* OFFICE OF INSPECTOR GENERAL, USDA, REPORT NO. 09601-8-TE, AUDIT REPORT: RURAL UTILITIES SERVICE BROADBAND LOAN AND LOAN GUARANTEE PROGRAM 4, 10 (2009), *available at* http://www.usda.gov/oig/webdocs/09601-8-TE.pdf. We understand that the 2008 Farm Bill addressed six of these recommendations, and RUS is taking responsive action to the extent it can, regarding the remaining two recommendations. *See id.* at 4.

[117] The Community Connect Grant Program provides financial assistance to unserved areas to connect critical community facilities, such as schools, libraries, hospitals, law enforcement, emergency services, and public safety organizations. Funds may be used to finance the construction and acquisition of facilities to deploy broadband and to purchase end-user equipment. At a minimum, a project must deploy basic broadband to critical community facilities free of charge for two years; offer basic broadband to all residential and business customers within the service area; and provide free access at a community center for at least two years. *See* Broadband Grant Program, 69 Fed. Reg. 44,896, 44,897 (Jul. 28, 2004); 7 C.F.R. § 1739.11–. 12.

[118] The Distance Learning and Telemedicine programs provide a combination of loans and grants to improve educational and health care opportunities. The grant program focuses primarily on connecting students and teachers or medical providers and patients at separate locations, while the loan and combination loan/grant program seeks to fund additional resources to improve medical care and education. Funds generally are used to finance broadband infrastructure, purchase land and buildings, acquire end-user and other equipment, and provide technical assistance and instruction. *See* 7 C.F.R. §§ 1703.121, 1703.130, 1703.140; USDA Telecommunications Program: Loans and Grants, http://www.usda.gov/rus/telecom/RDtelecom-loansandgrants.htm (last visited May 19, 2009).

[119] Recovery Act, Division A, Title I, Rural Utilities Service (RUS Appropriations).

[120] *See, e.g.*, Press Release, IBEC (Feb. 2, 2004), http://www.ibec.net/pdf/IBEC%20Achieves%20Conditional %20RUS%20approval%20for%20its%20BPL%20Solut ion.pdf.

[121] *See* Joan Engebretson, *Funding Fiber to the Farm*, TELEPHONY ONLINE, Feb. 5, 2007, http://telephonyonline.com/mag/telecom_funding (quoting Larry Sevier, Rural Telephone CEO and General Manager, "'As we grew from one rural exchange to 29, we put in the latest technology,' he said. 'All through the process, we used the RUS program. This area is extremely sparsely populated, with an average of two households per square mile.'").

[122] *Id.*

[123] *See generally* Presentation of Scott Zimmer, President of Air Advantage, Using Broadband To Make Rural Michigan A Better Place to Live and Work, http://wireless.fcc.gov/outreach/presentations/saginaw2008/ SuccessStories/SAGINAW%20-%20Air%20Advantage%20Success%20Story.pdf (last visited May 8, 2009).

[124] *See* Press Release, Motorola Corporation (May 7, 2009), http://news.prnewswire.com/ DisplayReleaseContent.aspx?ACCT=1 04&STORY=/www/story/05-07-2009/0005021 331 &EDATE.

[125] *Id.*

[126] *Id.*

[127] *Id.*

[128] Recovery Act, Division A, Title II, National Telecommunications and Information Administration (NTIA Appropriations); Recovery Act § 6001. The BTOP has five enumerated purposes in the Recovery Act: "(1) provide access to broadband service to consumers residing in unserved areas of the United States; (2) provide improved access to broadband service to consumers residing in underserved areas of the United States; (3) provide broadband education, awareness, training, access, equipment, and support to [organizations including schools, libraries, health care providers, and outreach organizations]; (4) improve access to, and use of, broadband service by public safety agencies; and (5) stimulate the demand for broadband, economic growth, and job creation." Recovery Act § 6001(b); *see also* United States Department of Commerce, Information Related to the American Recovery and Reinvestment Act of 2009, http://www.commerce (last visited May 19, 2009).

[129] NTIA may award competitive grants to: "(1) acquire equipment, instrumentation, networking capability, hardware and software, digital network technology, and infrastructure for broadband services; (2) construct and deploy broadband service related infrastructure; (3) ensure access to broadband service by community anchor institutions; (4) facilitate access to broadband service by low-income, unemployed, aged, and otherwise vulnerable populations in order to provide educational and employment opportunities to members of such populations; (5) construct and deploy broadband facilities that improve public safety broadband communications services; and (6) undertake such other projects and activities as the Assistant Secretary finds to be consistent with the purposes for which the program is established." Recovery Act § 6001(g).

[130] Specifically, the Recovery Act states, "To be eligible for a grant under the program, an applicant shall—(1)(A) be a State or political subdivision thereof, the District of Columbia, a territory or possession of the United States, an Indian tribe (as defined in section 4 of the Indian Self-Determination and Education Assistance Act (25 U.S.C. 45 0(b)) or native Hawaiian organization; (B) a nonprofit—(i) foundation, (ii) corporation, (iii) institution, or (iv) association; or (C) any other entity, including a broadband service or infrastructure provider, that the Assistant Secretary finds by rule to be in the public interest. In establishing such rule, the Assistant

Secretary shall to the extent practicable promote the purposes of this section in a technologically neutral manner." *Id.* § 6001(e).

[131] *Id.* § 6001(h)(1).

[132] *Id.* § 6001(h)(2).

[133] *Id.* § 600 1(h)(3).

[134] *Id.* § 6001(j). Section 6001(j) of the Recovery Act states, "Concurrent with the issuance of the Request for Proposal for grant applications pursuant to this section, the Assistant Secretary shall, in coordination with the Commission, publish the non-discrimination and network interconnection obligations that shall be contractual conditions of grants awarded under this section, including, at a minimum, adherence to the principles contained in the Commission's broadband policy statement (FCC 05-15, [1] adopted August 5, 2005)." *Id.*

[135] Recovery Act § 6001(l).

[136] *Id.*

[137] Appendix B provides a non-exhaustive list of these programs.

[138] *See National Broadband Plan NOI* at paras. 113–22.

[139] *See, e.g.*, AFBF Comments at 2; NASUCA Comments at 6; NATOA Comments at 5; Nebraska Commission Comments at 5–6; Pennsylvania Comments at 2; USTA Comments at 2; WISPA Comments at 4.

[140] The terms "Tribal Nation," "Indian Tribe[s]," or "Tribes" refer to any Indian or Alaska Native tribe, band, nation, pueblo, village, or community that is acknowledged by the federal government to constitute a governmental entity necessary to enter into a government-to-government relationship with the United States and thereby be eligible for the programs and services established by the United States for Indians. *See* The Federally Recognized Indian Tribe List Act of 1994, Pub. L. 103-454, 108 Stat. 4791 (1994) (Indian Tribe Act) (requiring the Secretary of the Interior to publish in the Federal Register an annual list of all Indian Tribes which the Secretary recognizes to be eligible for the special programs and services provided by the United States to Indians because of their status as Indians).

[141] *See supra* note 54 (concerning the definition of Indian Country).

[142] For example, Sacred Wind explains that some federal rules and practices fail to recognize the "special characteristics" of Tribal lands. *See* Sacred Wind Comments at 1–2. Sacred Wind states, for example, that the definition of a rural community for the USDA's RUS Community Connect Grant Program "excludes communities that are not registered as Census Designated Places" and only a handful of the 111 chapters of the Navajo Nation are registered. Sacred Wind further explains that federal law requires environmental and archeological surveys before any federally-financed construction on Tribal lands, even if the lands already have been surveyed or are within utility easements. *See* Sacred Wind Comments at 2.

[143] *See Statement of Policy on Establishing a Government-to-Government Relationship with Indian Tribes*, Policy Statement, 16 FCC Rcd 4078 (2000) (*Tribal Policy Statement*).

[144] *See infra* Part VI.C (providing a detailed discussion of Tribal Land bidding credits).

[145] *See* FCC, ITI, http://www.fcc.gov/indians/iti.html (last visited May 19, 2009). Since its inception, ITI has organized informational workshops to provide Tribes and Tribal organizations with information about Commission rules and policies, such as cellular tower siting procedures and broadband deployment that affect the deployment of telecommunications infrastructure and services on Tribal lands. The ITI seeks to offer clear, practical, solution- oriented information the Tribes can use to benefit their communities and enhance their economic and social development regarding deployment and subscribership challenges. In addition, the Commission's Consumer and Governmental Affairs Bureau, through its Liaison to Tribal Governments and the ITI, seeks to: (1) educate and inform Tribes about telecommunications; (2) confer with Tribal representatives about the Commission's rules, regulations, and policies; (3) assist Tribes in networking with each other to develop viable telecommunications systems; and (4) engage in dialogue with Tribes how to overcome about barriers to obtaining telecommunications technologies. The Commission also conducts outreach through various other mechanisms, including publications or advisories tailored specifically to consumers living on Tribal lands, attendance and participation at Tribal conferences and events, and periodic meetings with Tribal representatives. In response to feedback received directly from the Tribes, the Commission has conducted targeted outreach regarding telecommunications issues of interest to Tribes, such as the Lifeline/Link Up programs and broadband deployment.

[146] *See supra* Part III.B (noting a lack of broadband service on Tribal lands); *see also* GAO, CHALLENGES TO ASSESSING AND IMPROVING TELECOMMUNICATIONS FOR NATIVE AMERICANS ON TRIBAL LANDS, GAO-06-189, at 10–13 (2006) http://www.gao.gov/new.items/d06513t.pdf (2006 GAO TRIBAL LANDS REPORT) (explaining that according to the 2000 census, the telephone subscribership rate for Native American households on Tribal lands in the lower 48 states was 68.6%, while for Alaska Native Villages it was 87.0%—both substantially below the national rate of 97.6%).

[147] *See, e.g.*, *Sacred Wind Communications, Inc. and Qwest Corporation, Joint Petition for Waiver of the Definition of "Study Area" Contained in Part 36, Appendix-Glossary of the Commission's Rules, Sacred Wind Communications, Inc., Related Waivers of Parts 36, 54, and 69 of the Communication's Rules*, CC Docket No. 96- 45, Order, 21 FCC Rcd 9227, 9231, para. 9 (2006); *see also Federal-State Joint Board on Universal Service*, CC Docket No. 96-45, Memorandum Opinion and Order, and Further Notice of Proposed

Rulemaking, 15 FCC Rcd 12208, 12217–18, para. 16 (2000) (amending Lifeline and Link-Up assistance rules applicable to eligible residents of Tribal lands, consisting of qualifying low-income consumers living on or near reservations, as defined in 25 C.F.R. § 20.1(r), (v)); *Federal-State Joint Board on Universal Service; Promoting Deployment and Subscribership in Unserved and Underserved Areas, Including Tribal and Insular Areas*, CC Docket No. 96-45, Order and Further Notice of Proposed Rulemaking, 15 FCC Rcd 17122 (2000) (seeking additional comment on extending the enhanced Lifeline and Link-Up measures to qualifying low-income consumers living in areas near reservations to target support to underserved, geographically isolated, and impoverished areas that are characterized by low subscribership).

[148] Consultation and Coordination with Indian Tribal Governments, Exec. Order No. 13175, 65 Fed. Reg. 67249 (Nov. 9, 2000).

[149] *See infra* Part V.B (discussing state initiatives regarding mapping and data collection); *see also* Massachusetts Commission Comments at 4–5; Michigan Commission Comments at 5; NGA CTR. FOR BEST PRACTICES, STATE EFFORTS TO EXPAND BROADBAND ACCESS 4 (2008), http://www.nga.org/Files/pdf/0805BROADBANDACCESS.PDF (NGA, STATE EFFORTS TO EXPAND BROADBAND) (explaining that California, Missouri, Hawaii, and Maryland have used state task forces to evaluate the current state of broadband deployment and identify possible regulatory changes to increase service availability in their states).

[150] For example, the Commonwealth of Virginia's Office of Telework and Broadband Assistance has produced a community guidebook or "toolkit" containing specific tools and resources that communities can leverage when they are assessing how to deploy broadband in their communities. *See, e.g.,* Virginia.gov, Office of Telework Promotion and Broadband Assistance, Community Broadband Toolkit, http://www.otpba.vi.virginia.gov/roundtable_toolkit.shtml (last visited May 19, 2009) (Virginia.gov Toolkit); *see also* Karen Jackson, Director of the Telework Promotion and Broadband Access for the Commonwealth of Virginia, Address at the National Governors Association: State Efforts to Expand Broadband Access (Dec. 18, 2008), (transcript available at http://www.apt.org/events/2008/1218broadband.txt (last visited May 19, 2009)) (describing the efforts of the Commonwealth of Virginia's Center for Innovative Technology, and addressing related broadband issues such as "broadband-friendly zoning," along with expediting the permitting process and waiving or reducing fees, as well as mapping and engaging local communities).

[151] The Joint Conference serves as a forum for an ongoing dialogue among the Commission, state regulators, and local and regional entities regarding the deployment of advanced telecommunications capabilities. It was convened in 1999 as an early step in the Commission's efforts to ensure that advanced services are deployed as rapidly as possible to all Americans, and reconstituted by the Commission on March 19, 2008. *See Commission Seeks Nominations for Federal-State Joint Conference on Advanced Services,* CC Docket No. 99-294, News Release (rel. Mar. 19, 2008); *Federal-State Joint Conference on Advanced Telecommunications Services*, CC Docket No. 99-294, Order, 14 FCC Rcd 17622 (1999). The Joint Conference is comprised of commissioners from state public utilities commissions and from the Commission, and is chaired by the Commission Chairman or his designee.

[152] 2008 Farm Bill § 61 12(a)(1)(C).

[153] Joint Conference Comments at 5.

[154] Connected Nation Comments at 13.

[155] The key to many success stories was the formation of a core local group or team comprised of individuals from both the public and private sectors. For example, a local businessman who partnered with the community and local government succeeded in bringing to Stevenson, Washington, a municipality with a population of 1,300, a Wi-Fi network that blankets the entire downtown area. *See, e.g.*, City of Stevenson, Washington, Wi-Fi Project, http://www.cityofstevenson.com/wifi.html (last visited May 19, 2009). The Blacksburg Electronic Village, an online community center serving the needs of a diverse population, began with a collaboration among the Town of Blacksburg, Virginia Tech, Bell Atlantic (now Verizon), and other interested members of the community. *See* Blacksburg Electronic Village, http://www.bev.net/ (last visited May 19, 2009). In addition, Connected Nation's local leadership teams, comprised of community leaders from key sectors such as healthcare, education, security, and the local private and public sectors, were integral in creating a large public wireless broadband project in a very rural area, the Green River Area Development District of Kentucky. Connected Nation Comments at 13.

[156] *See* Virginia.gov Toolkit, http://www.otpba.vi.virginia.gov/roundtable_toolkit.shtml.

[157] The Connecting Rural Communities project was developed by the Penn State University and University of Minnesota Extension and was funded by the Southern Rural Development Center/Mississippi State University, in partnership with Cooperative State Research, Education, and Extension Service (CSREES)/USDA, as part of "The Rural e-Commerce Extension Initiative: A National Demonstration Project." *See* Connecting Rural Communities, Welcome to Connecting Rural Communities, http://www.connectingcommunities.info/index.cfm (last visited Apr. 29, 2009). This guidebook simplifies the process to several discrete steps: (1) create a team, either an individual or small group, that realizes how connectivity can help future economic and social well-being; (2) learn about broadband; (3) assess what one has, such as what infrastructure and people skills one already has in the community; (4) design a network that can accommodate future growth; (5) create

an action plan; (6) implement and evaluate; and (7) tell the story. *See* Connecting Rural Communities, Getting Started, http://www.connectingcommunities.info/article.cfm?id=198 (last visited Apr. 29, 2009).

[158] *See* MICH. DEP'T OF INFO. TECH., ACTION PLAN FOR DEPLOYING BROADBAND INTERNET TO MICHIGAN LOCAL GOVERNMENTS, http://www.michigan.gov/documents/dit/Broadband_Reference_Guidebook_FINAL_212166_7.pdf (last visited May 19, 2009).

[159] *See, e.g.*, Lone Eagle Consulting Comments at 2 (suggesting the creation of an Office of Broadband Innovation Best Practices to focus on the identification and dissemination of best practices as they emerge); *see also supra* Part IV.A.2 (discussing coordination with state and local authorities).

[160] *See, e.g.*, NMMLP Comments at 1; Mimi Pickering Comments at 1–2; Rural Broadband Policy Group Comments at 7; Pennsylvania Comments at 6; New America Foundation Comments, Attach. at 3; *see also* Broadband Conduit Deployment Act of 2009, H.R. 2428, 111th Cong. § 2 (2009) (requiring the installation of broadband conduit in highway construction projects).

[161] *See* Benton Foundation Comments, Attach. A at 4.

[162] *See* Access Humboldt Comments at 3; CTEF Comments at 1.

[163] *See, e.g.*, ONE ECONOMY CORPORATION, ANNUAL REPORT 2007, at 4, *available at* http://www.one-economy.com/sites/all/files/report2007-low-res.pdf (last visited May 14, 2009) (stating that One Economy has worked to change the affordable housing finance policies in 42 states to foster the inclusion of affordable broadband into the homes of low-income individuals).

[164] *See* John Cromartie & Shawn Bucholtz, *Defining the "Rural" in Rural America*, 28 AMBER WAVES 28, 29 (2008), *available at* http://www.ers.usda.gov/AmberWaves/June08/Features/RuralAmerica.htm (stating that there are more than two dozen rural definitions currently used by Federal agencies); *see also, e.g.*, AT&T Comments at 2– 3; NCTA Comments at 7–8; NRTC Comments at 8; Nebraska Commission Comments at 4; SBA Comments at 7; WISPA Comments at 5.

[165] *See National Broadband Plan NOI* at paras. 116–18.

[166] *See* 2008 Farm Bill § 61 12(a)(2).

[167] The workshops provide communities and organizations in rural America seeking to bring the benefits of broadband to their communities with an opportunity to learn about the resources, programs, and policies of the Commission and USDA. *FCC and USDA to Conduct Regional Educational Workshops on Rural Broadband During 2008*, Public Notice, 23 FCC Rcd 1263 (2008).

[168] *FCC and USDA Launch Web Site Focused on Rural Broadband Opportunities*, News Release, FCC, Feb. 1, 2008. The site provides information on the different technology platforms that can be used to provide broadband service, how to access spectrum necessary for delivery of wireless broadband services, government funding for broadband services, relevant Commission and USDA proceedings and initiatives, including the programs and initiatives created as a result of the recent broadband legislation, and data on broadband deployment. In addition, the site provides instructions on how to locate companies already licensed to provide wireless services in or near specific rural communities, as well as helpful links to other government and private resources related to encouraging broadband opportunities in rural America. *Id.*

[169] 2008 Farm Bill § 61 12(a)(1)(C).

[170] *Id.* § 61 12(a)(1)(D).

[171] *Id.* § 61 12(a)(1)(C).

[172] These include "wired" technologies, using fiber, coaxial cable, copper wires, or BPL; a variety of terrestrial wireless technologies; and also satellite technologies.

[173] We also note that there are a wide variety of broadband applications requiring different functionalities. Some of the next-generation broadband applications require functionalities such as, for example, dramatically faster file transfer speeds for both uploads and downloads, and the ability to transmit streaming video. *See* INFORMATION TECHNOLOGY & INNOVATION FOUNDATION, THE NEED FOR SPEED: THE IMPORTANCE OF NEXT-GENERATION BROADBAND NETWORKS (March 5, 2009), *available at* http://www.itif.org/files/2009-slides-needforspeed.pdf (last visited Apr. 30, 2009).

[174] *See generally* Letter from Michele C. Farquhar, Counsel, LEMKO Corporation, to Marlene H. Dortch, Secretary, FCC, GN Docket No. 09-29 (filed Apr. 16, 2009) (describing advanced wireless network architectures for rural deployments); *see also* Letter from Thomas Cohen, Counsel, Calix, to Marlene H. Dortch, Secretary, FCC, GN Docket No. 09-29 (filed Apr. 7, 2009) (describing new approaches to fiber deployments, particularly in rural areas).

[175] Rural broadband networks are typically in locations that are geographically removed from Internet peering points or nodes. As a consequence, there may be no dedicated, high-capacity middle-mile line available to connect the local ISP with a peering point, and a rural ISP may therefore not be capable of providing robust broadband Internet access to its customers. *See* NECA Comments at 5–6 (finding that 55% of rural telephone companies are located more than 70 miles from a node and 10% are more than 200 miles away).

[176] *See* DigitalBridge Comments ("The lack of middle-mile infrastructure is one of the greatest obstacles to building sustainable rural broadband networks. Many middle-mile facilities were originally built by telephone and cable companies for ordinary telecommunications or cable television services. Rural communities are often

still reliant upon these antiquated copper telephone and cable infrastructures, which lack the capabilities to deliver high-speed, broadband access.").

[177] *See* Letter from Stephen L. Goodman, Counsel, ADTRAN, Inc., to Marlene H. Dortch, Secretary, FCC, WC Docket No. 09-40, app.1 at 3 (filed Apr. 13, 2009) (discussing allowable latency requirements for networks to retain an interactive experience for applications) (ADTRAN April 13, 2009 *Ex Parte* Letter); STUART CHESHIRE, VOLPE, WELTY, ASSET MGMT, LLC, LATENCY AND THE QUEST FOR INTERACTIVITY (Nov. 1996) (finding in 1996 that a signal can move through the Internet backbone from Stanford to Boston and back in less than 80 milliseconds (ms) and discussing theoretical speeds as a matter of physics), http://www.stuartcheshire.org/papers/LatencyQuest.html, *cited in* ADTRAN April 13, 2009 *Ex Parte* Letter, app. 1 at 6 n.5. *See also* Letter from C. Douglas Jarrett, Counsel, API, to Marlene H. Dortch, Secretary, FCC, GN Docket No. 09-51 at 1 (filed May 12, 2009) (stating that the petroleum industry needs "more robust broadband infrastructure . . . in rural areas" and that "the industry's operational/critical infrastructure industry requirements" "operate most efficiently with a maximum latency of 20 ms, and that latency approximating 100 ms adversely impacts these applications").

[178] Interconnected VoIP providers made $514 million in 2006, the first year that the Commission has data for that type of provider. *See* 2008 TRENDS IN TELEPHONE at tbl. 15.4.

[179] VRS is a form of telecommunications relay service that allows individuals with a hearing disability to communicate with voice telephone users using sign language that is transmitted through video equipment connected to a broadband Internet connection. The video link allows a relay agent to view and interpret the user's signed conversation and relay the conversation back and forth with a voice telephone user. *See* 47 C.F.R. 64.601(a)(26); *see generally* 47 C.F.R. 64.60 1 *et seq.*

[180] Educational industries have emerged that are capable of supplying interactive educational experiences, such as for music lessons, over broadband connections. *See, e.g.*, Internet2, Internet2 Member Community Education Initiatives, http://www.internet2.edu/arts/member-education.html (last visited May 20, 2009).

[181] *See, e.g.*, Cisco, WebEx, How it Works for Meetings, http://www.webex.com/how-it-works/for-meetings.html (last visited Apr. 23, 2009) (describing interactive video conferencing that allows users to share files virtually, show presentations, mutually browse, and record the meeting, among other features).

[182] Youtube, Cisco Telepresence Magic, http://www.youtube.com/watch?v=rcfNC_x0VvE (last visited Apr. 23, 2009) (demonstrating telepresence and claiming $1 billion in sales in 3 years for this "green" product).

[183] *See, e.g.*, American Telemedicine Association, Home Telehealth & Remote Monitoring SIG, http://www.americantelemed.org/i4a/pages/index.cfm?pageID=3320 (last visited Apr. 24, 2009) (discussing its home telemonitoring program and providing links that list various abstracts of peer reviewed articles involving interactive home telemedicine applications).

[184] Interactive online gaming is a billion dollar per year industry for one American-based video game developer alone. Seth Schiesel, *An Online Game, Made in America, Seizes the Globe*, N.Y.TIMES, Sept. 5, 2006, at A1 (noting that Blizzard Entertainment, Inc. made over $1 billion the previous year mostly for ongoing subscriptions from the interactive massive multiplayer online game, World of Warcraft). *See also* Alex Pham, *The Work of Play: Video Games Grow Up*, L.A.TIMES, Oct. 19, 2008 (reporting that interactive online gaming is expected to make $8 billion over the next ten years and that the video game industry employs thousands in California); Hiawatha Bray, *Computer vs. The Console: Downloads Give PC Games a Boost Battle for Buyers*, BOSTON GLOBE, Jun. 17, 2008.

[185] *See* CISCO SYSTEMS, INC., CISCO VISUAL NETWORKING INDEX—FORECAST AND METHODOLOGY, 2007–2012, at 4, tbl. 3 (2008), http://www.cisco.com/en/US/solutions/collateral white_paper_c1 1-48 1360.pdf (last visited May 15, 2009).

[186] *See, e.g.*, *supra* notes 178–184 (discussing current and emerging revenue-generating network applications).

[187] *See* ROBERT D. ATKINSON *et al.*, INFO. TECH. & INNOVATION FOUND., EXPLAINING INTERNATIONAL BROADBAND LEADERSHIP app. D at D1, D3 n.5 (May 2008), http://www.itif.org/files/ExplainingBBLeadership.pdf (last visited May 19, 2009) (providing the average data speeds for Japan based on advertised speeds of several major carriers in the country).

[188] *See* Amit Roy Choudhury, *Platform for Next Level of Growth: The Next Generation National Broadband Network Will Change the Way Singaporeans Work, Play and Live*, BUS. TIMES. (SINGAPORE), Mar. 30, 2009 (discussing Singapore's plans to have 100 Mbps to 95% of homes and businesses by 2012 with consideration to upgrade to 1 Gbps shortly thereafter); BBC Monitoring Service, *South Korea: Super-Speed Internet Planned by 2012*, BBC INT'L REP. (MEDIA), Feb. 2, 2009 (reporting that South Korea has committed to have 1 Gbps to the home with 10 Mbps wireless overlay by 2012); Richard Mumford, *RF and Microwaves in Asia: Economies of Scale*, 51 MICROWAVE J., INT'L ED. 118 (2008) (noting that Japan and Hong Kong already have 1 Gbps services to many homes and that South Korea is not far behind); *see also* Leslie Cauley, *FCC Pursues Goal of a Nationwide Affordable, Fast Internet*, USA TODAY, Apr. 8, 2009, at 5B (noting that Australia recently committed to 100 Mbps nationwide).

[189] *Korea to get 1Gbps Wired Internet by 2012*, ELECTRONISTA, Feb. 2, 2009, http://www.electronista.com/articles/09/02/02/korea.to.get.1gbps.web (last visited May 15, 2009); *see also*

Sung So-young, *IT Plan Calls for Big Spending, Jobs*, JOONGANG DAILY, Feb. 2, 2009, http://joongangdaily.joins.com/article/view.asp?aid=2900490.

190 The International Telecommunication Union has published many recommendations concerning the availability and performance of various wireless services such as the fixed and mobile services, as well as satellite services. For example, ITU-R Recommendation F.1400 provides information on the performance and availability requirements and objectives for fixed wireless access to the public switched telephone network. ITU-R Recommendation S. 1783 provides similar data for the satellites. There are numerous other Recommendations available for each service.

191 *See* Kodiak-Kenai Cable Comments at 3–4 (discussing satellite's frequent service disruptions as justification for building the "all weather" Northern Fiber Link); JOHN S. SEYBOLD, INTRODUCTION TO RF PROPAGATION 257 (2005) (discussing satellite rain fade and rain attenuation); SAMI TABBANE, HANDBOOK OF MOBILE RADIO NETWORKS 32(2000) (discussing how wireless technologies are affected by rain).

192 *See* SELECT BIPARTISAN COMMITTEE TO INVESTIGATE THE PREPARATION FOR AND RESPONSE TO HURRICANE KATRINA, US HOUSE OF REPRESENTATIVES, A FAILURE OF INITIATIVE: THE FINAL REPORT OF THE SELECT BIPARTISAN COMMITTEE TO INVESTIGATE THE PREPARATION FOR AND RESPONSE TO HURRICANE KATRINA (Feb. 15, 2006), http://katrina.house.gov/full_katrina_report.htm (noting that Katrina's destruction of pole mounted backhaul facilities was a major contributing factor to the long-term disruption of communications in the wake of that hurricane).

193 *See, e.g.*, Ellen Nakashima & R. Jeffrey Smith, *Electric Utilities May Be Vulnerable to Cyberattack*, WASH. POST, Apr. 9, 2009, at A4.

194 JOHN S. SEYBOLD, INTRODUCTION TO RF PROPAGATION 2, 6 (2005). *See* Valerie Fast Horse Comments at 3 (urging the Commission to allow the use of new low-frequency spectrum that can penetrate trees and make non-lineof-site communication possible). For example, as compared to providing service at higher frequencies, the unique propagation characteristics of the 700 MHz band mean that fewer towers will be needed to serve a given license area, thereby lowering infrastructure costs. *See Service Rules for the 698-746, 74 7-762 and 777-792 MHz Bands*; *Revision of the Commission's Rules to Ensure Compatibility with Enhanced 911 Emergency Calling Systems*; *Section 68.4(a) of the Commission's Rules Governing Hearing Aid-Compatible Telephones*; *Biennial Regulatory Review—Amendment of Parts 1, 22, 24, 27, and 90 to Streamline and Harmonize Various Rules Affecting Wireless Radio Services*; *Former Nextel Communications, Inc. Upper 700 MHz Guard Band Licenses and Revisions to Part 27 of the Commission's Rules*; *Implementing a Nationwide, Broadband, Interoperable Public Safety Network in the 700 MHz Band*; *Development of Operational, Technical and Spectrum Requirements for Meeting Federal, State and Local Public Safety Communications Requirements Through the Year 2010*, WT Docket Nos. 06-150, 96-86, 03- 264, 01-309, 06-169, CC Docket No. 94-102, PS Docket No. 06-229, Report and Order and Further Notice of Proposed Rulemaking, 22 FCC Rcd 8064 (2007) (*700 MHz Report and Order*); Second Report and Order, 22 FCC Rcd 15289, 15348, para. 154 (2007), *recon. pending* (*700 MHz Second Report and Order*).

195 *See* GEORGE OU, THE INFORMATION TECHNOLOGY AND INNOVATION FOUNDATION, MANAGING BROADBAND NETWORKS: A POLICYMAKER'S GUIDE 10(2008).

196 *See id.*

197 *See supra* Part III.B (discussing the current dearth of data regarding the state of broadband in America).

198 ACS Comments at 4; Connected Nation Comments at 4 ("[I]dentifying and mapping unserved and underserved areas—is a necessary factual guide for any rural broadband strategy."); NASUCA Comments at 5 ("Another key to a national rural broadband strategy is knowing where broadband service is available, and at what speeds and at what prices."); NATOA Comments at 6.

199 *See Development of Nationwide Broadband Data to Evaluate Reasonable and Timely Deployment of Advanced Services to All Americans, Improvement of Wireless Broadband Subscribership Data, and Development of Data on Interconnected Voice over Internet Protocol*, WC Docket No. 07-3 8, Report and Order and Further Notice of Proposed Rulemaking, 23 FCC Rcd 9691, 9709–11, para. 36 (2008) (*2008 Broadband Data Gathering Order*).

200 *See id.* at 9710, para. 37.

201 *See infra* Part V.C.

202 *See* Microsoft Comments at 2 (arguing that schools need capacity sufficient to deliver high-quality video to the classroom, but noting that currently, there is no reliable data on the number or type of schools that have such capacity).

203 As discussed above, this backbone is part of the foundation needed to make broadband services available in all areas. *See supra* Part V.A (regarding technological considerations).

204 *See supra* note 103 (regarding the Commission's section 706 Reports).

205 *See, e.g.*, *Section 706 Fifth Report*, 23 FCC Rcd at 9685 (Commissioner Copps, dissenting).

206 *See 2008 Broadband Data Gathering Order*, 23 FCC Rcd at 9695–99, paras. 10–15.

207 The Commission updated the broadband reporting tiers to include upload and download speeds of: (1) greater than 200 kbps but less than 768 kbps; (2) equal to or greater than 768 kbps but less than 1.5 Mbps; (3) equal to

or greater than 1.5 Mbps but less than 3.0 Mbps; (4) equal to or greater than 3.0 Mbps but less than 6.0 Mbps, (5) equal to or greater than 6.0 Mbps but less than 10.0 Mbps; (6) equal to or greater than 10.0 Mbps but less than 25.0 Mbps; (7) equal to or greater than 25.0 Mbps but less than 100.0 Mbps; and (8) equal to or greater than 100 Mbps—for a total of 72 tiers. *2008 Broadband Data Gathering Order*, 23 FCC Rcd at 9700–01, para. 20. The previous five "speed tiers" were based on the transfer rate of the connection's faster direction and were divided as follows: (1) greater than 200 kbps but less than 2.5 Mbps; (2) greater than or equal to 2.5 Mbps but less than 10 Mbps; (3) greater than or equal to 10 Mbps but less than 25 Mbps; (4) greater than or equal to 25 Mbps but less than 100 Mbps; and (5) greater than or equal to 100 Mbps. *Local Telephone Competition and Broadband Reporting*, WC Docket No. 04-141, Report and Order, 19 FCC Rcd 22340, 22347–48, para. 14 (2004).

208 *See 2008 Broadband Data Gathering Order*, 23 FCC Rcd at 9698–99, para. 16.

209 *See id.* at 9703–04, para. 23.

210 These maps are provided through a contract with American Roamer. *See supra* note 47.

211 *Id.*

212 *See 1998 Biennial Regulatory Review—Annual Report of Cable Television Systems, Form 325, filed Pursuant to Section 76.403 of the Commission's Rules*, CS Docket No. 98-61, Report and Order, 14 FCC Rcd 4720 (1999).

213 *E.g., Annual Report and Analysis of Competitive Market Conditions with Respect to Domestic and International Satellite Communications Services*, IB Docket No. 07-252, Second Report, 23 FCC Rcd 151570 (2008).

214 *See supra* Part III.D (regarding recent legislative developments).

215 BDIA § 103(a).

216 *Id.* § 103(c).

217 Specifically, section 103(b) of the BDIA states: "As part of the assessment and report required by section 706 of the Telecommunications Act of 1996 (47 U.S.C. 157 note), the Federal Communications Commission shall include information comparing the extent of broadband service capability (including data transmission speeds and price for broadband service capability) in a total of 75 communities in at least 25 countries abroad for each of the data rate benchmarks for broadband service utilized by the Commission to reflect different speed tiers." *Id.* § 103(b)(1).

218 *See Comment Sought on International Comparison and Consumer Survey Requirements in The Broadband Data Improvement Act*, GN Docket No. 09-47, Public Notice, 24 FCC Rcd 3908 (2009).

219 CALIFORNIA BROADBAND REPORT at 7.

220 *Id.*

221 XO/Nextlink Comments at 2.

222 *See supra* at 65; *see also* Virginia.gov Toolkit.

223 COMMONWEALTH'S BROADBAND ROUNTABLE, FINAL REPORT (Sept. 9, 2008) *available at* http://www.otpba.vi.virginia.gov/pdf/Governor_report.pdf (last visited May 18, 2009) (VIRGINIA BROADBAND ROUNDTABLE FINAL REPORT); *see also* XO/Nextlink Comments at app. A.

224 *See, e.g.*, Connected Nation Comments at 1–3.

225 Connected Nation at Comments at 1. ConnectKentucky, a pilot program that led to Connected Nation, worked with the private sector to create an online map of broadband availability in Kentucky and to conduct consumer surveys to identify barriers to broadband adoption in that state. Connected Nation Comments at 1, 12–13.

226 Connected Nation Comments at 8; *see* Harris Corporation Comments at 4 (supporting public-private partnerships because private companies can contribute "managerial efficiencies and technological proficiency," while government can "provide the economic incentive for private companies to deploy broadband in rural areas, especially where such an incentive has never previously existed").

227 *See infra* Part V.C (regarding stimulating demand).

228 *See, e.g.*, NGA, STATE EFFORTS TO EXPAND BROADBAND (stating that the California Public Utilities Commission allocated $100 million to provide matching funds of up to 40% of the total project cost of broadband infrastructure deployment projects in California); *id.* (stating that the Vermont Telecommunications Authority has the authority to issue up to $40 million in state-backed bonds to finance the construction of broadband infrastructure in Vermont).

229 *See* Pennsylvania Comments at 2. *See generally* BROADBAND INVESTMENT FOR ECONOMIC RECOVERY: PERSPECTIVES OF AN AD-HOC GROUP OF STATE BROADBAND ENTITIES 5–6 (Feb. 9, 2009), *attached to* Massachusetts Comments (describing a variety of approaches that states have used to encourage broadband deployment, including creating specific agencies to focus on broadband; fostering public-private cooperation and co-investment; funding access with grants or loans; streamlining rights of way; mapping broadband facilities; promoting education efforts; encouraging inclusion of telecommunications infrastructure in the planning and deployment of capital improvement projects; and setting goals for broadband deployment and use).

230 HierComm Comments at 2–3.

231 Connected Nation Comments at 8.

232 Recovery Act § 6001(l).

233 *Id.*

[234] *See, e.g.*, Connected Nation Comments at 8 (identifying Alabama, Arkansas, California, Colorado, Hawaii, Illinois, Kansas, Kentucky, Maine, Massachusetts, Minnesota, North Carolina, Ohio, Tennessee, Virginia, and West Virginia); *see* Broadband Access in Illinois, Institute for Regulatory Policy Studies, Illinois State University, August 2007, *available at* http://www.irps.ilstu.edu/broadband (last visited March 20, 2009); e-NC, Broadband Access in North Carolina, http://e-ncbroadband.org/ (last visited April 1, 2009).

[235] *See, e.g.*, Connected Nation Comments at 1, 7 (explaining that Connected Nation is a non-profit organization with operations in nine states that grew out of the ConnectKentucky pilot initiative).

[236] *See, e.g.*, CALIFORNIA BROADBAND REPORT; VIRGINIA BROADBAND ROUNDTABLE FINAL REPORT.

[237] *See, e.g.*, VT DEP'T OF PUB. SERV., APPROXIMATE BROADBAND AVAILABILITY IN VERMONT—2006 (2007), http://publicservice.vermont.gov/cable/broadband (providing a map of broadband availability).

[238] *See, e.g.*, e-NC Authority, Who We Are, http://www.e-nc.org/WhoWeAre.asp (last visited April 1, 2009).

[239] *See, e.g.*, Comments of Connected Nation, WC Docket No. 07-3 8 at 35.

[240] *See, e.g.*, American Public Power Ass'n. Comments, WC Docket No. 07-38, at 5; Consumers Union Comments, WC Docket No. 07-38, at 6–7 n.4, 17; Kentucky Municipal Utilities Reply, WC Docket No. 07-38, at 6, 13. *But see* Connected Nation Comments, WC Docket No. 07-38, at 6, 10, 12-13 (stating that the vast majority of Connected Nation's funding is from public sources and it is unbiased and that its ability to keep data confidential increases voluntary disclosures by providers); Connected Nation Reply, WC Docket No. 07-38, at ii–iii, 12–18.

[241] *See supra* Part IV.A (discussing interagency coordination).

[242] *See supra* Part V.B. 1 (discussing broadband data collection).

[243] *See* Connected Nation Comments at 9 (stating that according to a Connected Nation study, approximately 90% of households in the U.S. have access to some form of broadband service, but only 50% of households choose to subscribe); NCTA Comments at 15, Attach. 1 at 2 (estimating that there are approximately 35 million households in the U.S. that have access to broadband but do not currently use it).

[244] *See, e.g.*, Rural Broadband Policy Group at 5 (noting the diversity of rural America in "terrains, cultures, foods, peoples, and knowledge" and opining that there is no "one size fits all solution"); Access Humboldt Comments at 1 (agreeing with the Rural Broadband Policy Group).

[245] For example, one estimate of the demand for Internet access services indicates that approximately 29% of the U.S. population in 2007 did not use the Internet. Internet use is defined as a household with a subscription to either broadband or dial-up or use of a terminal outside the home to access the Internet. *See* U.S. CENSUS BUREAU, TABLE 1118. HOUSEHOLD INTERNET USAGE IN AND OUTSIDE OF THE HOME, BY SELECTED CHARACTERISTICS: 2007, http://www.census.gov/compendia/statab/tables/09s1118.pdf (last visited May 19, 2009) (TABLE 1118). TIA, citing to the Pew Internet and American Life Project, claims that approximately one quarter of the population does not use the Internet. TIA Comments at 4.

[246] *See, e.g.*, NCTA Comments at 15; TIA Comments at 4.

[247] *See, e.g.*, NCTA Comments, Attach. at 10 (citing John B. Horrigan, *Obama 's Online Opportunities II: If You Build It, Will They Log On?*, at iii, 12 (2009))

[248] *See* CONNECTED NATION, CONSUMER INSIGHTS TO AMERICA'S BROADBAND CHALLENGE 11 (Oct. 13, 2008), *available at* http://www.connectednation.com/_documents/ConsumerInsightsBroadbandChallenge_20081013.pdf (CONNECTED NATION REPORT), *cited by* TIA Comments at 4. According to NCTA, nearly half of the population that does not subscribe to broadband says it does not need such a connection. NCTA Comments, Attach. at 10 (citing CONNECTED NATION REPORT at 2).

[249] *See* 2008 PEW BROADBAND ADOPTION STUDY at 12–13.

[250] According to TIA, 34% of rural consumers do not have Internet access due to the lack of a computer. TIA Comments at 4.

[251] In the *National Broadband Plan NOI*, the Commission sought comment on the extent to which it should encourage or subsidize broadband subscription in areas where service is already available. *National Broadband Plan NOI*at paras. 27, 39–41, 54.

[252] *See* NCTA Comments, Attach. A at 11 (citing 2008 PEW BROADBAND ADOPTION STUDY at ii, 11, to show that 35 percent of consumers surveyed responded that the price of broadband would have to fall for them to subscribe).

[253] *See* CFA/CU Comments at 2. Another estimate shows that for households with an income in the $20,000 to $24,999 range, 28.9% had broadband subscriptions, whereas 77% of the households with incomes in the $75,000 to $99,999 range subscribed to broadband services. *See* NTIA, HOUSEHOLDS USING THE INTERNET. The Benton Foundation also provides an estimate of decreasing broadband adoption by income. For Americans with incomes under $20,000 annually, broadband penetration has fallen from 29% in 2007 to 25% in 2008. Benton Foundation Comments at 6.

[254] *See, e.g.*, Benton Foundation Comments at 8, Attach. at 7, 10; Rural Broadband Policy Group Comments at 1; Connected Nation Comments at 3, 9; NASUCA Comments at 5–6; NCTA Comments at 15, Attach. at 1–2, 4.

[255] Recovery Act § 600 1(b).

[256] Recovery Act, Division A, Title II, National Telecommunications and Information Administration (Broadband Technology Opportunities Program Appropriations).

[257] *National Broadband Plan NOI* at paras. 5 5–57.

[258] *Id.*

[259] *Id.*

[260] For example, Connected Nation operates community-based organizations that sponsor computer training and education to increase the value of accessing the Internet for businesses. Connected Nation Comments at 10. Connected Nation also sponsors grassroots "eCommunity Leadership Teams" comprised of community leaders from key sectors that develop and implement technology promotion plans within their communities. Connected Nation Comments at 11–12. However, we note that these programs are not necessarily specifically directed to rural areas.

[261] *See* ALA Comments at 3.

[262] *See* ALA Comments at 3, 6.

[263] Public computing facilities can provide numerous benefits to users in the community. *See generally* U.S. Dep't of Hous. and Urban Dev., Multifamily Housing – Neighborhood Networks, http://www.hud.gov/offices/hsg/mfh/nnw/nnwaboutnn.cfm (last visited May 19, 2009) (describing HUD's Neighborhood Networks initiative, which encourages property owners and managers to open onsite, multiservice technology centers, which has resulted in more than 1,400 Neighborhood Networks centers across the U.S. that offer job-training classes in computer skills, resume writing, and interviewing techniques; job placement services; General Educational Development (GED) preparation; English as a Second Language (ESL) classes; programs that allow seniors to become familiar with computers and use them to better their lives, whether through staying in touch with family and friends via e-mail or searching for healthcare benefits online; and computer access for children whose families cannot afford their own).

[264] *See* Pennsylvania Comments at 2.

[265] *See* PA. DEP'T OF CMTY. AND ECON. DEV., BROADBAND OUTREACH AND AGGREGATION FUND, PROGRAM GUIDELINES 1 (June 2006), http://www.newpa.com/download.aspx?id=989.

[266] For example, Pennsylvania has established a special program that aggregates demand in communities looking to achieve an agreed upon deployment threshold. *See* Pennsylvania Comments at 2.

[267] *See, e.g.*, NCTA Comments at 15. For example, recognizing that customer premises equipment (CPE) costs can be a barrier to broadband implementation and adoption, the National Rural Telecommunications Cooperation (NRTC) recently announced a subsidized leasing program for its members to pass on to customers in their rural markets. Thus, WildBlue satellite service customers can now gain access to the service for just $99, which includes CPE and installation. Under this program, the actual CPE cost of $400 is subsidized by NRTC and its members. This program has been launched as a test to help expand access to broadband. NRTC Comments at 6.

[268] *See High-Cost Universal Service Support; Federal-State Joint Board on Universal Service; Lifeline and Link Up; Universal Service Contribution Methodology; Numbering Resource Optimization; Implementation of the Local Competition Provisions in the Telecommunications Act of 1996; Developing a Unified Intercarrier Compensation Regime; Intercarrier Compensation for ISP-Bound Traffic; IP-Enabled Services*, CC Docket Nos. 01-92, 99-200, 99-68, 96-98, 96-45, WC Docket Nos. 06-122, 05-337, 04-36, 03-109, Order on Remand and Report and Order and Further Notice of Proposed Rulemaking, FCC 08-262, para. 40; app. A, paras. 64–91; app. C, paras. 60–87 (rel. Nov. 5, 2008) (*November 2008 Further Notice*). Specifically, the Commission sought comment on making available $300 million each year for three years to enable eligible telecommunications carriers to provide discounts on broadband Internet access service and the necessary access devices to low-income consumers. *See id.* The Commission's existing Low-Income program, discussed *infra*, helps ensure that the costs of starting and continuing telephone service do not prevent low-income consumers from receiving telephone service. *See infra* Part VI.A (discussing Universal Service).

[269] *See* HierComm Comments, Attach. 3, tbl. 18 (listing equipment and costs for wireless broadband deployers).

[270] *See generally* Fibertech/KDL Comments (discussing pole make ready costs and rents).

[271] *See generally* HierComm Comments, Attach. 1 (discussing assessments made by deployers).

[272] *See* GAO, BROADBAND DEPLOYMENT IS EXTENSIVE THROUGHOUT THE UNITED STATES, BUT IT IS DIFFICULT TO ASSESS THE EXTENT OF DEPLOYMENT GAPS IN RURAL AREAS, GAO-06-426, at 19 (May 2006) (GAO BROADBAND REPORT); Embarq Comments at 3 (stating that the high cost of deployment remains the "principal barrier to rural advanced services"); MSS/ATC Coalition Comments at 3 (asserting that both wireline access and terrestrial wireless networks are economically impractical in remote areas).

[273] *See, e.g.*, Embarq Comments at 12 n.18 (noting the cost differences per line between rural customers that live in a cluster and those in outlying areas); Stephouse Networks Comments at 2 (describing the low population densities and mountainous terrain of its service areas).

[274] For example, commenters point out that deployment in rural areas can be especially expensive and complicated because of relatively inaccessible, rugged terrain. *See* Valerie Fast Horse Comments at 3 (describing low and thickly forested areas where there are "no possible means to deliver broadband"); PBIA Comments at 6 (stating that because of their remoteness and challenging terrain, the communities of western Alaska have not been accessible to fiber networks). This terrain often includes mountains or forests that can obstruct wireless

technologies that require line-of-sight transmission. GAO BROADBAND REPORT at 19; *see also supra* Part V.A (discussing Technological Considerations).

[275] *See, e.g.*, NRECA Comments at 1 (stating that "[l]ow population densities coupled with the issue of traversing vast expanses of remote and often rugged topography" create a financial barrier to rural broadband deployment); Embarq Comments at 13 ("In rural America, however, high-costs and low population densities make investment and provision of broadband service uneconomic (just as has been and still is the case with basic voice service pursuant to carrier-of-last-resort obligations).").

[276] *See supra* Part V.A (discussing Technological Considerations).

[277] *See, e.g.*, VT. DEP'T OF PUB. SERV., UNDERSTANDING BROADBAND DEPLOYMENT IN VERMONT 14(2007), http://publicservice.vermont.gov/Broadband/Broadband%20Deployment%20in%20Vermont%20Final.pdf (VT. BROADBAND REPORT) (stating that in many rural areas of Vermont the average build-out cost for fiber-to-thepremise networks would likely be close to $4,000 per subscriber).

[278] NATIONAL EXCHANGE CARRIER ASSOCIATION, THE PACKET TRAIN NEEDS TO STOP AT EVERY DOOR 13 (2006) (estimating the cost of upgrading a traditional telephone line to DSL as follows: $988 per line for the telephone company's central office service area; $1,083 per line for the mid-range service area; and $4,865 per line for areas not served by central offices).

[279] *See, e.g.*, VT. BROADBAND REPORT at 13 (estimating the cost for cable line extensions in Vermont at $20,000 per mile).

[280] The local access network, either wireline (*e.g.*, cable, DSL, fiber) or wireless (*e.g.*, WiMAX, EV-DO), connects a local network in a particular area, which could range in size from a single house or farm in a rural area to a small town.

[281] The backhaul network, which can be either wireline or wireless, connects the local network to an Internet backbone connection point.

[282] *See* NECA Comments at 5–6 (noting from its 2001 study that 55% of rural switches are more than 70 miles from an Internet Backbone Provider node and 10% are more than 200 miles away).

[283] *See* DigitalBridge Comments at 8–9 ("Many middle-mile facilities were originally built by telephone and cable companies for ordinary telecommunications or cable television services. Rural communities are often still reliant upon these antiquated copper telephone and cable infrastructures, which lack the capabilities to deliver high-speed, broadband access."); *see also, e.g.*, City of Shafter Comments at 4 n.6.

[284] Commenters state that if an incumbent LEC does not build or tariff any high-capacity middle-mile lines, the ISP may have its data come to a crawl or stall as the data attempt to join the greater communications network using the LEC's low-capacity facilities. *See* DigitalBridge Comments at 8–9; GCI Comments at 3.

[285] *See* Microsoft Comments at 4 ("We understand from some network operators, for example, that the local cost of upgrading wires and distributing broadband is not a hurdle. The marginal cost of adding subscribers to broadband systems can be more than covered by subscriber fees. However, rural or remote providers cannot take advantage of those economics, because the cost of acquiring high-capacity facilities between the Internet backbone and the community is too high. Once this hurdle is overcome, we believe that in many instances the market can remedy the problem of delivering affordable broadband to Main Street and neighborhoods.").

[286] *See* NECA Comments at 6 n. 15 (quoting Verizon as stating that "[i]n some rural high-cost areas, however, the cost of the additional transport mileage is high enough to impinge on a rural broadband provider's ability to offer services in those areas"); *see also* OPASTCO Comments at 8–9 ("Another significant obstacle that rural ILECs face in deploying broadband to additional rural consumers and increasing the broadband speeds that they offer is the high price of access to the Internet backbone. The price of backbone access is based upon mileage, among other factors, and the further removed a carrier is from a backbone facility, the higher the price they must pay. Thus, rural ILECs who are on average much further away from these facilities face higher than average costs for Internet backbone access. In addition, rural ILECs that wish to upgrade their broadband networks to provide their customers with higher speeds must also upgrade their backbone access, incurring even higher costs. Moreover, the majority of rural ILECs have only one choice of provider for connecting to the Internet backbone.") (citations omitted).

[287] Several commenters have expressed concern that high backhaul costs present a barrier to broadband deployment in rural areas. *See* DBC Comments at 8–9 (urging the Commission to ensure that affordable backhaul is available for rural operations); Fibertech/KDL Comments at 9 (stating that backhaul costs can be "prohibitively expensive" in rural areas); GCI Comments at 3 (citing cost-effective middle-mile transport as essential to broadband deployment in rural Alaska); ACA Comments at 3 (arguing that grant and loan programs should allow operators to build middle- mile fiber backhaul infrastructure to increase speeds and lower operational costs).

[288] *See* NECA Comments at 6–7 (encouraging partnerships and consortia by interested groups); *see also generally* ALA Comments at 2, CFA/CU Comments at 4, Microsoft Comments at 1–4 (all recommending using government funds to rollout fiber to anchor institutions, thereby providing incentives for last mile providers to buildout later).

[289] *See* NECA Comments at 6 nn. 15–16.

[290] *See* Fibertech/KDL Comments at 2-3; New America Foundation Comments, Attach. at 3 (arguing that fiber should be deployed with highway construction); Letter from Michele C. Farquhar, Counsel to Sprint Nextel Corporation, Special Counsel to FiberTower Corporation and Rural Telecommunications Group, Inc. to Marlene H. Dortch, Secretary, FCC, GN Docket No. 09-29, ET Docket Nos. 04-186, 02-3 80 at 1-2 (filed Apr. 13, 2009) (Sprint Nextel *et al.* April 13, 2009 *Ex Parte* Letter) (asking the Commission to expand the amount of licensed spectrum that can be used for wireless backhaul services, such as White Spaces, which lie fallow in rural areas); City of Shafter Comments at 4 n.6 (noting that despite building a fiber network and having two Internet backbone providers nearby, the city had difficulty connecting directly to an Internet Backbone Provider).

[291] *See* NTCA Comments at 26–27; Sprint Comments at 8; OPASTCO Comments at 8–9.

[292] *See, e.g., Thirteenth CMRS Competition Report* at para. 100 (finding that the inability of new firms to borrow sums sufficient to finance efficient start-ups can be an entry barrier).

[293] *See id.*

[294] *See supra* Part III.B (The State of Broadband in Rural Areas).

[295] *See supra* Part III.A (Rural Broadband: Why it Matters).

[296] *See Federal-State Joint Board on Universal Service*, CC Docket No. 96-45, Report and Order, 12 FCC Rcd 8776, 8783, para. 8 (1997) (subsequent history omitted) (*Universal Service First Report and Order*) ("At the simplest level, increasing the number of people connected to the telecommunications network makes the network more valuable to all of its users by increasing its usefulness to them.").

[297] This "network effect" is a reason why the Commission has an explicit universal service program to ensure that people are connected to the telephone network. *See id.* A "network effect" may be illustrated by observing that all else being equal (with two non-interconnected networks), a customer is more likely to choose a network that serves 80% of the population, instead of one that serves 20% of the population, because the larger network is more likely to serve more people that the customer may want to call. *See* JONATHAN E. NUECHTERLEIN & PHILIP J. WEISER, DIGITAL CROSSROADS 4–5 (2005); Big Think Strategies Comments at 7.

[298] Aside from network effects, larger networks also have other indirect benefits, or positive externalities, that arise merely because the network reaches more users. In the *Universal Service First Report and Order*, the Commission noted that "[i]ncreasing subscribership also benefits society in ways unrelated to the value of the network per se. For example, all of us benefit from the widespread availability of basic public safety services, such as 911." *See Universal Service First Report and Order*, 12 FCC Rcd at 8783, para. 8.

[299] A broadband-connected Internet also produces network effects. Unlike dial-up connections, broadband connections can be used to create "two-way networks" for applications like VoIP and video conferencing. *See supra* notes 178–184 (listing interactive broadband applications). In addition to these network effects, ubiquitous broadband would produce other positive externalities. *See* Big Think Strategies Comments at 17. For instance, a vastly interconnected broadband network would allow Next Generation 911 to be deployed throughout the country to the benefit of those who live in or may visit areas that otherwise would lack access to broadband services. *See* DEP'T OF TRANS., NEXT GENERATION 9-1-1 SYSTEM PRELIMINARY CONCEPT OF OPERATIONS (2005), http://www.its.dot.gov/NG911/pdf/ConOps.pdf. Dial-up connections are increasingly impractical for many Internet applications; thus, dial-up users do not add as great a value to the network and cannot contribute to as many of the positive externalities as broadband-connected users. *See generally supra* Part III.A (describing the benefits of broadband).

[300] *See, e.g.,* MISS. CODE ANN. § 5 7-87-1 *et seq.* (providing tax credits to providers that deploy broadband equipment, ranging from five to 15% of the cost of the broadband equipment, with the higher tax credits going to those providers that deploy equipment in the least populous regions of the state); FLA. STAT. § 220.183 (providing receive income tax credits to broadband providers for any project designed to increase a community's access to broadband facilities, including those communities located in rural areas); HAW. REV. STAT. § 235-110.51 (providing that up to 4 % of a commercial building's renovations costs may be deducted if the building's telecommunications facilities are upgraded to high-speed telecommunications systems that can provide Internet access, direct satellite communications access, and videoconferencing facilities).

[301] *See* Chuck Huckaby, *Maryland Broadband Cooperative Connects Rural Counties*, WORK AT HOME BUSINESS OPPORTUNITIES, Oct. 11, 2007, http://work-at-home.business-opportunities.biz/2007/10/1 1/maryland-broadbandcooperative-connects-rural-counties/ (last visited May 13, 2009) (describing the Maryland Broadband Cooperative (MDBC)'s connection of nine counties via fiber optic cable and plans to lay fiber optic cable through Western Maryland); Ryan Bentley, *Cooperative Ready to Begin Making Broadband a Reality by Early Summer*, PETOSKEY NEWS REVIEW, Apr. 23, 2008, http://www.petoskeynews.com/articles/2008/04/23/news/ doc480f396d259ba6591 16224.prt (last visited May 13, 2009) (announcing plans to seek rural development loans from the USDA to fund broadband infrastructure deployment to residents and businesses in 12 rural counties in Northern Michigan); SCTC Comments at 2 (stating that it offers nearly ubiquitous DSL service, ranging in speeds from 1.5 Mbps download/512 Kbps upload to 6 Mbps download/1 Mbps upload, to a broad region that includes Salem, Oregon).

[302] A number of previously unserved or underserved communities have sponsored or deployed local broadband networks (consisting of either fiber, cable, or wireless networks). *See infra* notes 303–304 (listing communities

that have sponsored or deployed local broadband networks); *see also* Allan J. Hastings Comments at 2 (stating that public development of shared digital roads, managed just as traditional roads are managed, lowers the costs for private sector service providers and creates new opportunities for start-up companies and incumbent providers that have said they cannot afford to build fiber networks in rural areas); Big Think Strategies Comments at 7 (noting no strong objection to government owned and operated infrastructure since governmental units in the United States today typically provide roads, water and sewer service, and airports as well as a great deal of gas distribution, electrical service, and health care).

[303] Communities such as Minneapolis, Minnesota, Lafayette, Louisiana, and Fort Wayne, Indiana, have deployed high speed networks. *See* Benton Foundation Comments at 11, nn. 19–20; Esme Vos, *Muniwireless Updates List of Cities and Counties with Large Wi-Fi Networks*, MUNIWIRELESS, Mar. 28, 2009, http://www.muniwireless.com/ 2009/03/28/muniwireless-list-of-cities-with-wifi/ (article includes a link to the list of cities, *available at* http://www.muniwireless.com/reports/Mar-28-2009-list-of-cities.pdf).

[304] We note that counties in rural areas such as Allegheny County, Maryland; Craven County, North Carolina; and Cambria County, Pennsylvania, are just a few that have sponsored deployment of county-wide networks. AllCoNet.org, What is AllCoNet, http://www.allconet.org/ (last visited Apr. 29, 2009); *Craven County, North Carolina Goes Wireless: Rural Wi-Fi Connects Schools, Towns*, MUNIWIRELESS, June 6, 2008, http://www.muniwireless.com/2008/06/06/craven-county-north-carolina-goes-wireless-rural-wi-fi-connects-schoolstowns; Cambria Connected, Introducing a New Way to Communicate in Cambria County, http://www.cambriaconnected.net/ (last visited May 11, 2009); Melissa Block, *Widening the Internet Highway to Rural America*, NATIONAL PUBLIC RADIO, Dec. 14, 2005, http://www.npr.org/templates/story/ story.php?storyId=5053488 (last visited May 19, 2009).

[305] *See* Craig Dingwall, *Municipal Broadband: Challenges and Perspectives*, 56 FED. COMM. L.J. 92, 92 (2006).

[306] *See* FED. TRADE COMM'N STAFF, MUNICIPAL PROVISION OF WIRELESS INTERNET 28–35 (2006), http://www.ftc.gov/os/2006/10/V060021municipalprovwirelessinternet.pdf (FTC REPORT).

[307] *See* FTC REPORT at 30; Dingwall, *supra* note 305, at 92.

[308] *See, e.g.*, Art Menius Comments at 2 (urging adoption of a federal rural broadband policy that encourages local ownership of broadband infrastructure and public ownership of broadband infrastructure in areas where private entities are not forthcoming in building broadband infrastructure). We note that at least 35 states have considered limiting municipal broadband, and at least 19 states have enacted legislation specifically addressing municipal broadband. *See* A.H. Tapia & J.A. Ortiz, Municipal Responses to State-Level Broadband Internet Policy 2 (Sept. 2006), *available at* http://web *see also, e.g.*, ARK. CODE ANN. § 23-17-409(b)(1) (West 2009); COLO. REV. STAT. ANN. § 29-27-201 (West 2009); FLA. STAT. ANN. § 166.047 (West 2009); IOWA CODE ANN. § 388.10 (West 2009); LA. REV. STAT. ANN. § 45:884.41, *et seq.* (2009); MICH. COMP. LAWS ANN. § 484.2252 (West 2009); MINN. STAT. ANN. § 237.19 (West 2009); MO. ANN. STAT. § 392.410 (West 2009); NEB. REV. STAT. ANN. §§ 86-575, 594 (West 2009); NEV. REV. STAT. ANN. §§268.086, 7 10.147 (West 2009); OHIO REV. CODE. ANN. § 1332.01, *et seq.* (West 2009); 66 PA. CONS. STAT. ANN. § 3014(h) (West 2009); S.C. CODE ANN. § 5 8-9-2600, *et seq.* (2009); TENN. CODE ANN. § 7-52-601, *et seq.* (West 2009); TEX. UTIL. CODE ANN. § 54.20 1 (2009); UTAH CODE ANN. § 10-18-201, *et seq.* (West 2009); VA. CODE ANN. § 15.2-2160 (West 2009); WASH. REV. CODE ANN. §54.16.330 (West 2009); WIS. STAT. ANN. § 66.0422 (West 2009). The Supreme Court has upheld the legality of regulations restricting municipal provision of broadband service. *See Nixon v. Missouri Municipal League*, 541 U.S. 125 (2004) (determining that the Communications Act does not preempt states from restricting municipal provision of telecommunications services). Some states prohibit or strictly limit municipal provision of broadband, while others impose some or all of the following: hearings, local referendums, cost-benefit analyses, feasibility studies, and strict rules against cross-subsidization.

[309] *See* Esme Vos, *Australia Plans 100 Mbps to 90 Percent of Homes and Offices*, MUNIWIRELESS, Apr. 7, 2009, http://www.muniwireless.com/2009/04/07/australia-plans-100mbps-national-network

[310] *Id.* To accomplish this enormous task, the Australian government will establish a new company that will deploy the network in partnership with the private sector. The Australian government estimates that the network will cost $31 billion and take eight years to build. *Id.*

[311] Under Phase 1 of France's program, for which public funding of $44 million is allocated for passive infrastructure, the goal is to target coverage of some 1,800 towns with 1,250 sites. *See* AUTORITÉ DE RÉGULATION DES COMMUNICATIONS ÉLECTRONIQUES ET DES POSTES, ARCEP ANNUAL REPORT: 2007, at 338 (2007), http://www.arcep.fr/fileadmin/uploads/tx_gspublication/rap2007-eng.zip (last visited May 13, 2009).

[312] Big Think Strategies Comments at 14 (stating that more than thirty organizations in Stockholm have set up their own transmission equipment and built their facilities using the municipality's open fiber network).

[313] *See* FTC REPORT at 20.

[314] *See* FTC REPORT at 22–23.

[315] *See* FTC REPORT at 23.

[316] Big Think Strategies Comments at 17 ("On some level this is simple economics: there are strong positive externalities to having a robust and open communications infrastructure, which means that the benefits of such

an infrastructure exceed—we believe, far exceed—the returns that the operator of the network can actually extract from its users. This means that the economically and socially optimal open network will never be financed and built by private entities—again, they cannot capture the benefits that arise from such a network so they will have no incentive to build it.").

[317] 2008 Farm Bill § 61 12(a)(1)(D).

[318] *See supra* Part IV.A (discussing Promoting Interagency Coordination).

[319] *Inquiry Concerning High-Speed Access to the Internet Over Cable and Other Facilities; Internet Over Cable Declaratory Ruling; Appropriate Regulatory Treatment for Broadband Access to the Internet Over Cable Facilities*, CS Docket No. 02-52, Declaratory Ruling and Notice of Proposed Rulemaking, 17 FCC Rcd 4798, 4801, para. 4 (2002) (*Cable Modem Declaratory Ruling*), *aff'd Nat'l Cable & Telecomms. Ass 'n v. Brand X Internet Servs.*, 545 U.S. 967 (2005) (*NCTA v. Brand X*) (cable modem Internet access service); *Appropriate Framework for Broadband Access to the Internet over Wireline Facilities; Universal Service Obligations of Broadband Providers; Review of Regulatory Requirements for Incumbent LEC Broadband Telecommunications Services; Computer III Further Remand Proceedings: Bell Operating Company Provision of Enhanced Services; 1998 Biennial Regulatory Review—Review of Computer III and ONA Safeguards and Requirements; Conditional Petition of the Verizon Telephone Companies for Forbearance Under 47 U.S. C. § 160(c) with Regard to Broadband Services Provided via Fiber to the Premises; Petition of the Verizon Telephone Companies for Declaratory Ruling or, Alternatively, for Interim Waiver with Regard to Broadband Services Provided via Fiber to the Premises; Consumer Protection in the Broadband Era*, CC Docket Nos. 02-33, 95-20, 98-10, 01-337, WC Docket Nos. 04-242, 05-271, Report and Order and Notice of Proposed Rulemaking, 20 FCC Rcd 14853 (2005) (*Wireline Broadband Internet Access Services Order*), *aff'd Time Warner Telecom, Inc. v. FCC*, 507 F.3d 205 (3d Cir. 2007) (wireline broadband Internet access service); *United Power Line Council's Petition for Declaratory Ruling Regarding the Classification of Broadband over Power Line Internet Access Service as an Information Service*, WC Docket No. 06-10, Memorandum Opinion and Order, 21 FCC Rcd 13281 (2006) (*BPL -Enabled Internet Access Services Order*) (BPL-enabled Internet access service); *Appropriate Regulatory Treatment for Broadband Access to the Internet Over Wireless Networks*, WT Docket No. 07-53, Declaratory Ruling, 22 FCC Rcd 5901, 5909-10, para. 22 (2007) (wireless broadband Internet access service).

[320] *Wireline Broadband Internet Access Services Order*, 20 FCC Rcd at 14899, para. 86.

[321] *See, e.g., Petition of AT&T, Inc. for Forbearance Under 47 U.S. C. § 160(c) from Title II and Computer Inquiry Rules with Respect to Its Broadband Services and Petition of BellSouth Corporation for Forbearance Under 47 U.S. C. § 160(c) from Title II and Computer Inquiry Rules with Respect to Its Broadband Services*, WC Docket No. 06-125, Memorandum Opinion and Order, 22 FCC Rcd 18705 (2007) (*AT&T Enterprise Broadband Forbearance Order*) (retaining statutory and regulatory requirements designed to ensure the sufficiency of federal universal service support mechanisms, promote access to telecommunications services by individuals with disabilities, protect customer privacy, and increase the effectiveness of emergency services, among other objectives).

[322] *See National Broadband Plan NOI* at para. 49.

[323] The 2008 Farm Bill states that the "Chairman of the Federal Communications Commission, in coordination with the Secretary, shall update and evaluate the report described in subsection (a) during the third year after the date of enactment of this Act." 2008 Farm Bill § 6112(b).

[324] Telecommunications Act of 1996, Pub. L. No. 104-104, 110 Stat. 56 § 254 (1996) (codified at 47 U.S.C. § 254); *see also Federal-State Joint Board on Universal Service*, CC Docket No. 96-45, Notice of Proposed Rulemaking and Order Establishing Joint Board, 11 FCC Rcd 18092 (1996). Historically, the purpose of universal service support has been to promote universally available basic telephone service at reasonable and affordable rates. Before the 1996 Act, universal service was promoted largely through implicit support mechanisms.

[325] 47 U.S.C. § 254(a), (b)(5).

[326] 47 U.S.C. § 254(b)(1)–(7).

[327] 47 U.S.C. § 254(b)(1)–(3). Specifically, the Communications Act requires that universal service policies be based on the following principles: "(1) QUALITY AND RATES.—Quality services should be available at just, reasonable, and affordable rates. (2) ACCESS TO ADVANCED SERVICES.—Access to advanced telecommunications and information services should be provided in all regions of the Nation. (3) ACCESS IN RURAL AND HIGH COST AREAS.—Consumers in all regions of the Nation, including low-income consumers and those in rural, insular, and high cost areas, should have access to telecommunications and information services, including interexchange services and advanced telecommunications and information services, that are reasonably comparable to those services provided in urban areas and that are available at rates that are reasonably comparable to rates charged for similar services in urban areas. (4) EQUITABLE AND NONDISCRIMINATORY CONTRIBUTIONS.—All providers of telecommunications services should make an equitable and nondiscriminatory contribution to the preservation and advancement of universal service. (5) SPECIFIC AND PREDICTABLE SUPPORT MECHANISMS.—There should be specific, predictable and sufficient Federal and State mechanisms to preserve and advance universal service. (6) ACCESS TO ADVANCED

TELECOMMUNICATIONS SERVICES FOR SCHOOLS, HEALTH CARE, AND LIBRARIES.—Elementary and secondary schools and classrooms, health care providers, and libraries should have access to advanced telecommunications services as described in subsection (h). (7) ADDITIONAL PRINCIPLES.—Such other principles as the [Federal-State Joint Board on Universal Service] and the Commission determine are necessary and appropriate for the protection of the public interest, convenience, and necessity and are consistent with this Act." 47 U.S.C. § 2 54(b). The Commission adopted the additional principle that federal support mechanisms should be competitively and technologically neutral. *Federal-State Joint Board on Universal Service*, CC Docket No. 96-45, Report and Order, 12 FCC Rcd 8776, 8801–02, paras. 46–48 (1997) (*Universal Service First Report and Order*).

[328] In 2008, the Universal Service Administrative Company (USAC) disbursed approximately $7.1 billion in universal service support: approximately $4.4 billion for the High-Cost Program; approximately $1.7 billion for the E-rate Program; approximately $819 million for the Low-Income Program; and approximately $49 million for the Rural Health Care Program. USAC, 2008 ANNUAL REPORT (2008) (USAC 2008 ANNUAL REPORT).

[329] Section 2 54(d) of the Communications Act directs that "[e]very telecommunications carrier that provides interstate telecommunications services shall contribute, on an equitable and nondiscriminatory basis, to the specific, predictable, and sufficient mechanisms established by the Commission to preserve and advance universal service." 47 U.S.C. § 2 54(d). Section 254(d) further provides that "any other provider of interstate telecommunications may be required to contribute to the preservation and advancement of universal service if the public interest so requires." *Id.* The Commission has determined that any entity that provides interstate telecommunications to others for a fee must contribute, based on contributors' interstate and international end-user telecommunications revenues, to the universal service fund. *Universal Service First Report and Order*, 12 FCC Rcd at 9 183–84, para. 795. The Commission also requires certain other providers of interstate telecommunications to contribute to the universal service fund. *See, e.g.*, *Universal Service Contribution Methodology*, WC Docket Nos. 06-122, 04-36, CC Docket Nos. 96-45, 98-171, 90-571, 92-237,99-200, 95-116, and 98-170, Report and Order and Notice of Proposed Rulemaking, 21 FCC Rcd 7518 (2006) (*2006 Interim Contribution Methodology Order*) (requiring interconnected VoIP providers to contribute to the universal service fund), *aff'd in part and vacated in part on other grounds*, *Vonage Holdings Corp v. FCC*, 489 F.3d 1232 (D.C. Cir. 2007).

[330] *See November 2008 Further Notice*, apps. A, C (seeking comment on requiring recipients of high-cost support to offer broadband services, and seeking comment on a low-income support pilot program for broadband services).

[331] 47 C.F.R. Part 54, Subpart G.

[332] 47 C.F.R. § 54.623; *Universal Service First Report and Order*, 12 FCC Rcd at 9141, para. 705. [333]USAC 2008 ANNUAL REPORT.

[334] *See Rural Health Care Support Mechanism*, WC Docket No. 02-60, Order, 22 FCC Rcd 20360 (2007) (*RHC Pilot Selection Order*); *Rural Health Care Support Mechanism*, WC Docket No. 02-60, Order, 21 FCC Rcd 11111 (2006) (*2006 Rural Health Care Pilot Program Order*). The Rural Health Care Pilot Program also supports costs associated with connecting to nationwide backbone providers, Internet2 or National LambdaRail, and connecting to the public Internet. *See RHC Pilot Selection Order*, 22 FCC Rcd at 20361, para. 2.

[335] *See 2006 Rural Health Care Pilot Program Order*, 21 FCC Rcd 11111.

[336] *See RHC Pilot Selection Order*, 22 FCC Rcd 20360.

[337] Press Release, FCC, *FCC Update on Rural Healthcare Pilot Program Initiative* (Apr. 16, 2009), *available at* http://hraunfoss.fcc.gov/edocs_public/attachmatch/DOC-290141A1.pdf. Of the 67 projects, 29 have developed or posted requests for proposals (RFPs) to select vendors to build out their networks, while the remaining projects are preparing their RFPs as part of the competitive bidding process. *Id.*

[338] 47 C.F.R. Part 54, Subpart F; *see also* USAC, Schools and Libraries' Eligible Services List for Funding Year 2009, November 21, 2008, http://www.usac.org/_res/documents/sl/pdf/ESL EligibleServicesList_1 12108.pdf (Funding Year 2009 ESL). The eligible services list (ESL) indicates whether specific products or services are eligible to receive discounts under the E-rate Program.

[339] Schools and libraries can apply discounts toward broadband connections that include digital subscriber lines (DSL), cable modems, fiber optics, integrated services digital networks (ISDN, BRI, PRI), satellite services, T1 s, T2s, T3s, and fractional T1s. Funding Year 2009 ESL at 2, 3, 7.

[340] *Id.* at 6, 7, 10, 11.

[341] *See* USAC, Schools and Libraries Division, http://www.sl.universalservice.org/funding (1998– 2008 data).

[342] U.S. DEPT. OF EDUC., INSTITUTE OF EDUC. SCI., INTERNET ACCESS IN U.S. PUBLIC SCHOOLS AND CLASSROOMS: 1994-2005, at 4–5 (2006), http://nces.ed.gov/pubs2007/2007020.pdf. The survey also found that nearly 100% of large schools were connected to the Internet using broadband while only 94% of small schools reported using broadband connections. *Id.* at 5. Although it is likely that the percentage of small schools using dial-up has decreased since 2006, any school still using dial-up will be limited in its use of the Internet and its applications.

[343] *Id.* at 14, 16.

[344] INFO. USE MGMT. & POLICY INSTITUTE, COLLEGE OF INFO., FLA. STATE UNIV., PUBLIC LIBRARIES AND THE INTERNET 2006: STUDY RESULTS AND FINDINGS 1 (2006), http://www.ii.fsu.edu/projectFiles/plinternet/2006/2006_plinternet.pdf.

[345] *See* STATE EDUC. TECH. DIRECTORS ASS'N, HIGH-SPEED BROADBAND ACCESS FOR ALL KIDS: BREAKING THROUGH THE BARRIERS 22–23 (2008), http://www.setda.org/c/document_library/get_file?folderId=270&name=DLFE-211.pdf.

[346] *Id.* at 14–15.

[347] *Id.*

[348] *Id.*

[349] The program allows eligible carriers who serve these areas to recover some of their operating costs from the federal universal service fund. *See* 47 C.F.R. Part 54, Subpart D.

[350] The public switched network is not a single-use network, and modern network infrastructure can provide access not only to voice service, but also to data, graphics, video, and other services. The Commission's policies seek to promote the deployment of modern plant capable of providing access to advanced services. *See Federal-State Joint Board on Universal Service, Multi-Association Group (MAG) Plan for Regulation of Interstate Services of Non-Price Cap Incumbent Local Exchange Carriers and Interexchange Carriers,* Fourteenth Report and Order, Twenty- Second Order on Reconsideration, and Further Notice of Proposed Rulemaking in CC Docket No. 96-45, and Report and Order in CC Docket No. 00-256, 16 FCC Rcd 11244, 11322, para. 200 (2001) (*Rural Task Force Order*).

[351] 47 C.F.R. Part 54, Subpart E.

[352] The Lifeline program currently provides low-income consumers with discounts of up to $10.00 off of the monthly cost of telephone service for a single telephone line in their principal residence, though this amount adjusts, in part, to reflect the carrier's tariffed federal subscriber line charge. *See* 47 C.F.R. § 54.403. Link Up provides low-income consumers with discounts of up to $30.00 off of the initial costs of obtaining telephone service. *See* 47 C.F.R. § 54.411(a). Under the Commission's rules, there are four tiers of federal Lifeline support. All eligible subscribers receive Tier 1 support, which provides a discount equal to the eligible telecommunications carrier's subscriber line charge. Tier 2 support provides an additional $1.75 per month in federal support, available if all relevant state regulatory authorities approve such a reduction. (All fifty states have approved this reduction.) Tier 3 of federal support provides one half of the subscriber's state Lifeline support, up to a maximum of $1.75. Only subscribers residing in a state that has established its own Lifeline/Link Up program may receive Tier 3 support, and they may receive this support only if their service provider is an eligible telecommunications carrier that has all necessary approvals to pass on the full amount of this total support in discounts to subscribers. Tier 4 support provides eligible subscribers living on tribal lands up to an additional $25 per month towards reducing basic local service rates, but this discount cannot bring the subscriber's cost for basic local service to less than $1. *See* 47 C.F.R. § 54.403.

[353] 54 C.F.R. § 54.403.

[354] 54 C.F.R. § 54.409(a)(4).

[355] 54 C.F.R. §§ 54.411; 54.415. The discount is up to 50% of the first $60 of the installation fee. Qualified subscribers living on Tribal lands with installation or activation fees above $60 receive an additional discount of up to $70, bringing the total discount to a maximum of $100. 54 C.F.R. §§ 54.411. The program also allows subscribers to pay the remaining amount that they owe on an interest-free deferred schedule.

[356] *See November 2008 Further Notice* at para. 40; app. A, paras. 64–91; app. C, paras. 60–87.

[357] *See id.*

[358] 47 U.S.C. § 254(c).

[359] *High-Cost Universal Service Support, Federal-State Joint Board on Universal Service*, WC Docket No. 05-33 7, CC Docket No. 96-45, Recommended Decision, 22 FCC Rcd 20477, 20488–89, para. 47 (Fed-State Jt. Bd. 2007).

[360] *November 2008 Further Notice* at paras. 30–37.

[361] *Id.* at para. 40, app. A, paras. 19–3 1, app. C, paras. 19–3 1.

[362] *Id.*

[363] For example, the Commission sought comment on: (1) "what modifications [to universal service], if any should be considered as part of a national broadband plan;" (2) each program's "effectiveness and efficiency as a mechanism to help achieve national broadband goals;" (3) "[i]f broadband services become eligible to receive high- cost and low-income support, should we also require contributions to universal service form broadband providers;" and (4) how the programs can be modified "to encourage community broadband development." *National Broadband Plan NOI* at paras. 39–41.

[364] *See generally* Embarq Comments; CTIA Comments; USA Coalition Comments; Benton Foundation Comments; and ALA Comments.

[365] *Use of the Carterfone Device in Message Toll Service*, Decision, 13 FCC 2d 420 (1968), *recon. denied*, 14 FCC 2d 571 (1968).

[366] *Appropriate Framework for Broadband Access to the Internet over Wireline Facilities; Review of Regulatory Requirements for Incumbent LEC Broadband Telecommunications Services; Computer III Further Remand Proceedings: Bell Operating Company Provision of Enhanced Services; 1998 Biennial Regulatory Review – Review of Computer III and ONA Safeguards and Requirements; Inquiry Concerning High-Speed Access to the Internet Over Cable and Other Facilities Internet Over Cable Declaratory Ruling; Appropriate Regulatory Treatment for Broadband Access to the Internet Over Cable Facilities*, Policy Statement, 20 FCC Rcd 14986, 14987–88, para. 4 (2005).

[367] *National Broadband Plan NOI.*

[368] *See* WIRELESS BROADBAND ACCESS TASK FORCE, FCC, CONNECTED & ON THE GO 2 (2005), *available at* http://hraunfoss.fcc.gov/edocs_public/attachmatch/DOC-256693A1.pdf (WIRELESS BROADBAND ACCESS TASK FORCE REPORT); *see also* APPALACHIAN REGIONAL COMM'N, EVALUATION OF THE APPALACHIAN REGIONAL COMMISSION'S TELECOMMUNICATIONS PROJECTS: 1994–2000, at v–vii (2003), http://www.arc.gov/images telecomeval/telecom_eval.pdf (discussing reducing infrastructure costs).

[369] *See* WIRELESS BROADBAND ACCESS TASK FORCE REPORT at 13–14; *see also see also* 2009 HIGH SPEED REPORT at 1 (explaining that information about subscribership to high-speed services, including advanced services, includes data concerning terrestrial wireless service providers, in addition to data concerning wireline telephone companies, cable system operators, and satellite service providers).

[370] For example, the Commission auctioned 90 megahertz of Advanced Wireless Service (AWS) spectrum in the 2 GHz band beginning in 2006 and 52 megahertz of commercial spectrum in the 700 MHz Band in 2008. In addition, beginning in 2002, the Commission also auctioned 18 megahertz of commercial spectrum in the Lower 700 MHz Band. *See Service Rules for Advanced Wireless Services in the 1.7 GHz and 2.1 GHz Bands*, WT Docket No. 02- 353, Report and Order, 18 FCC Rcd 25162 (2003) (*AWS-1 Service Rules Report and Order*), *modified by Service Rules for Advanced Wireless Services In the 1.7 GHz and 2.1 GHz Bands*, WT Docket No. 02-353, Order on Reconsideration, 20 FCC Rcd 14058 (2005) (*A WS-1 Recon Order*); *700 MHz Report and Order*, 22 FCC Rcd 8064; *700 MHz Second Report and Order*, 22 FCC Rcd 15289; http://wireless.fcc.gov/auctions/44/ (with links to later auctions of Lower 700 MHz Band licenses). In addition, beginning in 2004, the Commission restructured the 2.5 GHz Broadband Radio Service/Educational Broadband Service band plan and modified the rules for the use of that spectrum to enable the development and deployment of mobile broadband services, including WiMAX technologies. In restructuring the band plan, the Commission moved from interleaved spectrum to more cohesive, contiguous blocks of spectrum. *See Amendment of Parts 1, 21, 73, 74 and 101 of the Commission's Rules to Facilitate the Provision of Fixed and Mobile Broadband Access, Educational and Other Advanced Services in the 2150-2162 and 2500-2690 MHz Bands*, WT Docket No. 03-66, Report and Order and Further Notice of Proposed Rulemaking, 19 FCC Rcd 14165, 14270, 14271, 14272, paras. 281, 286 (2004); *Amendment of Parts 1, 21, 73, 74 and 101 of the Commission's Rules to Facilitate the Provision of Fixed and Mobile Broadband Access, Educational and Other Advanced Services in the 2150-2162 and 2500-2690 MHz Bands*, WT Docket No. 03-66, Third Memorandum Opinion and Order and Second Report and Order, 21 FCC Rcd 5606 (2006).

[371] As envisioned by the Commission, such a network would bring the promise of broadband to all public safety entities, including those in rural areas that may not otherwise have the resources or the ability to procure broadband capabilities for themselves. The Commission is continuing to examine opportunities to utilize the 700 MHz band to enhance the ability of the public safety community to gain access to a state of the art broadband network. *See Implementing a Nationwide, Broadband, Interoperable Public Safety Network in the 700 MHz Band*, PS Docket No. 06-229, Ninth Notice of Proposed Rulemaking, 21 FCC Rcd 14837 (2006); Report and Order and Further Notice of Proposed Rulemaking, 22 FCC Rcd 8064 (2007); Second Report and Order, 22 FCC Rcd 15289 (2007); Second Further Notice of Proposed Rulemaking, 23 FCC Rcd 8047 (2008); Third Further Notice of Proposed Rulemaking, 23 FCC Rcd 14301 (2008).

[372] The spectrum most commonly used by unlicensed devices in the United States is at 902-928 MHz, 2.4 GHz, and between 5.3 and 5.8 GHz. *See generally* 47 C.F.R. Part 15.

[373] *Unlicensed Operation in the TV Broadcast Bands*, ET Docket No. 04-186, Second Report and Order and Memorandum Opinion and Order, 23 FCC Rcd 16807 (2008). *See* Sprint Nextel *et al.* April 13, 2009 *Ex Parte* Letter at 1–2.

[374] *See Wireless Operations in the 3650-3 700 MHz Band*; *Rules for Wireless Broadband Services in the 3650-3 700 MHz Band*; *Additional Spectrum for Unlicensed Devices Below 900 MHz and in the 3 GHz Band*; *Amendment of the Commission's Rules with Regard to the 3650-3700 MHz Government Transfer Band*, ET Docket Nos. 04-151, 98- 237, 02-380, WT Docket No. 05-96, Report and Order and Memorandum Opinion and Order, 20 FCC Rcd 6502 (2005), *recon. granted in part*, Memorandum Opinion and Order, 22 FCC Rcd 10421 (2007) (*3650 Memorandum Opinion and Order*). *See* API Comments at 7 (stating the Commission's recent decision regarding the 3650-3700 MHz Band is a commendable example of efficient management in a new band with registration conditions placed on licensees in that band and the accompanying protocol restrictions helping to ensure the viability of that band to multiple users in the future). The rules for this band provide for nationwide, non-exclusive, licensing of terrestrial wireless operations, and licensees must use equipment with a contention-based protocol. *See 3650 Memorandum Opinion and Order*, 22 FCC Rcd at 10425, para. 11. This approach

combines the benefits of the unlicensed approach, in which the number of wireless services providers is not restricted, with other elements, namely that service providers in the band must register their operations and use equipment that incorporates a contention-based protocol.

[375] In the *Terrestrial Wireless Rural Report and Order*, the Commission found that the record supported finding, in principle, that "increasing power limits in rural areas can benefit consumers in rural areas by reducing the costs of infrastructure and otherwise making the provision of spectrum-based services to rural areas more economic." *Terrestrial Wireless Rural Report and Order*, 19 FCC Rcd at 19 126-27, para. 86.

[376] *See id.* at 19126–35 paras. 86–104; *700 MHz Report and Order*, 22 FCC Rcd at 8099–100, para. 93.

[377] In deciding to employ the use of power spectral density (PSD) in the 700 MHz band, the Commission adopted a general power limit of 1 kW/MHz ERP for 700 MHz base stations and a 2 kW/MHz ERP limit for 700 MHz base stations operating in rural areas. *Id.* In March 2008, the Commission extended the PSD model to the Broadband PCS and AWS-1 spectrum bands, thereby adopting a power limit of 1640 Watts EIRP/MHz in non-rural areas and 3280 Watts EIRP/MHz in rural areas. *See Biennial Regulatory Review—Amendment of Parts 1, 22, 24, 27, and 90 to Streamline and Harmonize Various Rules Affecting Wireless Radio Services*, Third Report and Order, 23 FCC Rcd 5319, 5330, para. 25 (2008) (*Streamlining and Harmonization Order*).

[378] *See 700 MHz Report and Order*, 22 FCC Rcd at 8103, para. 105; *Streamlining and Harmonization Order*, 23 FCC Rcd at 5330, para. 25.

[379] For example, in the AWS-1 Spectrum, the Commission increased the amount of spectrum licensed on a small geographic area basis (CMAs). *See A WS-1 Recon Order*, 20 FCC Rcd 14058. In addition, in the 700 MHz Band, the Commission adopted "keep-what-you-use" rules which provide that if a licensee fails to meet its end of term benchmark, its authorization to operate will terminate automatically for those geographic areas of its license authorization in which it is not providing service, and those unserved areas will become available for reassignment. *See 700 MHz Second Report and Order*, 22 FCC Rcd at 15353–54, paras. 170–74.

[380] These policies have led to the development and deployment of cutting-edge mobile broadband network technologies in many areas of the country, as well as innovative, powerful, and personalized mobile broadband applications and devices. *See generally Thirteenth CMRS Competition Report*, paras. 66–67, 126–27.

[381] "Partitioning" a license refers to transferring a portion of the geographic area of a spectrum license, such as a county within a Cellular Market Area (CMA), to another entity. "Disaggregation" refers to transferring a portion of the spectrum included in a license, such as 10 megahertz of a 30 megahertz license, to another entity.

[382] *See generally Promoting Efficient Use of Spectrum Through Elimination of Barriers to the Development of Secondary Markets*, First Report and Order and Further Notice of Proposed Rulemaking, 18 FCC Rcd 20604 (2003); *Promoting Efficient Use of Spectrum Through Elimination of Barriers to the Development of Secondary Markets*, Second Report and Order, Order on Reconsideration, and Second Further Notice of Proposed Rulemaking, 19 FCC Rcd 17503 (2004); 47 CFR §§ 1.9001 *et seq.*

[383] The Commission may provide rural telephone companies that qualify as small businesses with bidding credits. *See Amendment of Part 1 of the Commission's Rules—Competitive Bidding Procedures*, WT Docket No. 97-82, Order on Reconsideration of the Third Report and Order, Fifth Report and Order, and Fourth Further Notice of Proposed Rule Making, 15 FCC Rcd 15293, 153 19–20, para. 52 (2000).

[384] *See, e.g., Revision of Part 22 and Part 90 of the Commission's Rules to Facilitate Future Development of Paging Systems*; *Implementation of Section 309(j) of the Communications Act -- Competitive Bidding*, WT Docket No. 96- 18, PR Docket No. 93-253, Memorandum Opinion and Order on Reconsideration and Third Report and Order, 14 FCC Rcd 10030, 10091, para. 112 (1999). *But see also* David Mason Comments at 2 (urging the Commission to allow licensing of WiMAX spectrum to small ISPs across the nation for free -- instead of auctioning spectrum to one large nationwide service provider who bids the most -- because a single nationwide provider of service will never get to his small town since the market is too small).

[385] *Extending Wireless Telecommunications Services to Tribal Lands*, WT Docket No. 99-266, Report and Order and Further Notice of Proposed Rulemaking, 15 FCC Rcd at 11794, 11802, para. 22 (2000) (*TLBC First Report and Order*).

[386] In order to obtain a bidding credit in a particular market, a winning bidder must indicate on its long-form application (FCC Form 601) that it intends to serve Tribal lands in that market. *Id.* at 11805, para. 31. Following the long-form application filing deadline, the applicant provided an additional 90 calendar days beyond the deadline to amend its application to identify the Tribal lands to be served, and provide certification from the Tribal government(s). In particular, applicants must provide certification from the applicable Tribal government that: (1) it will allow the bidder to site facilities and provide service on its Tribal Land(s), in accordance with our rules; (2) it has not and will not enter into an exclusive contract with the applicant precluding entry by other carriers, and will not unreasonably discriminate against any carrier; and (3) its Tribal Land is a qualifying Tribal Land as defined in Commission rules. In addition, at the conclusion of the grace period, the applicant was required to amend its long-form application to file a certification that it would comply with the bidding credit buildout requirement, and that it would consult with the Tribal government regarding the siting of facilities and deployment of service on the Tribal Land. *Id.* at 11805–06, para. 32. The Commission, on its own motion, subsequently extended the grace period to 180 days. A winning bidder now

has 180 days to amend its long-form application to identify the Tribal Land it seeks to serve, and to provide the required certification from the Tribal government. The winning bidder also has 180 days to file a certification that it will comply with the Tribal lands buildout requirements, and consult with the Tribal government regarding the siting of facilities and deployment of service on Tribal lands. *See Extending Wireless Telecommunications Services to Tribal Lands*, WT Docket No. 99-266, Second Report and Order and Second Further Notice of Proposed Rulemaking, 18 FCC Rcd 4775, 4779 para. 10 (2003) (*TLBC Second Report and Order*); *see also* 47 C.F.R. § 1.211 0(f)(3)(ii)(A) (2003).

387 *Extending Wireless Telecommunications Services to Tribal Lands*, WT Docket No. 99-266, Third Report and Order, 19 FCC Rcd 17652, 17659, para. 18 (2004) (*TLBC Third Report and Order*).

388 *TLBC First Report and Order,* 15 FCC Rcd at 11,806–07, para. 35. The Commission also requires that, at the conclusion of the three-year period, licensees file a notification of construction indicating that they have met the 75% construction requirement on the Tribal lands for which the credit was awarded. If the licensee fails to comply with any condition, it is required to repay the bidding credit plus interest 30 days after the conclusion of the construction period. In the event the licensee fails to repay the amount, the license automatically cancels. *Id.* at 11807, para. 37. Additionally, in order to verify compliance with the Tribal lands construction requirement, the Commission requires licensees to file an attachment along with their notification of construction, stating that it covers 75% of the population of the Tribal area for which the credit was awarded and providing the data to support that statement. *TLBC Second Report and Order*, 18 FCC Rcd at 4780, para. 13. Finally, the Commission expressly codified penalties for failure to comply with buildout requirements, and failure to timely repay the bidding credit. *Id.* at 4781, para. 14; *see also* 47 C.F.R. § 1.2110(f)(3)(vii) (2003).

389 DigitalBridge Comments at 4–6; *see also, e.g.*, FiberTower *et al.* Comments at 1–2 & app. (urging the Commission to expand the amount of licensed spectrum that can be used for wireless backhaul services).

390 *See* API Comments at 5 (stating that secondary markets for suitable broadband spectrum are dysfunctional in that incentives for auction winners or spectrum lessors to lease, partition, or sublease spectrum in rural areas are not compelling).

391 *See, e.g.*, Big Think Strategies Comments at 8 (urging the Commission to consider the need for some sort of fixed, reasonable pricing on backhaul/middle mile costs to an upstream Internet access point); ACS Comments at 3 (stating that a component of broadband sustainability is actual access to the Internet based on economically available backhaul facilities); Fibertech/KDL Comments at 9 (stating that backhaul represents an enormous cost to wireless providers and can be prohibitively expensive in rural areas). We also note that roundtable participants at the March 19, 2009 NTIA/RUS Public Meeting about the Broadband Technology Opportunities Program explained that, in some areas, backhaul costs were the reason communities were unserved. *See, e.g.,* John Rose, President of OPASTCO, Remarks at the U.S. Dep't of Commerce Roundtable on Rural and Unserved Areas (Mar. 19, 2009) (transcript available at http://www.ntia.doc.gov/ broadbandgrants/090319/NTIA_031909_1445_1615_session.pdf).

392 *See, e.g.*, New America Foundation Comments at 3–5 (arguing that TV "whitespaces" rules should allow variable power limits for rural areas); WISPA Comments at 5 (stating that the Commission should be favorably disposed to quickly grant waiver requests seeking relief from power limits or other technical parameters in rural areas); and Main Street Broadband Comments at 2 (recommending that the Commission raise the power limits for 3.65 GHz operations in rural areas allowing service providers to reach more customers using less infrastructure).

393 The Commission has noted the potential for a "good faith" negotiation process in recent spectrum rulemaking proceedings. *See Service Rules for Advanced Wireless Services in the 2155-2175 MHz Band*, WT Docket No. 07- 195, Notice of Proposed Rulemaking, 22 FCC Rcd 17035, 17090 para. 131 (2007) (seeking comment on whether requiring licensees to make "good faith" efforts to negotiate with potential spectrum lessees could help increase access to spectrum, including in rural areas). *Cf. 700 MHz Report and Order*, 22 FCC Rcd at 8086–87 paras. 53–54 (declining to adopt rules requiring licensees to make "good faith" efforts to negotiate based, in part, on other incentives under the rules to enter into spectrum leasing arrangements).

394 *See supra* Part V.D (Addressing Network Costs).

395 *See generally* 47 C.F.R. Parts 61, 69.

396 *See, e.g.*, New America Foundation Comments at 1, 5; GCI Comments at 3; and Alaska Commission Comments at 5.

397 *Special Access Order and NPRM*, WC Docket No. 05-25, RM-10593, Order and Notice of Proposed Rulemaking, 20 FCC Rcd 1994, 1994, para. 1 (2005).

398 *Parties Asked to Refresh Record in the Special Access Notice of Proposed Rulemaking*, WC Docket No. 05-25, RM-10593, Public Notice, 22 FCC Rcd 13352 (2007).

399 *See, e.g., Petition of AT&T Inc. for Forbearance under 47 U.S. C. § 160(c) from Title II and Computer Inquiry Rules with Respect to its Broadband Services*, WC Docket No. 06-12 5, Memorandum Opinion and Order, 22 FCC Rcd 18705 (2007); *see also, e.g.,* Joint Statement of Chairman Kevin J. Martin and Commissioner Deborah Taylor Tate, *Petition of the Verizon Telephone Companies for Forbearance under 47 U.S. C. § 160(c) from Title II and Computer Inquiry Rules with Respect to their Broadband Services*, WC Docket No.

04-440 (Mar. 20, 2006) (discussing Verizon's "deemed granted" petition). Verizon's "deemed granted" petition was upheld in *Sprint Nextel Corp. v. FCC*, 508 F.3d 1129 (D.C. Cir. 2007).

[400] We note, however, that unfavorable weather conditions such as rain and wind can adversely affect the capacity of these transmissions for wireless backhaul using certain spectrum bands (such as 24 GHz and 39 GHz). *See* FCC Bulletin 70, Millimeter Wave Propagation Spectrum Management Implications (July 1997). In areas with irregular terrain, intermediate repeaters that ensure "line-of-sight" connectivity between transmitters and receivers are an indispensable part of any wireless backhaul system. *See supra* Part V.A (discussing technological considerations).

[401] *Unlicensed Operation in the TV Broadcast Bands; Additional Spectrum for Unlicensed Devices Below 900 MHz and in the 3 GHz Band*, ET Docket Nos. 04-186, 02-380, Second Report and Order and Memorandum Opinion and Order, 23 FCC Rcd 16807, 16847, paras. 104–05 (2008).

[402] *Id.* atpara. 105.

[403] *Id.* at para. 106. *See also* Statement of Commissioner Michael J. Copps, *id.* at 16928; Statement of Commissioner Jonathan S. Adelstein, *id.* at 16930; Statement of Commissioner Robert M. McDowell, *id.* at 16932; and Statement of Commissioner Deborah Taylor Tate, *id.* at 16935; *see also* Fibertech/KDL Comments at 9 (asserting that wireless backhaul can reduce cellular companies' special access costs by 90%). *See generally* Sprint Nextel *et al.* April 13, 2009 *Ex Parte* Letter.

[404] *November 2008 Further Notice* at paras. 3, 23–2 8.

[405] *Developing a Unified Intercarrier Compensation Regime*, CC Docket No. 01-92, Notice of Proposed Rulemaking, 16 FCC Rcd 9610 (2001) (*Intercarrier Compensation NPRM*).

[406] *Developing a United Intercarrier Compensation Regime*, CC Docket No. 01-92, Further Notice of Proposed Rulemaking, 20 FCC Rcd 4685 (2005) (*Intercarrier Compensation Further NPRM*) (seeking comment on a number of specific comprehensive reform proposals and alternative reform measures that would affect how intercarrier costs are recovered from end-users, carriers and universal service mechanisms).

[407] *Comment Sought on Amendments to the Missoula Plan Intercarrier Compensation Proposal to Incorporate a Federal Benchmark Mechanism*, CC Docket No. 01-92, Public Notice, 22 FCC Rcd 3362 (WCB 2007). *Comment Sought on Missoula Intercarrier Compensation Reform Plan*, CC Docket No. 01-92, Public Notice, 21 FCC Rcd 8524 (WCB 2006).

[408] *November 2008 Further Notice.*

[409] Pole Attachment Act of 1978, Pub. L. No. 95-234, 92 Stat. 33 (1978).

[410] Before 1996, section 224 did not guarantee cable television systems the right to attach to utility poles but merely provided that where cable television systems were able to obtain such attachments, the rates, terms, and conditions must be just and reasonable.

[411] 47 U.S.C. § 224(a)(4).

[412] *See* 47 U.S.C. § 224(d), (e).

[413] 47 U.S.C. § 224(f)(2).

[414] *Implementation of the Local Competition Provisions in the Telecommunications Act of 1996; Interconnection between Local Exchange Carriers and Commercial Mobile Radio Service Providers*, CC Docket Nos. 96-98, 95-185, First Report and Order, 11 FCC Rcd 15449, 16058–107, paras. 1119–240 (1996) (*Local Competition Order*) (Part XI.B. "Access to Rights of Way").

[415] *Id.* at 16067–68, para. 1143. *See generally* 47 C.F.R § § 1.1403–. 1418 (section 224 complaint rules). In addition, under current Commission rules, an attacher may execute a pole attachment agreement with a utility, and then later file a complaint challenging the lawfulness of a provision of that agreement. *See, e.g., S. Co. Svcs., Inc. v. FCC*, 313 F.3d 574, 582–84 (D.C. Cir. 2002).

[416] *Implementation of Section 224 of the Act; Amendment of the Commission's Rules and Policies Governing Pole Attachments*, WC Docket No. 07-245, RM-1 1293, RM-1 1303, Notice of Proposed Rulemaking, 22 FCC Rcd 20195 (2007) (*Pole Attachments NPRM*).

[417] *See Pole Attachments NPRM*, 22 FCC Rcd at 20209, para. 36. In the *Pole Attachments NPRM*, the Commission also tentatively concluded that this broadband rate should fall within the established statutory rates. *See* 47 U.S.C. § 224(d)–(e).

[418] *See Pole Attachments NPRM*, 22 FCC Rcd at 20205–06, para. 25. The Commission stated that "[s]ection 706 of the 1996 Act directs us to promote the deployment of broadband infrastructure, and this directive leads us to separate out those pole attachments that are used to offer broadband Internet access service from those used for other services." *Id* at 20209, para. 36.

[419] *See* CTIA Comments, WC Docket No. 05-337, Attach. A at 20 (filed Apr. 17, 2008) ("States with lower population density require more new site investment rather than augmentation of existing network assets. More than 90% of the estimated investment for Alaska, Idaho, Montana, Nevada, and Wyoming is Greenfield or new site investment.").

[420] For instance, in July 2008, CTIA filed a petition for declaratory ruling requesting that the Commission clarify provisions of the Communications Act regarding state and local review of wireless facility siting applications. *Petition for Declaratory Ruling to Clarify Provisions of Section 332(c) (7) (B) to Ensure Timely Siting Review and to Preempt under Section 253 State and Local Ordinances that Classify All Wireless Siting Proposals as*

Requiring a Variance, Petition for Declaratory Ruling, WT Docket No. 08-165 (filed July 11, 2008). In 2006, the Commission sought comment on whether it should impose additional requirements on the construction of new communications towers to reduce the number of migratory bird collisions with such towers. *Effects of Communications Towers on Migratory Birds*, WT Docket No. 03-187, Notice of Proposed Rulemaking, 21 FCC Rcd 13241, 13241, 13256, paras. 1, 32 (2006). The Commission also is presently reviewing its rules and procedures regarding the process for environmental impact assessment of registered antenna structures pursuant to a remand from the United States Court of Appeals for the District of Columbia Circuit. *See Am. Bird Conservancy, Inc. v. FCC*, 516 F.3d 1027 (D.C. Cir. 2008); *see also, Wireless Telecommunications Bureau Seeks Comment on Petition for Expedited Rulemaking and Other Relief Filed on Behalf of American Bird Conservancy, Defenders of Wildlife and National Audubon Society Regarding Commission Implementation of the National Environmental Policy Act, the Endangered Species Act, and the Migratory Bird Treaty Act*, WT Docket Nos. 08-61, 03-187, Public Notice, DA 09-904 (WTB rel. Apr. 29, 2009).

[421] *See Annual Assessment of the Status of Competition in the Market for the Delivery of Video Programming*, MB Docket No. 07-269, Notice of Inquiry, 24 FCC Rcd 75 (2009); Supplemental Notice of Inquiry, 24 FCC Rcd 4401 (2009).

[422] *See Implementation of the Cable Television Consumer Protection Act of 1992, Development of Competition and Diversity in Video Programming Distribution: Section 628(c)(5) of the Communications Act: Sunset of Exclusive Contract Prohibition; Review of the Commission's Program Access Rules and Examination of Program Tying Arrangements*, MB Docket Nos. 07-29, 07-198, Report and Order and Notice of Proposed Rulemaking, 22 FCC Rcd 17791 (2007).

In: Broadband and Rural America
Editor: Bethany J. Lenz
ISBN: 978-1-61668-637-6

Chapter 2

Broadband Deployment is Extensive throughout the U.S., but it is Difficult to Assess the Extent of Deployment Gaps in Rural Areas

United States Government Accountability Office

Why GAO Did This Study

Both Congress and the President have indicated that access to broadband for all Americans is critically important. Broadband is seen as a critical economic engine, a vehicle for enhanced learning and medicine, and a central component of 21st century news and entertainment. As part of our response to a mandate included in the Internet Tax Nondiscrimination Act of 2004, this report examines the factors that affect the deployment and the adoption of broadband services. In particular, this report provides information on (1) the current status of broadband deployment and adoption; (2) the factors that influence the deployment of broadband networks; (3) the factors that influence the adoption, or purchase, of broadband service by households; and (4) the options that have been suggested to spur greater broadband deployment and adoption.

What GAO Recommends

GAO recommends that FCC develop information regarding the cost and burden that would be associated with various options for improving the information available on broadband deployment and report this information to the relevant Senate and House committees to help them determine what actions, if any, are necessary. FCC provided technical comments on this report, but did not comment on this recommendation.

What GAO Found

About 30 million American households have adopted broadband service, but the Federal Communications Commission's (FCC) data indicating the availability of broadband networks has some weaknesses. FCC conducts an extensive data collection effort using its Form 477 to assess the status of advanced telecommunications service in the United States. For its zip-code level data, FCC collects data based on where *subscribers* are served, not where providers have deployed broadband infrastructure. Although it is clear that the deployment of broadband networks is extensive, the data may not provide a highly accurate depiction of local deployment of broadband infrastructures for residential service, especially in rural areas.

A variety of market and technical factors, government efforts, and access to resources at the local level have influenced the deployment of broadband infrastructure. Areas with low population density and rugged terrain, as well as areas removed from cities, are generally more costly to serve than are densely populated areas and areas with flat terrain. As such, deployment tends to be less developed in more rural parts of the country. Technical factors can also affect deployment. GAO also found that a variety of federal and state efforts, and access to resources at the local level, have influenced the deployment of broadband infrastructure.

A variety of characteristics related to households and services influence whether consumers adopt broadband service. GAO found that consumers with high incomes and college degrees are significantly more likely to adopt broadband. The price of broadband service remains a barrier to adoption for some consumers, although prices have been declining recently. The availability of applications and services that function much more effectively with broadband, such as computer gaming and file sharing, also influences whether consumers purchase broadband service.

Stakeholders identified several options to address the lack of broadband in certain areas. Although the deployment of broadband is widespread, some areas are not served, and it can be costly to serve highly rural areas. Targeted assistance might help facilitate broadband deployment in these areas. GAO found that stakeholders have some concerns about the structure of the Rural Utilities Service's broadband loan program. GAO was also told that modifications to spectrum management might address the lack of broadband infrastructure in rural areas. Also, because the cost of building land-based infrastructure is so high in some rural areas, satellite industry stakeholders noted that satellite broadband technology may be the best for addressing a lack of broadband in those regions. While several options such as these were suggested to GAO, each has some challenges to implementation. Also, a key difficulty for analyzing and targeting federal aid for broadband is a lack of reliable data on the deployment of networks.

ABBREVIATIONS

3G	third generation
ADS	Lasymmetric digital subscriber line
BPL	broadband over power lines
CTIA	Cellular Telecommunications and Internet Association
DBS	direct broadcast satellite
DSL	digital subscriber line
FCC	Federal Communications Commission
FTTH	fiber to the home
HFC	hybrid-fiber coaxial
IEEE	Institute of Electrical and Electronics Engineers
IG	inspector general
ILEC	incumbent local exchange carrier
Kbps	kilobits per second
Mbps	megabits per second
MSA	metropolitan statistical area
NARUC	National Association of Regulatory Utility Commissioners
NATOA	National Association of Telecommunications Officers and Advisors
NCTA	National Cable and Telecommunications Association
NTCA	National Telecommunications Cooperative Association
NTIA	National Telecommunications and Information Administration
RUS	Rural Utilities Service
SIA	Satellite Industry Association
UNE	unbundled network element
USF	Universal Service Fund
USIIA	US Internet Industry Association
USTA	United States Telecom Association
VoIP	voice over Internet protocol
Wi-Fi	wireless fidelity
WiMAX	Worldwide Interoperability for Microwave Access
WISP	wireless Internet service provider
WISPA	Wireless Internet Service Providers Association

May 5, 2006

The Honorable Ted Stevens
Chairman
The Honorable Daniel K. Inouye
Co-Chairman
Committee on Commerce, Science, and Transportation
United States Senate

The Honorable Joe L. Barton
Chairman

The Honorable John D. Dingell
Ranking Minority Member
Committee on Energy and Commerce House of Representatives

The universal availability of high speed Internet access over broadband technologies—commonly referred to as broadband Internet access—has become a national goal.[1] The Telecommunications Act of 1996 directed the Federal Communications Commission (FCC) and state commissions to encourage the deployment of advanced telecommunications capability. Similarly, in 2004, the President stated that there should be a national goal for universal, affordable access to broadband technology by 2007. The importance placed on access to broadband correlates to its many benefits for individuals and society. Broadband is seen as a critical economic engine, a vehicle for enhanced learning and medicine, and a central component of 21st century news and entertainment.

As part of our response to a mandate included in the Internet Tax Nondiscrimination Act of 2004, this report examines the factors that affect the deployment—that is, the building of infrastructure over which broadband services can be provided—and the adoption of broadband services. We focus particularly on the deployment and adoption of broadband to households, as opposed to businesses or institutions. In particular, this report provides information on (1) the current status of broadband deployment and adoption; (2) the factors that influence the deployment of broadband networks; (3) the factors that influence the adoption, or purchase, of broadband service by households; and (4) the options that have been suggested to spur greater broadband deployment and adoption. In January 2006, we released a report that examined the impact of the Internet tax moratorium on state and local tax revenues, as also mandated by the law.[2]

To respond to the objectives of this report, we selected eight states and conducted case studies on the status of broadband deployment and adoption. For each of the states—Alaska, California, Kentucky, Massachusetts, North Dakota, Ohio, Texas, and Virginia—we interviewed state and local officials, including local franchising authorities, state public utility regulators, and representatives from governors' offices; state industry and government associations; private cable and telephone providers; wireless Internet service providers; and municipal and cooperative telecommunications providers. We also spoke with a variety of individuals and organizations knowledgeable about broadband services, such as national industry associations and experts. We spoke with representatives from FCC, the National Telecommunications and Information Administration of the Department of Commerce, and the Rural Utilities Service (RUS) of the Department of Agriculture. To assess the status of broadband deployment and to understand the factors affecting the deployment and adoption of broadband, we used survey data from Knowledge Networks/SRI's *The Home Technology Monitor™: Spring 2005 Ownership and Trend Report*. Knowledge Networks/SRI interviewed approximately 1,500 randomly sampled households, asking questions about each household's purchase of Internet services and the availability of cable television service. Using these data, we estimated two econometric models: One model examined the factors affecting broadband deployment and the second examined the factors affecting households' adoption of broadband services. We combined the household survey data with information from FCC's Form 477 filings, which contain information on companies' provision of broadband services by zip codes. This enabled us to develop information about what options for broadband a particular household would have. To assess the impact of Internet taxes on broadband deployment and adoption, we contacted officials in 48 states and the District of Columbia to determine whether the state, or local governments in the state, imposed taxes on Internet access in 2005; we did not evaluate the level of taxation. We concluded that information from Knowledge Networks/SRI and FCC (with modifications discussed later in this report) was sufficiently reliable for the purpose of this report. All percentage estimates from the Knowledge Networks/SRI survey have margins of error of plus or minus 7 percentage points or less, unless otherwise noted. See appendix I for a more detailed discussion of the overall scope and methodology for this report, including a discussion of how we selected the case-study states; appendix II for an assessment of the data reliability of the Knowledge Networks/SRI survey; and appendix III for a more detailed explanation of, and results from, our deployment and adoption models.

We conducted our work from April 2005 through February 2006 in accordance with generally accepted government auditing standards.

RESULTS IN BRIEF

About 30 million American households purchase, or have adopted, broadband service, but it is difficult to assess the extent of gaps in the availability of broadband in local markets. Using a survey of American households, we found that 28 percent—or about 30 million—subscribed to broadband service in 2005. In addition, 30 percent of surveyed households subscribed to a dial-up Internet service, and 41 percent did not access the Internet from their home. Among households subscribing to broadband service, we found roughly an equal share taking cable modem and digital subscriber line (DSL) service, the two primary broadband services at this time. Households in rural areas were less likely to subscribe to broadband service, compared with households in urban and suburban areas. On a semiannual basis, FCC conducts an extensive data collection effort using its Form 477 to assess the availability of advanced telecommunications service in the United States. As of July 2005, FCC has found that 99 percent of Americans live in the 95 percent of zip codes that have at least one broadband provider reporting to be serving at least one subscriber. These data clearly indicate that deployment of broadband networks has been extensive. However, for its zip-code level data, FCC collects data based on where *subscribers* are served, not where providers have deployed broadband infrastructure. Based on our analysis is appears that these data may not provide a highly accurate depiction of deployment of broadband infrastructures for residential service in some areas.[3]

A variety of market and technical factors, as well as federal and state government efforts and access to resources at the local level have influenced the deployment of broadband infrastructure. Most importantly, companies contemplating the deployment of broadband infrastructure consider both the cost to deploy and operate a broadband network and the expected demand for broadband service. We found it is more costly to serve areas with low population density and rugged terrain with terrestrial facilities than it is to serve areas that are densely populated and have flat terrain. It also may be more costly to serve locations that are a significant distance from a major city. As such, these important factors have caused deployment to be less developed in more rural parts of the country. Firms also consider the extent of existing competition in the broadband market when making deployment decisions:

New entrants are more likely to enter markets with no competitors, but at the same time, we found that incumbent cable and telephone companies may respond to entry by new companies by rolling out broadband in markets where they had not yet provided service. Even when cost and demand factors are favorable, technical factors can limit the deployment of broadband service in certain contexts. For example, DSL—the primary broadband service provided by telephone companies—can generally extend only 3 miles[4] from the central office with copper plant, which precludes many households from obtaining DSL service.[5] Finally, we found that a variety of federal and state government efforts as well as access to resources at the local level have influenced the deployment of broadband infrastructure. At the federal level, one of the programs of the Universal Service Fund (USF)—known as the High Cost Fund—has indirectly facilitated broadband service in more rural areas. Similarly, the Department of Agriculture's Rural Utilities Service (RUS) provides grants and loans to promote broadband service in rural areas. At the local level, access to rights-of-way, pole attachments, wireless-tower sites as well as the video franchising process can influence the pace of deployment. We also found that strong leadership within a community can help promote broadband deployment by, for example, enhancing the likely market success of companies' entry into rural markets. Finally, using our econometric model, we found that the imposition of taxes was not a statistically significant factor influencing the deployment of broadband.

A variety of characteristics related to households and services influence whether consumers purchase (or adopt) broadband service. Based on our econometric model, we found that several characteristics of households influence the adoption decision. Our model showed that households with high incomes were 39 percentage points more likely to adopt broadband than lower-income households, and those with a college-educated head of household were 12 percentage points more likely to purchase broadband than households headed by someone who did not graduate from college. While rural households are less likely to adopt broadband, our findings indicate that this difference may be related in part to the lower availability of broadband in rural areas. In addition, based on discussions with stakeholders, we identified several characteristics of broadband service that influence whether a consumer purchases the service. The price of broadband service remains a barrier to adoption of broadband service for some consumers, although prices have been declining recently. The availability of applications and services that either require or function much more effectively with broadband—such as computer gaming and file sharing—also influences whether a particular consumer purchases broadband service. Using our model, we found that the imposition of the tax was not a statistically significant factor influencing the adoption of broadband service at the 5 percent level. It was statistically significant at the 10 percent level, perhaps suggesting that it is a weakly significant factor. However, giving the nature of our model, it is unclear whether this finding is related to the tax or other characteristics of the states in which the households resided.

Targeted government assistance might help facilitate the deployment of broadband service, and stakeholders we spoke with identified several options to spur greater deployment of broadband service in rural America. However, each of the policy options that stakeholders discussed with us had challenges to their implementation. For example, a few of the stakeholders we spoke with expressed concerns about the structure of the Rural Utilities Service's broadband loan program. Also, several of the stakeholders suggested that modifications to spectrum management might address the lack of broadband infrastructure in rural areas. Finally, because the cost of building land-based infrastructure is so high in some rural areas, satellite industry stakeholders noted that satellite broadband technology may be the best option for addressing a lack of broadband in those regions. Ultimately, we found that a key difficulty for analyzing and targeting any federal aid for broadband is a lack of reliable data on the deployment of networks.

We provided a draft of this report to the Department of Agriculture, the Department of Commerce, and FCC for their review and comment. The Department of Agriculture had no comments on the draft. The Department of Commerce and FCC provided technical comments that we incorporated, as appropriate.

In the draft, GAO recommended that FCC identify and evaluate strategies for improving the 477 data such that the data provide a more accurate depiction of residential broadband deployment throughout the country. In oral comments regarding this recommendation, FCC staff noted that the commission had recently determined that it would be costly and could impose large burdens on filers—particularly small entities—to require any more detailed filings on broadband deployment. As such, we recommend that FCC develop information regarding the degree of cost and burden that would be associated with various options for improving the information available on broadband deployment and should provide that information to the Senate Committee on Commerce, Science, and Transportation and the House Energy and Commerce Committee in order to help them determine what actions, if any, are necessary going forward. FCC did not comment on our final recommendation.

We also provided a draft of this report to several associations representing industry trade groups and state and local government entities for their review and comment. Specifically, the following associations came to GAO headquarters to review the draft: Cellular Telecommunications and Internet Association (CTIA), National Association of Regulatory Utility Commissioners (NARUC), National Association of Telecommunications Officers and Advisors (NATOA), National Cable and Telecommunications Association (NCTA), National Telecommunications Cooperative Association (NTCA), Satellite Industry Association (SIA), US Internet Industry Association (USIIA), United States Telecom Association (USTA), and Wireless Internet Service Providers Association (WISPA). Officials from CTIA, NARUC, and NTCA did not provide comments. Officials from NATOA, NCTA, SIA, and USIIA provided technical comments that were incorporated, as appropriate. USTA and WISPA provided comments that are discussed in appendix V.

BACKGROUND

Internet access became widely available to residential users by the mid 1990s. For a few years, the primary mechanism to access the Internet was a dial-up connection, in which a standard telephone line is used to make an Internet connection. A dial-up connection offers data transmission speeds up to 56 kilobits per second (Kbps). Broadband, or high-speed, Internet access became available by the late 1990s. Broadband differs from a dial-up connection in certain important ways. First, broadband connections offer a higher-speed Internet connection than dial-up—for example, some broadband connections offer speeds exceeding 1 million bits per second (Mbps) both upstream (data transferred from the consumer to the Internet service provider) and downstream (data transferred from the Internet service provider to the consumer).[6] These higher speeds enable consumers to receive information much faster and thus enable certain applications to be used and content to be accessed that might not be possible with a dial- up connection. Second, broadband provides an “always on” connection to the Internet, so users do not need to establish a connection to the Internet service provider each time they want to go online.

Consumers can receive a broadband connection to the Internet through a variety of technologies. These technologies include, but are not limited to, the following:

- **Cable modem.** Cable television companies first began providing broadband service in the late 1990s over their hybrid-fiber coaxial networks. When provided by a cable company, broadband service is referred to as cable modem service. Cable providers were upgrading their infrastructure at that time to increase their capacity to provide video channels in response to competition from direct broadcast satellite (DBS) providers such as DirecTV® and Dish Network. By also redesigning their networks to provide for two-way data transmission, cable providers were able to use their systems to provide cable modem service. Cable modem service is primarily available in residential areas, and although the speed of service varies with many factors, download speeds of up to 6 Mbps are typical. Cable providers are developing even higher speed services.

- **DSL.** Local telephone companies provide digital subscriber line (DSL) service, another form of broadband service, over their telephone networks on capacity unused by traditional voice service. Local telephone companies began to deploy DSL service in the late 1990s—some believe, in part, as a response to the rollout of cable modem service. To provide DSL service, telephone companies must install equipment in their facilities and remove devices on phone lines that may cause interference. While most residential customers receive asymmetric DSL (ADSL) service with download speeds of 1.5 to 3 Mbps, ADSL technology can achieve speeds of up to 8 Mbps over short distances. Newer DSL technologies can support services with much higher download speeds.

- **Satellite**. Currently, three providers of satellite service can offer nearly ubiquitous broadband service in the United States. These providers use geosynchronous satellites that orbit in a fixed position above the equator and transmit and receive data directly to and from subscribers. Signals from satellites providing broadband service can be accessed as long as the user’s reception dish has a clear view of the southern sky. Therefore, while the footprint of the providers’ transmission covers most of the country, a person living in an apartment with windows only facing north, or a person living in house in a heavily wooded area might not be able to receive Internet access via satellite. Earlier Internet services via satellite could only receive Internet traffic downstream— that is, from the satellite to the subscriber—and upstream Internet traffic was transmitted through a standard telephone line connection. Currently, however, satellite companies provide both upstream and downstream connections via satellite, eliminating the need for a telephone line connection and speeding the overall rate of service. Transmission of data via satellite typically adds one-half to three-fourths of a second, causing a slight lag in transmission and rendering this service less well-suited for certain applications over the Internet. While satellite broadband service may be available throughout the country, the price for this service is generally higher than most other broadband modes; both the equipment necessary for service and recurring monthly fees are generally higher for satellite broadband service, compared with most other broadband transmission modes.

- **Wireless.** Land-based, or terrestrial, wireless networks can offer a broadband connection to the Internet from a wide variety of locations and in a variety of ways. Some services are provided over unlicensed spectrum and others over spectrum that has been licensed to particular companies.[7] In licensed bands, some companies are offering fixed wireless broadband throughout cities. Also, mobile telephone carriers—such as the large companies that provide traditional cell phone service—have begun offering broadband mobile wireless Internet service over licensed spectrum—a service that allows subscribers to access the Internet with their mobile phones or laptops as they travel across cities where their provider supports the service. Such services are becoming widely deployed and are increasingly able to offer high-speed services. A variety of broadband access technologies and services are also provided on unlicensed spectrum— that is, spectrum that is not specifically under license for a particular provider's network. For example, wireless Internet service providers generally offer broadband access in particular areas by placing a network of antennae that relay signals throughout the network. Subscribers place necessary reception equipment outside their homes that will transmit and receive signals from the nearest antenna. Also, wireless fidelity (Wi-Fi) networks—which provide broadband service in so-called "hot spots," or areas up to 300 feet—can be found in cafes, hotels, airports, and offices. Some technologies, such as Worldwide Interoperability for Microwave Access (WiMAX), can operate on either licensed or unlicensed bands, and can provide broadband service up to approximately 30 miles in a line-of-sight environment.

Under section 706 of the Telecommunications Act of 1996, Congress directs FCC to encourage deployment of advanced telecommunications capability, which includes broadband, to all Americans. In implementing the act, FCC has treated the two most widely available broadband services—cable modem and DSL service—as information services having a telecommunications component. FCC's approach of not treating such services as telecommunications services has important legal implications because a service defined as a telecommunications service could be subject to regulation under Title II of the Communications Act, which imposes substantial common carrier regulations unless the commission choose to forebear from their enforcement. As part of its responsibilities, FCC periodically issues a report to Congress on the status of advanced telecommunication capability in the United States. To prepare this report, FCC developed a periodic reporting requirement using Form 477. In November 2004, FCC modified its rules regarding the filing of the 477 form, which went into effect for the companies' second filing in 2005. Specifically, FCC removed existing reporting thresholds,[8] and companies are now required to report their total state subscribership by technology.[9]

About 30 Million American Households Purchase Broadband Service; Despite Evidence of Substantial Broadband Deployment throughout the United States, It Is Difficult to Assess Deployment Gaps in Some Areas

We found that in 2005, about 30 million American households—or 28 percent—subscribed to broadband, although households in rural areas were less likely to subscribe to broadband service than were households in urban and suburban areas. Households appear to subscribe to cable modem and DSL services—the two primary broadband services—in approximately equal numbers. FCC requires providers of broadband service to report on the geographic areas in which they serve *subscribers*, but these data are sometimes used to infer the status of *deployment* of companies' Internet infrastructure. Some stakeholders find FCC data collection efforts useful for comparison of adoption of broadband across states, but we found that the data may not be as useful for understanding the status of broadband deployment across the country.

About 30 Million American Households Purchase Broadband Service

Based on survey data from 2005,[10] we found that 28 percent of American households subscribe to broadband service. Figure 1 illustrates how American households access the Internet, by various technologies, and also shows the percentage of households that do not own a computer. As shown, 30 percent of American households subscribe to dial-up access, and about 41 percent of American households do not have an Internet connection from home. Of those households that do not access the Internet, more than 75 percent do not have a computer in the home, while the remaining households own a computer but do not have online access.

Among online households, we found 50 percent subscribe to dial-up service, and 48 percent subscribe to a broadband service.[11] Additionally, we found that of those households subscribing to a broadband service, roughly half purchase DSL service and half purchase cable modem service. (See figure 2 for the types of connections purchased by online households.)

Finally, we found that households residing in rural areas were less likely to subscribe to broadband service than were households residing in suburban and urban areas.[12] Seventeen percent of rural households subscribe to broadband service, while 28 percent of suburban and 29 percent of urban households subscribe to broadband service. (See figure 3 for the percentage of urban, suburban, and rural households purchasing broadband service.)

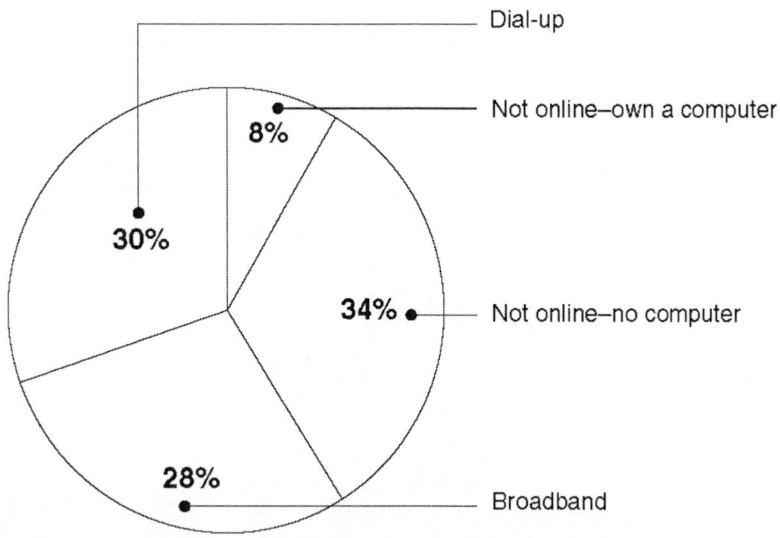

Source: GAO analysis of Knowledge Networks/SRI's The Home Technology MonitorTM: Spring 2005 Ownership and Trend Report.

Figure 1. Status of Household Computer Ownership and Internet Connection

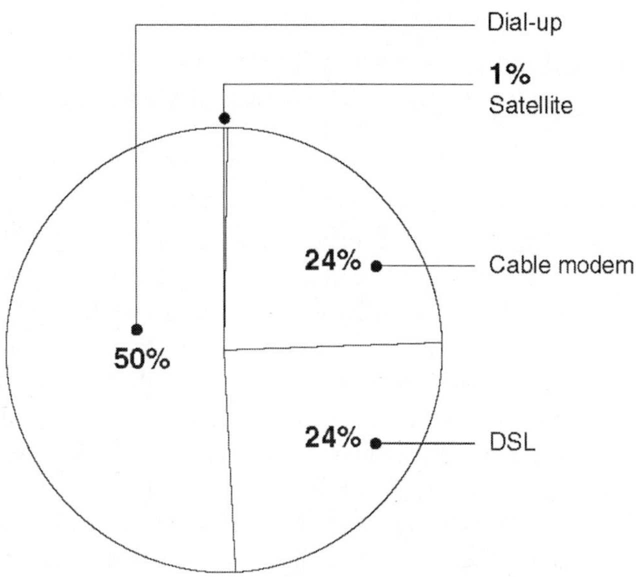

Source: GAO analysis of Knowledge Networks/SRI's The Home Technology Monitor™: Spring 2005 Ownership and Trend Report.

Figure 2. Household Online Connection

We also found that rural households were slightly less likely to connect to the Internet, compared with their counterparts in suburban areas.[13]

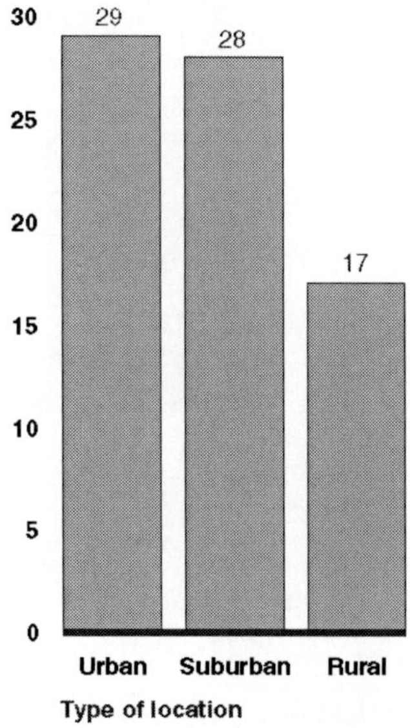

Source: GAO analysis of Knowledge Networks/SRI's The Home Technology Monitor™: Spring 2005 Ownership and Trend Report.

Figure 3. Percentage of Households Subscribing to Broadband, by Type of Location.

Deployment of Broadband Appears to Be Extensive, but FCC's Form 477 Data May Not Provide an Accurate Depiction of Gaps in Broadband Deployment

In order to fulfill its responsibility under section 706 of the Telecommunications Act, FCC collects data on companies' broadband operations. In early 2004, FCC initiated a proceeding to examine whether it should collect more detailed information for its broadband data gathering program than had previously been collected. Specifically, FCC asked whether it should do several things to enhance the broadband data including (1) requiring providers to report the speeds of their broadband services, (2) eliminating the reporting threshold such that all providers of broadband—no matter how small—must report to FCC on its services, and (3) requiring that providers report the number of connections by zip code. In late 2004, FCC released an order in which it decided to require all providers—no matter how small—of broadband to report in the 477 data collection effort on broadband and also required providers to report information about their services within speed tier categories. The commission decided not to require providers to report the number of connections (or subscribers) that they serve within each zip code or the number of connections in speed tiers or by technology within each zip code, finding that finding that such a requirement would impose a large burden on filers (particularly smaller entities), and would require significant time to implement. In particular, several providers commented in the 2004 proceeding that it would be costly and burdensome to develop the software and systems to generate the detailed zip code-level data and that the cost and burden of further reporting requirements would likely outweigh the benefits of more substantial information on broadband deployment in the United States. On the other hand, 3 state utility commissions asked FCC to gather more information within zip codes or by some other Census boundary because such information is, in their view, important for tracking broadband availability.

Based on the modifications to the filing requirements FCC implemented, FCC collects, through its Form 477 filings, information on several aspects of each company's provision of broadband at the state level, such as the total number of subscribers served, the breakdown of how those subscribers are served by technology, and estimates within each technology of the percentage of subscribers that are residential. For each technology identified in the state reporting, providers also submit a list of the zip codes in which they serve at least one customer. As discussed above, companies do not report the number of subscribers served or whether subscribers are business or residential within each zip code; they also do not report information on the locations within the zip code that they *can* serve.

In July 2005, FCC found that 99 percent of the country's population lives in the 95 percent of zip codes where at least one provider reported to FCC that it serves at least one high-speed subscriber as of December 31, 2004. In 83 percent of the nation's zip codes, FCC noted that subscribers are served by more than one provider, and the commission noted that for roughly 40 percent of zip codes in the United States, there are five or more providers reporting high-speed lines in service. Although these data indicate that broadband availability is extensive, we found that FCC's 477 data may not be useful for assessing broadband deployment at the local level.[14] While FCC, in general, notes that the 477 zip-code data are not meant to measure deployment of broadband, in its July 2005 report,[15] the commission states that in order to be able to evaluate *deployment*, the commission "instituted a formal data collection program to gather standardized information about *subscribership to high speed services. . . .*" (Emphasis added.) Based on our analysis, we found that collecting data about where companies have *subscribers* may not provide a clear depiction of their *deployment*, particularly in the context of understanding the availability of broadband for *residential* users.[16]

One quandary in analyzing broadband deployment is how to consider the availability of satellite broadband services. Even though broadband over satellite may not be seen by some as highly substitutable for other broadband technologies because of certain technical characteristics or because of its higher cost, satellite broadband service is *deployed*: Three companies have infrastructure in place to provide service to most of the country.[17] The actual *purchase* of satellite broadband is scattered throughout the country and shows up in FCC's 477 zip-code data only where someone actually purchases the service. It is not clear how satellite service should be judged in terms of deployment. Since it is available throughout the entire country, one view could be that broadband is near fully deployed. Alternatively, it could be viewed that satellite broadband—while available in most areas—does not reflect *localized* deployment of broadband infrastructure and should therefore not be counted as a deployed broadband option at all. In either case, FCC's zip-code data on satellite broadband—which is based on the pattern of scattered subscribership to this service—does not seem to be an appropriate indicator of its availability.

Aside from the question of how to view satellite deployment, other issues arise in using subscribership indicators for wire or wireless land-based providers at the zip-code level as an indicator of deployment. These issues include the following:

- Because a company will report service in a zip code if it serves just one person or one institution in that zip code, stakeholders told us that this method may overstate deployment in the sense that it can be taken to imply that there is deployment throughout the zip code even if deployment is very localized. We were told this issue might particularly occur in rural areas where zip codes generally cover a large geographic area. Based on our own analysis, we found, for example, that in some zip codes more than one of the large established cable companies reported service. Because such providers rarely have overlapping service territories, this likely indicates that their deployment was not zip-code-wide and that the number of providers reported in the zip code overstates the level of competition to individual households. We also found that a nontrivial percentage of households lie beyond the 3-mile radius of their telephone central office, indicating that DSL service was unlikely to be available to these homes.

- Companies report service in a zip code even if they only serve businesses. One academic expert we interviewed expressed a concern about this issue. Based on our own analysis, we found that many of the companies filing 477 data indicating service in particular zip codes only served business customers. As such, the number of providers reported as serving many zip codes is likely overstated in terms of the availability of broadband to residences.
- FCC requires that companies providing broadband using unbundled network elements (UNE)[18] report their broadband service in the zip code data. When a provider serves customers using UNEs, they purchase or lease underlying telecommunications facilities from other providers—usually incumbent telephone companies—to serve their customers. Having these providers report their subscribers at the state level is important to ensure that correct numbers on the total subscribers of broadband service is obtained. However, while UNE providers may make investments in infrastructure, such as in collocation equipment, they do not generally own or provide last-mile connectivity for Internet access. Thus, counting these providers in the zip-code-level data may overstate the extent of local infrastructure deployment in the sense that several reporting providers could be relying on the same infrastructure, owned by the incumbent telephone company, to provide broadband access.

Based on our analysis, we believe that the use of subscriber indicators at the zip-code level to imply availability, or deployment, may overstate terrestrially based deployment. We were able to check these findings for one state—Kentucky—where ConnectKentucky, a state alliance on broadband, had done an extensive analysis of its broadband deployment. ConnectKentucky officials shared data with us indicating that approximately 77 percent of households in the state had broadband access available as of mid-2005. In contrast, we used population data within all zip codes in Kentucky, along with FCC's 477 zip-code data for that state, and determined that, according to FCC's data, 96 percent of households in Kentucky live in zip codes with broadband service at the end of 2004. Thus, based on the experience in Kentucky, it appears that FCC's data may overstate the availability and competitive deployment of nonsatellite broadband.

Additionally, to prepare our econometric models, we relied on FCC's 477 data to assess the number of providers serving the households responding to Knowledge Networks/SRI's survey. Based on FCC's data, we found that the median number of providers reporting that they serve zip codes where the households were located was 8; in 30 percent of these zip codes, 10 or more providers report that they provide service. Only 1 percent of respondents lived in zip codes for which no broadband providers reported serving at least one subscriber, according to FCC's data. To better reflect the actual number of providers that each of the survey respondents had available at their residence, we made a number of adjustments to FCC's provider count based on our analysis of the providers, certain geographic considerations, and information provided by the survey respondents.[19] After making these adjustments, the median number of providers for the respondents fell to just 2, and we found that 9 percent of respondents likely had no providers of broadband at all.

CONNECTKENTUCKY

Source: ConnectKentucky.

The purpose of ConnectKentucky's Geographic Information Systems (GIS) mapping project is to produce an inventory of existing broadband infrastructure and service availability. The tool can produce maps at the state and census block level. Some of the items mapped include water towers, wireless towers, proposed sewer lines, roads, and population density. The maps also plot which areas are served by municipal, local exchange carrier, cable, and wireless broadband providers.

Despite these concerns about FCC's 477 data, several stakeholders, including a state regulatory office and a state industry association, said they found FCC's data useful. An official at a state governor's office also noted that analysis of FCC data allowed them to make conclusions about the extent of deployment in their state. Similarly, an official in another governor's office said that they use FCC's data to benchmark the accessibility of broadband in their state because it is the only data available. A few academic experts also told us that they use FCC's data.

A Variety of Market and Technical Factors, in Addition to Government Involvement and Access to Resources at the Local Level, Have Influenced the Deployment of Broadband

Several market characteristics appear to influence providers' broadband deployment decisions. In particular, factors related to the cost of deploying and providing broadband services, as well as factors related to consumer demand, were critical to companies' decisions about whether to deploy broadband infrastructure. At the same time, certain technical factors related to specific modes of providing broadband service influence how and where this service can be provided. Finally, a variety of federal and state government activities, as well as access to resources at the local level, have influenced the deployment of broadband infrastructure.

Several Key Market Factors Related to the Cost of Service and Demand Influence Deployment Decisions

As companies weigh investment decisions, they consider the likely profitability of their investments. In particular, because broadband deployment requires substantial investment, potential providers evaluate the cost to build and operate the infrastructure, as well as the likely demand—that is, the expected number of customers who will purchase broadband service at a given price—for their service. Based on our interviews, we found that several cost and demand factors influence providers' deployment decisions.

Cost factors

The most frequently cited cost factor affecting broadband deployment was the population density of a market. Many stakeholders, including broadband providers, state regulators, and state legislators, said population density—which is the population per square mile—was a critical determinant of companies' deployment decisions. In particular, we were told that the cost of building a broadband infrastructure in areas where people live farther apart is much higher than building infrastructure to serve the same number of people in a more urban setting. As such, some stakeholders noted that highly rural areas—which generally have low population density—can be costly to serve. Results from our econometric model confirm the views of these stakeholders. We found that densely populated and more urbanized locations were more likely to receive broadband service than were less densely populated and rural locations. For example, we found that urban areas were 9 percentage points more likely to have broadband service available than were rural areas.

Terrain was also frequently cited as a factor affecting broadband deployment decisions. In particular, we were told that infrastructure build- out can be difficult in mountainous and forested areas because these areas may be difficult to reach or difficult on which to deploy the required equipment. Conversely, we were told that flat terrain constitutes good geography for telecommunications deployment. For wireless providers, we were told that terrain concerns can present particular challenges because some wireless technologies require "line-of-sight," meaning that radio signals transmitted from towers or antennas need an unobstructed pathway—with no mountains, trees, or buildings—from the transmission site to the reception devices at users' premises. Terrain can also affect satellite broadband availability in rural areas that have rolling hills or many trees that can obstruct a satellite's signal.

BACKHAUL

Source: GAO.

In Alaska, backhaul from rural villages requires the use of satellites. This type of backhaul is costly because of the need for terrestrial infrastructure to send and receive signals from satellites as well as the need to either own or lease satellite transmitters. The high cost can affect whether providers deploy broadband service in a village. To help defray this cost, providers often look to serve an "anchor tenant" in a village, such as schools or health clinics that receive federal funding.

Some stakeholders also said costs for what is known as "backhaul" are higher for rural areas and can affect the deployment of broadband networks in these areas. Backhaul refers to the transmission of information—or data—from any of a company's aggregation points to an Internet backbone provider that will then transmit that data to any point on the Internet. This is also sometimes referred to as the "middle mile." Internet traffic originating from rural areas may need to travel a long distance to a larger city to connect to a major Internet backbone provider. Because the cost of transmitting over this distance—that is, the backhaul— can be high, one stakeholder noted that backhaul costs are another barrier to deployment in rural areas. However, using our econometric model, we did not find that the distance to a metropolitan area with a population of 250,000 or more—our proxy for backhaul—was associated with a lower likelihood of broadband deployment.

Demand factors

Based on our interviews with stakeholders, we found that certain demand factors influence providers' deployment decisions. In particular, because stakeholders noted that potential providers seek to deploy in markets where demand for their service will be sufficient to yield substantial revenues, certain elements of markets were identified as affecting the demand for broadband:

- **Ability to aggregate demand**. For rural locations, officials we spoke with stressed the importance of aggregating sufficient demand. For example, officials in one state told us that to justify the cost of deployment in rural areas where population density is low, telecommunications providers need to be able to aggregate all of the possible demand to make a business case. We were also told that aggregation is furthered by ensuring that a large "anchor tenant" will subscribe to the service. Possible anchor-tenant customers described to us included large businesses, government agencies, health-care facilities, and schools. Because the revenues from such customers will be significant, two stakeholders noted that the anchor tenant alone will help to cover a significant portion of the providers' expenses.

- **Degree of competition**. We found that the degree of existing broadband competition in a local market can inhibit or encourage deployment, depending on the circumstances. Some new entrants—companies not already providing a telecommunications service in an area—reported that they avoid entering markets with several existing providers and seek out markets where incumbent telephone and cable companies do not provide broadband service. The lack of existing service enables the entrant company to have the potential to capture many customers. At the same time, stakeholders told us that deployment by a new entrant often spurred incumbent telephone or cable providers to upgrade their infrastructures so as to compete with the entrant in the broadband market.

- **Technological expertise.** A few stakeholders noted that demand will be greater in areas where potential customers are familiar with computers and broadband, as these individuals are more likely to purchase broadband service.

Stakeholders we spoke with rarely mentioned the per-capita income of a service area as a factor determining deployment. In fact, a few stakeholders credited cable franchising requirements with ensuring deployment to low- income areas; in some cases, cable franchise agreements require cable providers to build out to all parts of the service territory. However, a 2004 study did find that areas with higher median incomes were more likely to have competitive broadband systems.[20] Similarly, results from our econometric analysis indicate that areas with higher per-capita income are more likely to receive broadband service than are areas with lower per- capita income.

Taxation of internet access

Using our econometric model, we did not find that taxation of Internet access by state governments influenced the deployment of broadband service. Taxes can raise consumer prices and reduce revenues and impose costs on providers, and thereby possibly reduce the incentive for companies to deliver a product or service. To assess the impact of Internet taxes on broadband deployment, we contacted officials in 48 states and the District of Columbia[21] to determine whether the state, or local governments in the state, imposed taxes on Internet access. To incorporate this analysis into our model, we used a binary variable to indicate the presence of the tax; that is, each state was placed into one of two groups, states with a tax and states without a tax. As such, this binary variable could potentially capture the influence of other characteristics of the states, in addition to the influence of the tax. While the parameter estimate in our model had the expected sign—indicating that the imposition of taxes may reduce the likelihood of broadband deployment—it was not statistically significant.

Certain Critical Technical Factors Affect Broadband Deployment

Many stakeholders we spoke with commented on issues related to technical characteristics of networks that provide broadband. In particular, many noted that certain technical characteristics of methods used to deliver broadband influence the extent of its availability. In terms of issues discussed for established modes of broadband delivery, we were told the following:

- DSL service can generally be provided over telephone companies' copper plant to residences and businesses that are within approximately 3 miles from the telephone company's facility, known as a central office. However, if the quality of the telephone line is not good, the distance limit can be reduced—that is, it may only be possible to provide DSL for locations within some lesser distance—perhaps 2 miles—from a central office. We were told, for example, that in some rural areas, DSL is more limited by distance because the telephone lines may be older. While the distance limit of DSL can be addressed by deploying certain additional equipment that extends fiber further into the network, this process can be expensive and time consuming.
- Across spectrum bands used to provide terrestrial wireless broadband service, technical characteristics vary: In some cases, signals may travel only a short distance, and in other cases, they may travel across an entire city; in some cases there may be a need for line-of-sight from the transmission tower to the user, but in other cases, the signals may be able to travel through walls and trees. Some stakeholders mentioned that wireless methods hold great promise for supporting broadband service.
- Satellite technology can provide a high-speed Internet service throughout most of the United States. However, the most economical package of satellite broadband service generally offers, at this time, upstream speeds of less than 200 kilobits per second, and therefore this service does not necessarily meet FCC's definition of *advanced telecommunications services*, while it does meet FCC's definition of high-speed service. Despite the near universal coverage of satellite service, consumers need a clear view of the southern sky to be able to receive transmissions from the satellites. Additionally, transmission via satellite introduces a slight delay, which causes certain applications, such as VoIP (i.e., telephone service over the Internet), and certain computer gaming to be ill-suited for use over satellite broadband.

Some emerging or expanding broadband technologies do not currently have significant subscribership, but have the potential to be important means of broadband service in the coming years. These technologies include deep fiber deployment (e.g., fiber to the home), WiMAX, broadband over power lines (BPL), and third-generation (3G) cellular. Each of these technologies has technical considerations that are influencing the nature of deployment. See appendix IV for a discussion of these technologies.

Federal and State Government Efforts, and Access to Resources at the Local Level, Have Impacted the Deployment of Broadband

We found that government involvement in several venues, and access to resources at the local level, have affected the deployment of broadband networks throughout the nation. In particular, we found that (1) certain federal programs have provided funding for broadband networks; (2) some state programs have assisted deployment; (3) state and local government policies, as well as access to resources at the local level, can influence broadband deployment; and (4) broadband deployment—particularly in more rural settings—is often influenced by the extent of involvement and leadership exercised by local government and community officials.

Federal programs have funded broadband infrastructure

We found that several federal programs have provided significant financial assistance for broadband infrastructure.

- The Universal Service Fund (USF) has programs to support improved telecommunications services. The high-cost program of the USF provides eligible local telephone companies with funds to serve customers in remote or rural areas where the cost of providing service is higher than the cost of service in more urbanized areas. The high-cost funds are distributed to providers according to formulas based on several factors, such as the cost of providing service, with funds distributed to small rural incumbent local exchange carriers (ILEC) and larger ILECs serving rural areas based on different formulas. Competitive local exchange carriers can also qualify to receive high-cost funds. While high-cost funds are not specifically targeted to support the deployment of broadband infrastructure, these funds do support *telecommunications* infrastructure that is also used to provide broadband services. We were told by some stakeholders in certain states that high-cost support has been very important for the upgrade of telecommunications networks and the provision of broadband services. In particular, some stakeholders in Alaska, Ohio, and North Dakota told us that high-cost support has been critical to small telephone companies' ability to upgrade networks and provide broadband services. Additionally, the e-rate program of the USF has provided billions of dollars in support of Internet connectivity for schools and libraries. Another USF program, the Rural Health Care Program, provides assistance for rural health facilities' telecommunications services.
- Some programs of the U.S. Department of Agriculture's Rural Utilities Service (RUS) provide grants to improve rural infrastructures providing broadband service. The Community Connect Program provides grants to deploy transmission

infrastructures to provide broadband service in communities where no broadband services exist, and requires grantees to wire specific community facilities and provide free access to broadband services in those facilities for at least 2 years. Grants can be awarded to entities that want to serve a rural area of fewer than 20,000 residents. Approximately $9 million was appropriated in 2004 as well as in 2005 for this purpose.

- RUS's Rural Broadband Access Loan and Loan Guarantee program provides loans[22] to eligible applicants to deploy infrastructures that provide broadband service in rural communities that meet the program's eligibility requirements. A wide variety of entities are eligible to obtain loans to serve small rural communities. To obtain a 4 percent loan, the applicant must plan on serving a community with no previously available broadband service, but loans at the Treasury interest rate do not have such a requirement.
- The Appalachian Regional Commission's Information Age Appalachia program focuses on assisting in the development and use of telecommunications infrastructure. The program also provides funding to assist in education and training, e-commerce readiness, and technology-sector job creation. We were told that in Kentucky, funding from the commission assisted the development and operations of ConnectKentucky, a state alliance that focuses on broadband deployment and adoption. The Appalachian Regional Commission also provided some funding for projects in Ohio and Virginia.

UNIVERSAL SERVICE FUND

Source: GAO.

The Universal Service Fund facilitates education and health care in rural Alaska by providing broadband Internet connections. Access to broadband connections provides learning opportunities for children, such as higher-level curriculum, and teacher certification programs. Additionally, it facilitates health care by allowing more immediate care in local villages rather than requiring long trips to a larger clinic or hospital.

Various state programs assist the deployment of broadband services

A number of states we visited have had programs to assist the deployment of broadband services, including the following:

- The Texas Telecommunications Infrastructure Fund began in 1996 and according to an official of the Texas Public Utility Commission committed to spend $1 billion on telecommunications infrastructure in Texas. Public libraries, schools, nonprofit medical facilities, and higher education institutions received grants for infrastructure and connectivity to advanced communications technology. The program is no longer operational.
- ConnectKentucky is an alliance of technology-focused businesses, government entities, and universities that work together to accelerate broadband deployment in the state. ConnectKentucky focuses on three goals: (1) raising public awareness of broadband services, (2) creating market-driven strategies to increase demand—particularly in rural areas, and (3) initiating policy to reduce regulatory barriers to broadband deployment.
- The Virginia Tobacco Indemnification and Community Revitalization Commission partially funded Virginia's Regional Backbone Initiative. The backbone initiative is designed to stimulate economic development opportunities by encouraging the creation of new technology-based business and industry in southern Virginia, which has historically relied heavily on tobacco production.

Local issues and access to resources impact the deployment of broadband services

The ability of a company to access local rights-of-way, telephone and electric poles, and wireless-tower sites can influence the deployment of broadband service. In particular, a few stakeholders we spoke with said difficulty in gaining access to these resources can serve as a barrier to the rapid deployment of broadband service because accessing these resources was a time-consuming and expensive process. Companies often require access to rights-of-way—such as areas along public roads—in order to install infrastructure for broadband service. In some instances, companies can face challenges gaining access to rights-of-way, which can hinder broadband deployment. For example, we were told that in one California community, providers had difficulty bringing wires across a highway, which limited their ability to provide service in all areas of the community. Some companies also require access to telephone and electric poles to install their broadband infrastructure. Depending on the entity owning the pole, we were told that acquiring access to poles could be costly and time consuming. For example, one BPL provider told us that it encountered difficulty accessing poles owned by the telephone company. Finally, wireless companies need access to towers or sites on which they can install facilities for their broadband infrastructure. A few stakeholders we spoke with told us that gaining this access can be a difficult process. For example, one company said providers are often challenged by the need to learn each town's tower-siting rules. While some stakeholders identified problems gaining access to these resources, other stakeholders did not identify access to rights-of-way, poles, and other resources as issues in deploying broadband services.

We found that the video-franchising process can also influence the deployment of broadband service because companies may be building infrastructure to simultaneously provide both video and broadband services. To provide video service, such as cable television, companies usually must obtain a franchise agreement from a local government. Some stakeholders assert that the video-franchising process can delay the deployment of broadband service because providers must negotiate with a large number of local jurisdictions. Further, these negotiations can be time consuming and costly. As a result, these stakeholders believe that local franchising can hinder their ability to deploy broadband infrastructures. Alternatively, some stakeholders believe that the video-franchising process is important because it helps promote deployment of broadband service to all areas of a community. For example, some jurisdictions have ubiquity requirements mandating deployment to all areas of a community, including those that are less affluent. These stakeholders argue that without the local ubiquity requirement, service providers could "cherry pick" and exclusively provide broadband services to more economically desirable areas.

BERKSHIRE CONNECT ATTRACTS DEPLOYMENT TOWESTERN MASSACHUSETTS

BERKSHIRE CONNECT, INC.

Source: Berkshire Connect, Inc.

Berkshire Connect, Inc. was founded to respond to the lack of reliable and affordable T-1 service for businesses in the Berkshire region of Massachusetts. The organization aggregates demand for broadband service among users in the region, such as hospitals, schools, large and small businesses, and nonprofit organizations. Through a request for proposal process, the organization selected "preferred providers" that offer members of Berkshire Connect an array of advanced telecommunications services at competitive prices. Berkshire Connect's primary goals are to provide and ensure competitive pricing, equal pricing throughout Berkshire County, sustained competition, and a community focus regarding the provision of telecommunications services.

In some instances, municipal governments provide broadband infrastructure and service. For example, we spoke with officials in five municipal governments that provide wire-based broadband service, often in conjunction with the government's electric utility. We also spoke with one municipal government that provided wireless broadband service. A few of these municipal government officials told us that their municipality had undertaken this deployment because they believe that their communities either do not have, or do not have adequate, private broadband service. A significant number of stakeholders we interviewed support a municipality's right to provide broadband services and believe that broadband service is a public utility, such as water and sewer. Some support municipal deployment of broadband, regardless of whether other providers are available in that area, while other stakeholders support a municipality's right to deploy broadband service only if there are no other broadband providers serving the area. However, other stakeholders we spoke with oppose municipal government deployment of broadband service. These stakeholders believe that municipal governments are not prepared to be in the business of providing broadband and that municipal deployment can hinder private-sector deployment.

Community leadership encourages the deployment of broadband services

We found that the cost of serving rural areas presents a challenge to the nationwide goal of universal access to broadband. One of the ways that some communities have addressed the lack of market entry into rural areas has been through initiatives wherein community leaders have worked to enhance the likely market success of private providers' entry into rural broadband markets. For example, some community leaders have worked to aggregate demand—that is, to coordinate the Internet needs of various users so that a potential entrant would be able to support a business plan. We were told that this leadership—sometimes by key government officials, sometimes through partnerships—was seen as critical in helping to spur the market in some unserved areas.[23] The following examples illustrate this point:

- In Massachusetts, several regional coalitions that have been called "connect" projects focus on demand aggregation as a tool to encourage further deployment of telecommunications backbone and broadband networks in more rural parts of the state that were not well served by broadband providers. In particular, three such regional groups said their demand aggregation model is designed to maximize the purchase of broadband services in their region by working with local hospitals, schools, home businesses, small business, and residents to demonstrate the full extent of the demand for broadband and thus encourage private investment in infrastructure. For the one project that was the most developed, a few stakeholders told us that the group had been critical in helping to spur infrastructure development in the area, and that leadership by State government was important to the development of the initiative.
- ConnectKentucky, as discussed earlier, is an example of a state coalition taking a leadership role to develop information on state deployment levels, educate citizens about the benefits of broadband service, and advocate broadband-friendly policies with the state legislature. Throughout our meetings in Kentucky, the work of ConnectKentucky was stated to have been instrumental in the development of a common understanding of the state of broadband deployment and adoption as well as in instigating new initiatives to advance the market. The key element of ConnectKentucky that was cited as crucial to its success was leadership from state government, in particular from the governor's office.
- In Alaska, we found that in one remote area—Kotzebue, a community 26 miles above the Arctic Circle—strong local leadership was important to the development of a public-private partnership that provides improved medical care to the region. The local leadership from the health cooperative brought together parties in the community and worked with them to develop a plan to provide enhanced health service throughout the community's villages. The Maniilaq Health Center uses a wireless "telecart" with a video camera that can send high-quality, real-time sound and video between the center and Anchorage. The center's physicians are able to

perform procedures under the guidance of experts in Anchorage who can "remotely" look over the physicians' shoulders. In addition, there are village clinics staffed by trained village health aides. These village clinics are connected to the main health center via a broadband link that allows them to share records and diagnoses via the telecart.

A VARIETY OF HOUSEHOLD AND SERVICE CHARACTERISTICS INFLUENCE THE ADOPTION OF BROADBAND

We developed an econometric model to assess the many factors that might influence whether a household purchases broadband service. The model examined two types of factors: the tax status of states in which respondents live, and the characteristics of households. We also discussed these issues, as well as the influence of characteristics and uses of broadband service, with stakeholders.

Based on our model and interviews with stakeholders, we identified several characteristics of households that influence broadband adoption.

First, our model indicated that high-income households are 39 percentage points more likely to purchase broadband service than are low-income households.[24] Similarly, some stakeholders we spoke with stated that adoption of broadband service is more widespread in communities with high income levels. A key underlying factor may be that computer ownership is substantially higher among higher-income households, according to a survey conducted by the Census Bureau. Second, our model results showed that households with a college graduate are 12 percentage points more likely to subscribe to broadband services compared with households without a college graduate. In fact, when discussing the effects of education on the demand for broadband, we were told that some college graduates see broadband as a necessity and would be less likely to choose to live in a rural area that did not have adequate broadband facilities. Third, we found that households headed by young adults are more likely to purchase broadband than are households headed by a person 50 or older.[25] Similarly, a few stakeholders we spoke with said that older adults are less likely to purchase broadband. This may be the case because older Americans generally have lower levels of computer ownership and computer familiarity. We also were told that households with children in school are more likely to have broadband service. Figure 4 provides some descriptive statistics to illustrate the relationship between several demographic characteristics and the adoption of broadband.

We also examined whether households residing in rural areas were less likely to purchase broadband service than those living in urban areas. As noted earlier, we found that only 17 percent of rural households subscribe to broadband service. Our model indicated, however, that when the availability of broadband to households, as well as demographic characteristics, are taken into account, rural households no longer appear less likely than urban households to subscribe to broadband. That is, the difference in the subscribership to broadband among urban and rural households appears to be related to the difference in availability of the service across these areas, and not to a lower disposition of rural households to purchase the service.

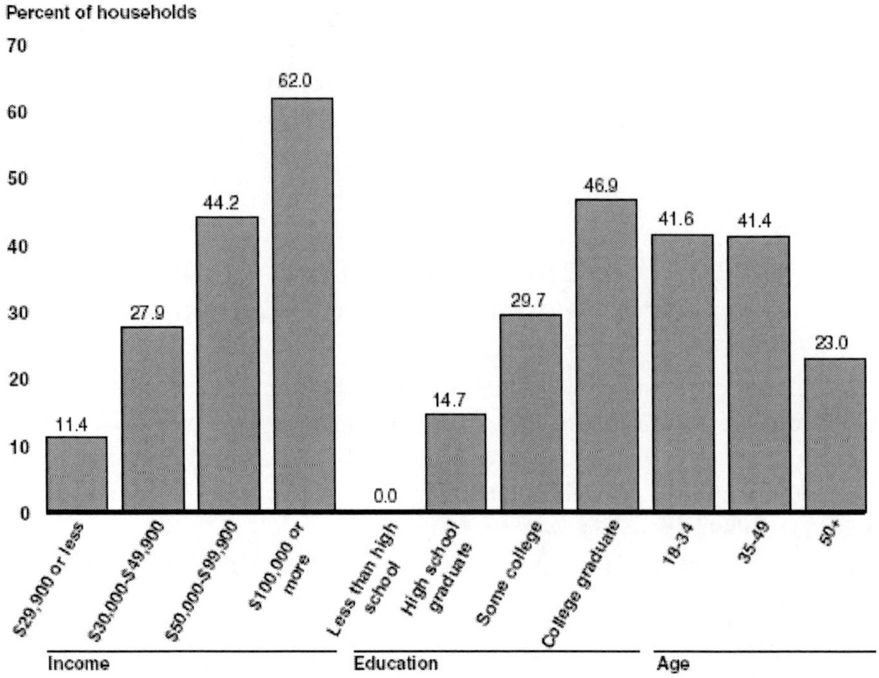

Source: GAO analysis of Knowledge Networks/SRI's The Home Technology MonitorTM: Spring 2005 Ownership and Trend Report.

Figure 4. Factors Influencing Subscription to Broadband

In addition to household characteristics, we also found that characteristics and uses of broadband service available to consumers can also influence the extent to which households purchase broadband service.

- Some stakeholders we spoke with mentioned that the price of broadband service is an important factor affecting a household's decision to purchase this service. Some stakeholders mentioned, for example, that one of the key reasons for the recent surge in DSL subscribership is due to recent price declines for the service: Some providers are now offering DSL for less than $15 per month. Conversely, because satellite broadband service is expensive and also requires the upfront purchase of expensive equipment needed to receive the satellite signal, several of those we spoke with said that the expense of satellite broadband deters its purchase. In fact, a recent study suggests that areas served by multiple providers, where prices may tend to be lower, may have higher rates of broadband adoption.[26] However, because we lacked data on the price of broadband service, we were unable to include this variable in our econometric model.[27] We did not find that the number of companies providing broadband service affected the likelihood that a household would purchase broadband service.
- Some stakeholders also told us that the availability of applications and content not easily accessible through dial-up, as well as the degree to which consumers are aware of and value this availability, contribute to a household's decision to adopt broadband. For example, some functions, applications, and content—such as gaming,

VoIP, and music and video downloads—either need or function much more effectively with broadband service than with dial-up service and, as such, make broadband a major attraction for households that value these types of services and content. Alternatively, some applications, such as e-mail, function adequately with dial-up service, and for households that primarily use the Internet for e-mail, there may be little need to upgrade to broadband service. Several of those we spoke with noted that a "killer application"—one that nearly everyone would view as essential and might entice more American households to adopt broadband—has not yet emerged.

- We also examined whether the tax status of the state in which each survey respondent lived influenced their likelihood to adopt broadband service. As mentioned earlier, we used a binary variable to represent the presence of Internet taxation. As such, the variable may capture the influence of other characteristics of the states in which the households resided, in addition to the influence of the tax. Further, lacking a variable for the price of broadband service, we cannot assess how the imposition of the tax influenced the price of the service. Using our model, we found that the parameter estimate had the expected sign— indicating that the imposition of the tax may have reduced the likelihood that a household would purchase broadband service. While the estimate was not statistically significant at the 5 percent level, it was statistically significant at the 10 percent level, perhaps suggesting that it is a weakly significant factor. However, given the nature of our model, it is unclear whether this finding is related to the tax or other characteristics of the states in which households resided.

Stakeholders Identified Several Options to Address the Lack of Broadband in Certain Areas, but Challenges Exist with Implementation

Stakeholders we spoke with identified several options to facilitate greater broadband service in unserved areas; however, each option poses special challenges. RUS broadband programs provide a possible means for targeted assistance to unserved areas, but stakeholders raised concerns about the effectiveness of the loan program and its eligibility criteria. USF programs have indirectly facilitated broadband deployment in rural areas, but it is unclear whether the program should be expanded to directly support broadband service. Finally, wireless technologies could help overcome some of the cost and technological limitations to providing service in remote locations, but congestion and the management of the spectrum remain possible barriers.

RUS Broadband Programs Could Provide a Source of Targeted Assistance, but Stakeholders Identified Several Concerns with the Programs

As mentioned earlier, RUS provides support through grants and loans to improve rural infrastructures providing broadband service. The Community Connect Broadband grant program provides funding for communities where no broadband service currently exists. One loan program, which provides loans at 4 percent, also requires that no existing broadband providers be present in a community, but loans at the Treasury interest rate are available to entities that plan to serve communities with existing broadband service. Several stakeholders with whom we spoke, as well as the findings of a recent report by the Inspector General (IG) of the Department of Agriculture, raised concerns about these programs:[28]

- **Effectiveness of loans**. It is not clear whether a loan program—such as the RUS loan program—is effective for helping rural areas gain access to broadband services. RUS requires applicants to submit an economically viable business plan—that is, applicants must show that their business will be sufficiently successful such that the applicant will be capable of repaying the loan. But developing a viable broadband business plan can be difficult in rural areas, which have a limited number of potential subscribers. As a result, RUS has rejected many applications because the applicant could not show that the business plan demonstrated a commercially viable and sustainable business. In fact, the agency has been unable to spend all of its loan program funds. Since the inception of the program in 2002, the agency has fallen far short of obligating the available funding in this program. For example, RUS officials told us that in 2004, they estimated that the appropriations for the broadband loan program could support approximately $2.1 billion in loans, but only 28 percent of this amount—or $603 million— was awarded for broadband projects. RUS officials also told us that its 2005 appropriations could support just over $2 billion in loans, but only 5 percent—or $112 million—was awarded to broadband projects. One stakeholder we spoke with suggested that a greater portion of RUS funds should be shifted from loans to grants in order to provide a more significant level of assistance for rural broadband deployment. RUS officials noted that they are currently evaluating the program and recognize that the program criteria limit the ability of the agency to utilize their full loan funding.

- **Competitive environment requirements**. During our interviews, some stakeholders expressed concerns about how the presence of existing broadband deployment was considered in evaluating RUS grant and loan applications. In the case of the grant program, RUS approves applications only for communities that have no existing broadband service. Some local government officials and a company we spoke with noted that this "unserved" requirement for RUS grants can disqualify certain rural communities that have very limited Internet access—perhaps in only one small part of a community.[29] Alternatively, regarding the Treasury rate loan program, a few providers and the IG's report criticized the program for supporting the building of new infrastructure where infrastructure already existed. In particular, we learned that loans were being let for deployment in areas that already had at least one provider and in some cases had several providers. As such, it is not clear whether these funds are being provided to communities most in need. RUS officials noted, however, that the statute specifically allows such loans. Additionally, the issue of how the status of existing service is gauged was a concern for one provider we spoke with. RUS obtains information about existing providers from applicants, and agency officials told us that agency field representatives review the veracity of information provided by applicants during field visits. However, RUS officials told us that FCC zip-code data is not granular enough for their needs in evaluating the extent of broadband deployment in rural areas.

- **Community eligibility**. A few local officials we spoke with criticized the community size and income eligibility requirements for the grant and loan programs. In Massachusetts, one stakeholder said that most small towns in part of that state exceed RUS's population requirements and thus do not qualify for grants or loans. The grant and loan programs also have per-capita personal income requirements. One service provider in Alaska said that the grant program income eligibility requirements can exclude Alaskan communities, while failing to take into account the high cost of living in rural Alaska.

- **Technological neutrality.** Satellite companies we spoke with said RUS's broadband loan program requirements are not readily compatible with their business model or technology. Once a company launches a satellite, the equipment that individual consumers must purchase is the remaining infrastructure expense. Because the agency requires collateral for loans, the program is more suited for situations where the *providers*, rather than individual consumers, own the equipment being purchased through the loan. Yet, when consumers purchase satellite broadband, it is common for them to purchase the equipment needed to receive the satellite signal, such as the reception dish. Additionally, broadband service must be provided at a speed of at least 200 kilobits *in both directions*—which is not necessarily the case for satellite broadband—for it to qualify for RUS loans. Moreover, RUS officials noted that for satellite broadband providers to be able to access RUS loans, they would have to demonstrate that each customer lives in a community that meets the community size eligibility requirement. As such, this program may not be easily utilized by satellite broadband providers. Yet for some places, satellite could be a cost-effective mechanism to provide broadband infrastructure into rural areas. For example, in 2005, the RUS Community Connect program provided grants to 19 communities that average 554 residents and 194 households. The total cost of these grants was roughly $9 million. Thus, RUS spent an average of $2,443 per covered household,[30] but the cost per household that adopted broadband would be even higher since only a subset of these households would choose to subscribe to broadband service. By contrast, two satellite providers we spoke with estimate that their consumer equipment and installation costs are roughly $600 per subscribing household. These figures might not fully represent the full nature of the services provided through the grant program and those available via satellite; for example, grantees of the RUS program are required to provide free Internet service to community centers.

USF Programs Indirectly Support Broadband Service, but Several Stakeholders Expressed Concerns

While the USF program does not directly fund broadband service, the funding provided to support telecommunications networks indirectly supports the development of infrastructure that can provide many communications services, including broadband. USF's high-cost program helps maintain and upgrade telecommunications networks in rural areas. Three stakeholders we spoke with in Alaska, Ohio, and North Dakota attributed the relative success of broadband deployment in rural areas to the USF program. Additionally, the Schools and Libraries Program and the Rural Health Care Program help facilitate broadband service to specific locations; according to two providers in Alaska, these programs have been very beneficial in bringing some form of broadband service to rural Alaskan villages that might have received no service without these government programs.

However, stakeholders we spoke with identified several concerns about the USF program:

- Large ILECs serving rural areas and rural ILECs receive high-cost fund support under different formulas. The two types of ILECs have different eligibility criteria under which they can qualify to receive high-cost support and more support is provided to rural companies than to nonrural companies serving rural areas.[31] Two stakeholders we spoke with suggested that the eligibility criteria should be modified, such that the criteria better reflect the cost to provide service in particular areas, rather than the type of company providing the service. Alternatively, two stakeholders we spoke with favor the current eligibility criteria and funding mechanism.
- Two stakeholders we spoke with expressed concerns about a lack of coordination across USF funding sources, which could lead to inefficient use of funds and inadequate leveraging of funds. For example, in Alaska, two stakeholders noted that governments and providers receive "silos" of funding for schools, libraries, and rural health centers. Because the programs are narrowly defined, multiple entities might be the recipient of funding for broadband service, which could lead to multiple broadband connections in relatively small rural communities. One stakeholder noted that since each entity might use only a fraction of its available broadband capacity, there can be capacity for Internet traffic available for other uses or users, but funding recipients are sometimes not allowed to share this capacity, either with other entities or with residents in the community. Thus, communities may be unable to leverage the available funding for other uses.
- While two stakeholders we spoke with suggested expanding the USF program to include broadband service, we found little support for this overall. Some stakeholders we spoke with expressed concern about funding the USF program at current levels of support. These stakeholders fear that expanding the USF program to include broadband service, which would increase program expenditures and thus require additional funding, could undermine support for the entire USF program.

Resolving Spectrum Congestion and Management Concerns Could Facilitate Greater Wireless Broadband Service

As mentioned previously, certain wireless technologies hold the potential for supporting broadband service in difficult-to-serve rural areas. In less densely populated areas, installing wire-based facilities for cable modem and DSL service represents a significant cost factor. Therefore, certain wireless technologies may be a lower-cost way to serve rural areas than wireline technologies.

While wireless technologies hold the promise of expanding the availability of broadband, some stakeholders we spoke with expressed concern about the degree of congestion in certain bands as well as the management of spectrum. For example, in some geographic areas, we heard that congestion in certain unlicensed spectrum bands makes providing wireless broadband Internet access more difficult, and a few stakeholders said that with more unlicensed spectrum, wireless providers could support greater broadband deployment. Additionally, wireless providers we spoke with also expressed concern about the management of spectrum, particularly the quality of certain bands and quantity of spectrum available for wireless broadband service. Two stakeholders mentioned that spectrum allocated to wireless broadband service is susceptible to having communications obstructed by interference from trees and buildings. In a 2005 report, we noted that experts agreed that the government should evaluate its allocation of spectrum between licensed and unlicensed uses.[32] But we also noted that these experts failed to agree on whether FCC should dedicate more or less spectrum to unlicensed uses. In June 2006, FCC will conduct an auction of spectrum dedicated to advanced wireless services, which will make available 90 MHz of spectrum for wireless broadband services. FCC staff also noted that the commission has other efforts underway to increase available spectrum for wireless broadband services.

CONCLUSION

In the past several years, the importance of broadband for Americans and for the American economy has been articulated by interested stakeholders, as well as by the President, Congress, and the last several FCC chairmen. Universal availability of broadband has been set forth as a policy goal for the near term—2007. And progress toward this goal has been substantial. The availability of broadband to residential consumers has grown from its nascent beginnings in the latter part of the 1990s to broad coverage throughout the country. In the last 10 years, providers in traditional communications industry segments—telephone and cable—have upgraded and redesigned miles of their networks in order to offer broadband services. The provision of broadband through various wireless means, as well as over the existing electricity infrastructure, have also been developed, and for many, if not most Americans, the burgeoning broadband marketplace is characterized by competitive choice in broadband access and creative and ever-expanding applications and content. Many would consider the rollout of broadband infrastructure as a success story of entrepreneurial initiative.

But not all places or people have experienced the full benefits of this rapid rollout of broadband services. As with many other technologies, the costs of bringing broadband infrastructure to rural America can be high. For private providers who must weigh the costs and returns of their investments, the feasibility of serving the most rural parts of our country may not work within a reasonable business model. While there are federal support mechanisms for rural broadband, it is not clear how much impact these programs are having or whether their design suggests a broad consideration of the most effective means of addressing the problem. And one of the difficulties of assessing the gaps in deployment and where to target any federal support is that it is hard to know exactly where broadband infrastructure has not been deployed. FCC does collect data on the geographic extent of providers' service, but these data are not structured in a way that accurately illustrates the extent of deployment to residential users. Without accurate, reliable data to aid in analysis of the existing deployment gaps, it will be difficult to develop policy responses toward gaps in broadband availability. This could hinder our country's attainment of universally available broadband. And as the industry moves quickly to even higher bandwidth broadband technologies, we risk leaving some of the most rural places in America behind.

RECOMMENDATION FOR EXECUTIVE ACTION

In a draft of this report provided to FCC for review and comment, GAO recommended that FCC identify and evaluate strategies for improving the 477 data such that the data provide a more accurate depiction of residential broadband deployment throughout the country. In oral comments regarding this recommendation, FCC staff acknowledged that the 477 data have some limitations in detailing broadband deployment, but also noted that there had recently been a proceeding examining its broadband data collection efforts and that some changes to the data collection had been implemented. In that proceeding, the commission also determined that it would be costly and could impose large burdens on filers—particularly small entities—to require any more detailed filings on broadband deployment. Although FCC staff told us that analysis of potential costs had been conducted, exact estimates of these costs and burdens have not yet been determined. Moreover, many have expressed concern about ensuring that all Americans—especially those in rural areas—have access to broadband technologies. Policymakers concerned about full deployment of broadband throughout the country will have difficulty targeting any assistance to that end without accurate and reliable data on localized deployment. As such, we recommend that FCC develop information regarding the degree of cost and burden that would be associated with various options for improving the information available on broadband deployment and should provide that information to the Senate Committee on Commerce, Science, and Transportation and the House Energy and Commerce Committee in order to help them determine what actions, if any, are necessary to employ going forward.

AGENCY COMMENTS AND OUR EVALUATION

We provided a draft of this report to the Department of Agriculture, the Department of Commerce, and the Federal Communications Commission for their review and comment. The Department of Agriculture provided no comments. The Department of Commerce and FCC provided technical comments that we incorporated, as appropriate. FCC did not comment on the final recommendation contained in this report.

We also provided a draft of this report to several associations representing industry trade groups and state and local government entities for their review and comment. Specifically, the following associations came to GAO headquarters to review the draft: Cellular Telecommunications and Internet Association (CTIA), National Association of Regulatory Utility Commissioners (NARUC), National Association of Telecommunications Officers and Advisors (NATOA), National Cable and Telecommunications Association (NCTA), National Telecommunications Cooperative Association (NTCA), Satellite Industry Association (SIA), US Internet Industry Association (USIIA), United States Telecom Association (USTA), and Wireless Internet Service Providers Association (WISPA). Officials from CTIA, NARUC, and NTCA did not provide comments. Officials from NATOA, NCTA, SIA, and USIIA provided technical comments that were incorporated, as appropriate. USTA and WISPA provided comments that are discussed in appendix V.

JayEtta Z. Hecker
Director, Physical Infrastructure Issues

APPENDIX I. SCOPE AND METHODOLOGY

The objectives of the report were to provide information on (1) the current status of broadband deployment and adoption, (2) the factors that influence the deployment of broadband networks, (3) the factors that influence the adoption of broadband service by households, and (4) the options that have been suggested to spur greater broadband deployment and adoption. To respond to the four objectives, we used a variety of approaches.

To gather opinions for all four objectives, we employed a case-study approach. This approach allowed us to identify issues at the state and local level that would not be apparent in nationwide data. We selected eight states for our case studies: Alaska, California, Kentucky, Massachusetts, North Dakota, Ohio, Texas, and Virginia. We selected these states based on Census Bureau data on statewide income, urbanization, population density, and percentage of households using the Internet. We also considered whether each state taxed Internet access. We sought to include states in diverse categories of each of our selection criteria. In each state, we interviewed state and local officials, including local franchising authorities, state public utility regulators, representatives from state governor's offices; associations; private cable and telephone providers; wireless Internet service providers; and municipal and cooperative telecommunications providers.

We also spoke with a variety of individuals and organizations knowledgeable about broadband services. In particular, we spoke with industry providers, trade associations, and academic experts. We also spoke with representatives from the Federal Communications

Commission (FCC), the National Telecommunications and Information Administration of the Department of Commerce, and the Rural Utilities Service of the Department of Agriculture.

To assess the factors influencing the deployment and adoption of broadband, we used survey data from Knowledge Networks/SRI's *The Home Technology Monitor™: Spring 2005 Ownership and Trend Report.* Knowledge Networks/SRI is a survey research firm that conducted a survey on household ownership and use of consumer electronics and media. Knowledge Networks/SRI interviewed approximately 1,500 randomly sampled telephone households, asking questions about the household's purchase of computers and Internet access. All percentage estimates from the Knowledge Networks/SRI survey have margins of error of plus or minus 7 percentage points or less, unless otherwise noted. See appendix II for a discussion of the steps we took to evaluate the reliability of Knowledge Networks/SRI's data. Using the data from Knowledge Networks/SRI, we estimated two econometric models. One model examined the factors affecting broadband deployment. We also developed a model to examine the factors affecting a household's adoption of broadband services. See appendix III for a more detailed explanation of, and results from, our deployment and adoption models.

To assess the status of broadband deployment, we used FCC's Form 477 data that identified companies providing broadband service by zip code. We used FCC's data to identify the companies reporting to provide broadband service in the zip codes where respondents to Knowledge Networks/SRI's survey resided. To assess the reliability of FCC's Form 477 data, we reviewed documentation, interviewed knowledgeable officials, and performed electronic testing of the data elements used in our analyses. We made several adjustments to these data, such as excluding satellite companies and companies only providing service to businesses. See appendix III for more on our methodology concerning adjustment to FCC's 477 data. With these adjustments to the data, we determined that they were sufficiently reliable for the purposes of this report.

We conducted our work from April 2005 through February 2006 in accordance with generally accepted government auditing standards.

Appendix II. Data Reliability

To obtain information on the types of Internet access purchased, or adopted, by U.S. households, we purchased existing survey data from Knowledge Networks Statistical Research (Knowledge Networks/SRI). Their survey was completed with 1,501 of the estimated 3,127 eligible sampled households for a response rate of 48 percent. The survey was conducted between February 22 and April 15, 2005.

The study procedures yielded a sample of members of telephone households in the continental United States using a national random-digit dialing method. Survey Sampling Inc. (SSI) provided the sample of telephone numbers, which included both listed and unlisted numbers and excluded blocks of telephone numbers determined to be nonworking or business-only. At least five calls were made to each telephone number in the sample to attempt to interview a responsible person in the household. Special attempts were made to contact refusals and convert them into interviews; refusals were sent a letter explaining the purpose of the study and an incentive. Data were obtained from telephone households and are weighted to the total number of households in the 2005 Current Population Survey adjusted for multiple phone lines.

As with all sample surveys, this survey is subject to both sampling and nonsampling errors. The effect of sampling errors due to the selection of a sample from a larger population can be expressed as a confidence interval based on statistical theory. The effects of nonsampling errors, such as nonresponse and errors in measurement, may be of greater or lesser significance but cannot be quantified on the basis of available data.

Sampling errors arise because of the use of a sample of individuals to draw conclusions about a much larger population. The study's sample of telephone numbers is based on a probability selection procedure. As a result, the sample was only one of a large number of samples that might have been drawn from the total telephone exchanges from throughout the country. If a different sample had been taken, the results might have been different. To recognize the possibility that other samples might have yielded other results, we express our confidence in the precision of our particular sample's results as a 95 percent confidence interval. We are 95 percent confident that when only sampling errors are considered each of the confidence intervals in this report will include the true values in the study population. All percentage estimates from the survey have margins of error of plus or minus 7 percentage points or less, unless otherwise noted. The 95 percent confidence interval for the estimate of the total number of U.S. households that subscribed to broadband service in 2005 is 28.5 million to 33.7 million households.

In addition to the reported sampling errors, the practical difficulties of conducting any survey introduce other types of errors, commonly referred to as nonsampling errors. For example, questions may be misinterpreted, some types of people may be more likely to be excluded from the study, errors could be made in recording the questionnaire responses into the computer-assisted telephone interview software, and the respondents' answers may differ from those who did not respond. Knowledge Networks/SRI has been fielding versions of this survey for over 20 years. In addition, to reduce measurement error, Knowledge Networks/SRI employs interviewer training, supervision, and monitoring, as well as computer- assisted interviewing to reduce error in following skip patterns.

For this survey, the 48 percent response rate is a potential source of nonsampling error; we do not know if the respondents' answers are different from the 52 percent who did not respond. Knowledge Networks/SRI took steps to maximize the response rate—the questionnaire was carefully designed and tested through deployments over many years, at least five telephone calls were made at varied time periods to try to contact each telephone number, the interview period extended over about 8 weeks, and attempts were made to contact refusals and convert them into interviews.

Because we did not have information on those contacted who chose not to participate in the survey, we could not estimate the impact of the nonresponse on our results. Our findings will be biased to the extent that the people at the 52 percent of the telephone numbers that did not yield an interview have different experiences with Internet access than did the 48 percent of our sample who responded. However, distributions of selected household characteristics (including presence of children, race, and household income) for the sample and the U.S. Census estimate of households show a similar pattern.

To assess the reliability of these survey data, we relied on a prior GAO report that made use of the Knowledge Networks/SRI 2004 survey for a similar purpose. In that prior assessment, we determined that the data were sufficiently reliable for our purposes. For this report we reviewed Knowledge Networks/SRI's documentation of survey procedures for 2005 and compared them to the procedures used in their 2004 survey. We determined that their survey methodology was substantively unchanged.

Additionally, we performed electronic testing of the 2005 survey data elements used in this report. We determined that the data were sufficiently reliable for the purposes of this report.

APPENDIX III. BROADBAND DEPLOYMENT AND ADOPTION MODELS

This appendix describes our models of broadband deployment and adoption. Specifically, we discuss (1) the design of our models, (2) the data sources, (3) our methodology for assessing broadband deployment, and (4) the estimation methodology and results.

Design of Our Broadband Deployment and Adoption Models

A company will deploy broadband service in an area only if the company believes that such a deployment will be profitable. Similarly, a household will purchase, or adopt, broadband service only if the value, or utility, to members of the household exceeds the price the household must pay to receive the service. In this section, we explain the two models we developed to examine the factors that influence the deployment and adoption of broadband service.

Deployment model

A company will deploy broadband service in an area only if the company believes that such a deployment will be profitable. Based on conversations with industry stakeholders, including companies deploying broadband service, we identified a number of factors that influence a company's decision to deploy broadband service. In particular, the following factors may influence the decision to deploy broadband service: population density, terrain, backhaul costs, existing or potential competition, the technical expertise of the population, the income of the population, and regulatory policies (such as rights-of-way policies). We also reviewed relevant studies, and noted the same and additional factors that may influence the deployment of broadband service.[33] Some of these factors, such as the population density and backhaul, will influence the cost of providing broadband service, while other factors, such as the income of the population, will influence the potential revenues that a company may hope to generate. Together, these revenue and cost factors will influence the potential profitability of providing broadband service, and ultimately the decision to deploy broadband service.

To empirically test these hypotheses, we estimated the following econometric model; since all the variables identified above were not available, we were unable to include some of the variables—such as terrain—in our model. The decision to deploy broadband service is a function of

- the population in the area;
- the population density in the area;

- the percentage of the population residing in an urban area;
- the per-capita income in the area;
- the educational attainment of the population in the area;
- the population teleworking in the area;
- the age of the population in the area;
- the distance to a metropolitan area with a population of 250,000 or more; and
- whether the state in which the area is located imposed a tax on Internet access in 2005.

Adoption model

Households will purchase, or adopt, broadband service only if the value, or utility, that members of the household receive from the service exceeds the price of the service. In conversations with industry stakeholders, we were told that several characteristics of households influence the extent to which households purchase broadband service; we also reviewed other studies, and noted characteristics of households that these studies associated with the purchase of broadband service.[34] In particular, the following characteristics of households may influence the decision to purchase broadband service: income, education, age of household members, presence of children in the household, and the technological knowledge of members of the household. These characteristics may be associated with the extent to which a household would benefit from, and therefore value, broadband service, such as using broadband to telework, conduct research for school, and playing games. Industry stakeholders also noted that price influences a household's decision to purchase broadband service.

To empirically test these hypotheses, we estimated the following econometric model; because we lacked data on the price of broadband service, we were unable to include this variable in our econometric model.[35] The decision to purchase, or adopt, broadband service is a function of

- the income of the household;
- the education attainment of the heads of the household;
- the age of the heads of the household;
- the presence of children in the household;
- the racial composition of the household;
- the occupation of the heads of the household;
- the number of people in the household;
- whether the household resides in an urban, suburban, or rural location;
- the number of companies providing broadband service in the area; and
- whether the state in which the household resides imposes a tax on Internet access.

Data Sources

We required several data elements to build the data set used to estimate our deployment and adoption models. The following is a list of our primary data sources. In addition, we list all of the variables, definitions, and sources for the deployment model in table 1 and the adoption model in table 2.

- We obtained data on a sample of households in the United States from Knowledge Networks/SRI, using Knowledge Networks/SRI's product *The Home Technology Monitor^{TM}: Spring 2005 Ownership and Trend Report*. From February through April 2005, Knowledge Networks/SRI interviewed a random sample of 1,501 households in the United States. Knowledge Networks/SRI asked participating households a variety of questions about their use of technology, including questions such as whether the household purchased broadband service, and about the household's demographic characteristics.
- From the Federal Communications Commission (FCC), we obtained information on the companies providing broadband service in zip codes throughout the United States in December 2004. For each zip code, FCC provided the names of companies reporting, through the agency's Form 477, that they provided broadband service to at least one residential or small business customer and the type of company providing the service (e.g., cable and satellite).
- We used the most recent information from the U.S. Census Bureau to obtain demographic information for the areas where the households responding to Knowledge Networks/SRI's survey were located.

Assessing Broadband Deployment

FCC's Form 477 data include information on companies providing broadband service to at least one residential or business customer in zip codes throughout the United States in December 2004. However, since zip codes can represent large geographic areas, companies providing broadband service in a zip code might not have facilities in place to serve all households in the zip code. Thus, while a household might reside in a zip code in which FCC's Form 477 indicates that broadband service is available, that service might not be available to the household. Additionally, as we note in the text, we identified other concerns with FCC's data. Therefore, we took additional steps to assess whether broadband service was available to households included in Knowledge Networks/SRI's survey. In particular, we took the following steps for each observation in our data set:

- removed firms providing only satellite service;
- removed firms that provided only broadband service to business customers, since residential households were the focus of our study;
- removed large incumbent local exchange carriers when the company was identified as providing service in areas that lay outside of its local exchange area, since these firms typically provide service only to business customers outside of their local exchange areas;[36]
- removed firms when 2 or more of the 10 largest cable operators reported providing broadband service, since large cable operators rarely have overlapping service territories;

- removed cable operators if the responding household indicated that cable service did not pass the residence; and
- removed companies providing telephone-based broadband service if the household's residence was greater than 2.5 miles from the central office facility, since DSL service is distance limited.

Table 1. Deployment Model: Definitions and Sources of Variables

Variable	Definition	Source
Deploy	A binary variable that equals 1 if broadband service is available to the household responding to Knowledge Networks/SRI's survey.	FCC 2004 Form 477 and GAO analysis
Internet taxation	A binary variable that equals 1 if the state where the household resides imposes a tax on Internet access.	GAO analysis
Population, in thousands	The number of residents in the area where the household resides, in thousands.	Census Bureau
Urbanization	The percentage of the population residing in an urban area.	Census Bureau
Distance	The distance to a metropolitan area with a population of 250,000 or more.	GAO analysis
Percentage of work-at-home residents	The percentage of the population working from home.	Census Bureau
Percentage of population under 16	The percentage of the population under the age of 16.	Census Bureau
Percentage of population 17 to 24	The percentage of the population 17 to 24 years old.	Census Bureau
Percentage of population 65 or older	The percentage of the population 65 or older.	Census Bureau
Percentage of population with a high-school degree	The percentage of the population with a high-school degree.	Census Bureau
Percentage of population with education beyond high school	The percentage of the population with education beyond high school.	Census Bureau
Per-capita income, in thousands	The per-capita income in the area, in thousands of dollars.	Census Bureau
Population density, in thousands	The ratio of population to square miles in the area, in thousands.	Census Bureau

Source: GAO.

Table 2. Adoption Model: Definitions and Sources of Variables

Variable	Definition	Source
Adopt	A binary variable that equals 1 if the household responding to Knowledge Networks/SRI's survey purchases broadband service.	Knowledge Networks/SRI
Internet taxation	A binary variable that equals 1 if the state where the household resides imposes a tax on Internet access.	GAO analysis
Number of broadband providers	The number of companies providing broadband service to the household.	FCC 2004 Form 477 and GAO analysis
Income between $30,000 and $49,900	A binary variable that equals 1 if the household's income is between $30,000 and $49,900.	Knowledge Networks/SRI
Income between $50,000 and $99,900	A binary variable that equals 1 if the household's income is between $50,000 and $99,900.	Knowledge Networks/SRI
Income $100,000 or more	A binary variable that equals 1 if the household's income is greater than or equal to $100,000.	Knowledge Networks/SRI
Race-white	A binary variable that equals 1 if the household reported its race as white.	Knowledge Networks/SRI
College graduate	A binary variable that equals 1 if either the man or woman of the household is a college graduate.	Knowledge Networks/SRI
Age 34 to 49	A binary variable that equals 1 if either the man or woman of the household is 34 to 49 years old, and neither is younger than 34.	Knowledge Networks/SRI
Age 50 or older	A binary variable that equals 1 if both the man and woman of the household are 50 years old or older.	Knowledge Networks/SRI
Children	A binary variable that equals 1 if a child age 17 or younger resides in the home.	Knowledge Networks/SRI
Household size	A binary variable that equals 1 if three or more people reside in the home.	Knowledge Networks/SRI
Occupation-professional	A binary variable that equals 1 if the man or woman of the household reports working in a professional position.	Knowledge Networks/SRI
Occupation-clerical, sales, or technical	A binary variable that equals 1 if the man or woman of the household reports working in a clerical, sales, or technical position, and neither reports working in a professional position.	Knowledge Networks/SRI
Occupation-blue collar	A binary variable that equals 1 if the man and woman of the household report working in a blue collar position.	Knowledge Networks/SRI
Occupation-other	A binary variable that equals 1 if the man or woman of the household reports working in a position other than a professional, clerical, sales, technical, or blue-collar position.	Knowledge Networks/SRI
Rural location	A binary variable that equals 1 if the household resides in an area outside a metropolitan statistical area (MSA).	GAO analysis
Suburban location	A binary variable that equals 1 if the household resides in an area inside an MSA but outside the central city of that MSA.	GAO analysis

Source: GAO.

Estimation Methodology and Results

For both the deployment model and adoption model, we are estimating a reduced-form, binary-choice model. That is, broadband service is either deployed in the area or it is not, and the household either purchases broadband service or it does not. Given the binary choice nature of the models, we employed the probit method to estimate the deployment and adoption equations.[37] In this section, we present descriptive statistics and estimation results for the two equations and discuss the results.

Deployment model

In table 3, we provide basic statistical information on all of the variables included in the deployment model, and in table 4, we provide the results from the probit estimation of the deployment model. Of the 1,501 respondents to Knowledge Networks/SRI's survey, we used 1,402 observations in the deployment model; we were unable to match the zip +4 code for all 1,501 observations with publicly available data, which was necessary to assess whether the residence was 2.5 miles from the serving central office facility.

Results from our model indicate that several factors related to the cost of providing broadband service and the demand for broadband service influence the likelihood that service will be available in a particular area. Regarding the cost factors, we found that urban areas and areas with greater population density are more likely to receive broadband service. For example, urban areas are about 9 percentage points more likely to receive broadband service than are similar rural areas.

Table 3. Deployment Model: Descriptive Statistics

Variable	Mean	Standard deviation	Minimum value	Maximum value
Deploy	0.911	0.285	0.000	1.000
Internet taxation	0.546	0.498	0.000	1.000
Population, in thousands	26.022	17.982	0.070	113.935
Urbanization	76.154	33.240	0.000	100.000
Distance	34.361	42.743	0.249	572.803
Percentage of work-at-home residents	3.257	2.064	0.000	33.333
Percentage of population under 16	24.018	4.651	6.225	41.219
Percentage of population 17 to 24	10.585	4.785	0.583	55.113
Percentage of population 65 or older	12.995	5.688	2.195	59.057
Percentage of population with a high-school degree	29.546	9.002	3.121	68.966
Percentage of population with education beyond high school	51.395	16.092	8.836	95.348
Per-capita income, in thousands	44.466	15.597	9.583	164.479
Population density, in thousands	2.976	6.876	0.002	74.814

Source: GAO.

Table 4. Deployment Model: Estimation Results

Variable	Parameter estimate and [p-value]
Intercept	-2.9299[0.0097][a]
Internet taxation	-0.1486[0.2275]
Population, in thousands	0.0099[0.1140]
Urbanization	0.0102[0.0001][a]
Distance	-0.0012[0.3115]
Percentage of work-at-home residents	-0.0600[0.0392][b]
Percentage of population under 16	0.0335[0.1192]
Percentage of population 17 to 24	0.0198[0.3027]
Percentage of population 65 or older	0.0468[0.0271][b]
Percentage of population with a high-school degree	0.0114[0.3260]
Percentage of population with education beyond high school	0.0121[0.1957]
Per-capita income, in thousands	0.0270[0.0074][a]
Population density, in thousands	0.1706[0.0159][b]
Number of observations	1,402
1-LogL/Log0	32.0077

Source: GAO.

[a]Significant at the 1 percent level.

[b]Significant at the 5 percent level.

These results are consistent with broadband service being less costly to deploy in densely populated, more urban environments, where a similar investment in facilities can serve a greater number of subscribers than is possible in rural areas. Regarding demand for broadband service, we found that areas with greater per-capita incomes are more likely to receive broadband service. Additionally, we found that areas with a greater number of people working from home are less likely to have broadband service and that areas with a greater percentage of people age 65 or older are more likely to have broadband service.

We did not find that taxation of Internet access by state governments influenced the deployment of broadband service. Taxes can raise consumer prices and reduce revenues and impose costs on providers, and thereby possibly reduce the incentive for companies to deliver a product or service. Since we used a binary variable to indicate the presence of taxes, this variable could also potentially capture the influence of other characteristics of the states, in addition to the influence of the tax. Results from our model indicate that Internet access taxes do not affect the likelihood that companies will deploy broadband service; while the parameter estimate has the expected sign, the estimate is not statistically significant.

Adoption model

In table 5, we provide basic statistical information on all of the variables included in the adoption model, and in table 6, we provide the results from the probit estimation of the adoption model. Since households can only chose to purchase, or adopt, broadband service where it is deployed, we only include households from Knowledge Networks/SRI's survey where we assessed that broadband service was available; based on our analysis, 133 respondents did not have broadband service available. Further, 355 respondents to

Knowledge Networks/SRI's survey did not answer one or more demographic questions and 29 did not answer, or did not know, what type of Internet connection their household purchased. Therefore, we excluded these respondents. Thus, we used 901 observations in the adoption model.[38]

Our model results indicate that four characteristics influence whether households purchase, or adopt, broadband service. First, we found that households with greater incomes are more likely to purchase broadband service than are lower-income households. For example, the 25 percent of households with the highest income levels were about 39 percentage points more likely to purchase broadband service than the 25 percent of households with the lowest income levels. Second, households with a college graduate are about 12 percentage points more likely to purchase broadband service than are households without a college graduate. We also found that white households are more likely to purchase broadband service than households of other races. Finally, older households are less likely to purchase broadband service than are younger households.

As with the deployment model, we did not find that taxation of Internet access by state governments influenced the adoption of broadband service. As mentioned earlier, we used a binary variable to represent the presence of Internet taxation. As such, the variable may capture the influence of other characteristics of the states in which the households resided, in addition to the influence of the tax. Further, lacking a variable for the price of broadband service, we cannot assess how the imposition of the tax influenced the price of the service and thus the household's adoption decision.

Table 5. Adoption Model: Descriptive Statistics

Variable	Mean	Standard deviation	Minimum value	Maximum value
Adopt	0.336	0.473	0.000	1.000
Internet taxation	0.553	0.497	0.000	1.000
Number of broadband providers	3.307	2.161	1.000	9.000
Income between $30,000 and $49,900	0.223	0.417	0.000	1.000
Income between $50,000 and $99,900	0.336	0.473	0.000	1.000
Income $100,000 or more	0.149	0.356	0.000	1.000
Race-white	0.858	0.349	0.000	1.000
College graduate	0.499	0.500	0.000	1.000
Age 34 to 49	0.378	0.485	0.000	1.000
Age 50 or older	0.424	0.494	0.000	1.000
Children	0.387	0.487	0.000	1.000
Household size	0.465	0.499	0.000	1.000
Occupation-professional	0.442	0.497	0.000	1.000
Occupation-clerical, sales, or technical	0.154	0.361	0.000	1.000
Occupation-blue collar	0.029	0.167	0.000	1.000
Occupation-other	0.244	0.430	0.000	1.000
Rural location	0.052	0.222	0.000	1.000
Suburban location	0.568	0.496	0.000	1.000

Source: GAO.

Table 6. Adoption Model: Estimation Results

Variable	Parameter estimate and [p-value]
Intercept	-1.4919[0.0001][a]
Internet taxation	-0.1683[0.0745]
Number of broadband providers	0.0118[0.6101]
Income between $30,000 and $49,900	0.4531[0.0024][a]
Income between $50,000 and $99,900	0.7429[0.0001][a]
Income $100,000 or more	1.1331[0.0001][a]
Race-white	0.2905[0.0405][b]
College graduate	0.3525[0.0009][a]
Age 34 to 49	-0.2239[0.0759]
Age 50 or older	-0.3316[0.0217][b]
Children	0.1318[0.3894]
Household size	0.1241[0.3894]
Occupation-professional	0.2610[0.1409]
Occupation-clerical, sales, or technical	0.2098[0.2867]
Occupation-blue collar	0.2638[0.3879]
Occupation-other	0.0212[0.9086]
Rural location	-0.3234[0.1892]
Suburban location	0.0983[0.3406]
Number of observations	901
1-LogL/Log0	16.2800

Source: GAO.

[a]Significant at the 1 percent level. [b]Significant at the 5 percent level.

Using our model, we found that the parameter estimate had the expected sign—indicating that the imposition of the tax may have reduced the likelihood that a household would purchase broadband service. While the estimate was not statistically significant at the 5 percent level, it was statistically significant at the 10 percent level, perhaps suggesting that it is a weakly significant factor. However, given the nature of our model, it is unclear whether this finding is related to the tax or other characteristics of the states in which households resided.

APPENDIX IV. ADDITIONAL COMMUNICATIONS TECHNOLOGIES

Based on our conversations with stakeholders, and our own research, we identified several emerging technologies that could further the deployment of broadband service.

Broadband over power lines. Broadband over power lines (BPL) is an emerging competitive source of broadband to the home. BPL transmits broadband by using existing electric distribution networks, such as the wires that deliver electricity to consumers. Although there are a few commercial deployments, most BPL efforts are currently at the trial stage. Trials and commercial deployments range across the urban-rural landscape, from Cullman County, Alabama, to Cincinnati.[39] Currently, BPL can provide upstream and downstream speeds of 3 million bits per second (Mbps), and next generation equipment is being developed to provide speeds of 100 Mbps.

Industry stakeholders have identified several concerns with BPL service. First, while traveling across the electric network, BPL can emit signals that interfere with other users of the spectrum, such as amateur radio and public safety. The Federal Communications Commission (FCC) has taken steps to document, mitigate, and alleviate this potential problem. Second, some stakeholders also expressed concern that, due to the age or condition of the electric network, providers in some areas would be unable to transmit Internet data at high speeds. Finally, some stakeholders expressed varied opinions about the feasibility of BPL to bring broadband service to rural areas. Some stakeholders were optimistic about BPL's ability to serve these communities, while others expressed skepticism, pointing out that overcoming BPL's distance limitations would require more equipment and additional costs.

Wireless Fidelity (Wi-Fi). Wi-Fi-enabled wireless devices, such as laptop computers, can send and receive data from any location within signal reach—about 300 feet—of a Wi-Fi-equipped access point. Wi-Fi provides data transmission rates, based on the current transmission standard, of up to a maximum of 54 Mbps,[40] which is shared by multiple users. Wi-Fi equipment and services are based on the 802.11 series standards developed by the Institute of Electrical and Electronics Engineers (IEEE) and operate on an unlicensed basis in the 2.4 and 5 GHz spectrum bands. Several stakeholders we spoke with said that Wi-Fi service complemented, rather than substituted for, other broadband services.

The number of areas that can access Wi-Fi service, known as "hot spots," has grown dramatically and, according to one equipment manufacturer, may exceed 37,000. Wi-Fi hot spots include such diverse entities as airports, colleges, retail establishments, and even entire towns. Increasingly, municipalities are planning or deploying larger area or citywide hot spots; some municipalities considering or deploying a Wi-Fi network include Atlanta, Philadelphia, San Francisco, and Tempe, Arizona. While Wi-Fi service is widely deployed in urban and suburban areas, some stakeholders identified a few problems with the service. Because Wi-Fi hot spots operate in unlicensed spectrum, interference can be a problem. Several stakeholders we spoke with mentioned congestion or limited distance capability in Wi-Fi as a potential limitation of the service.

Worldwide Interoperabiity for Microwave Access (WiMAX). With WiMAX service, the distance covered and data transmission speeds can exceed those found with Wi-Fi service. WiMAX can provide data transmission speeds of 75 Mbps with non-line-of-sight service—that is, the signal can pass through buildings, trees, or other obstructions—or up to 155 Mbps with line-of-sight service. In a non-line-of-sight environment, WiMAX can provide service in an area with a radius of 3 miles or more; in a line-of-sight environment, WiMAX can provide service up to approximately 30 miles. WiMAX equipment and services are based on the IEEE 802.16 series of standards and operate in unlicensed and licensed spectrum.

WiMAX networks are being deployed on a trial commercial basis, but some challenges remain for further deployment. More than 150 pilot and commercial deployments of WiMAX networks are currently in use. Because of its greater capabilities in terms of distance and speed, WiMAX can extend wireless broadband to less densely populated communities, where wired solutions may be more expensive to deploy. Stakeholders we spoke with serving smaller, less densely populated areas indicated that they were testing or interested in WiMAX to serve their communities. However, concerns have been raised about spectrum availability, interference, and the ability of different manufacturers' equipment to support the same level of broadband applications. FCC has several initiatives under way to increase the availability of spectrum for WiMAX services. While the WiMAX Forum Certification Lab certifies WiMAX equipment, the standard allows manufacturers of equipment various options, such as different levels of security protocols, and thus, not all equipment may support the same level of service, such as carrying voice over the Internet (VoIP) and security.

Third Generation (3G) cellular broadband. Recently, several major commercial wireless companies have introduced broadband service based on advances in cellular technology and data protocols. Focused primarily on the business customer and more expensive than cable modem and DSL services, 3G services permit consumers to receive broadband service while mobile. 3G services typically provide data transmission speeds of 400 to 700 kilobits per second (Kbps). There are two competing technologies: EVDO service, introduced by Verizon and Sprint; and HSDPA, introduced by Cingular. Currently, Verizon Wireless reports that its service is available nationally in 181 major metropolitan markets, covering approximately 150 million people. Sprint reports providing EV-DO service in major airports and business districts in 212 markets, covering approximately 140 million people. For HSDPA service, Cingular reports that its service is available to nearly 35 million people in 52 communities. Industry stakeholders expressed concerns about the ubiquity of service, data transmission speeds, and the monthly costs associated with 3G service. Opinions varied as to whether cellular broadband services would be a competitive threat, or a complementary service, for consumers of other broadband services.

Fiber To The Home (FTTH). FTTH provides a high-speed, wire-based alternative to traditional cable and telephone networks. According to the FTTH Council, as of September 2005, 2.7 million homes were passed by fiber and over 300,000 homes were connected to fiber in 652 communities in 46 states. Stakeholders expressed concerns about the high cost associated with deploying FTTH, and also that FTTH deployment was concentrated in urban and suburban communities, or in newly developed communities (known as "greenfields").

APPENDIX V. COMMENTS FROM INDUSTRY PARTICIPANTS

We provided a draft of this report to several associations representing industry trade groups and state and local government entities for their review and comment. The following associations came to GAO headquarters to review the draft: Cellular Telecommunications and Internet Association (CTIA), National Association of Regulatory Utility Commissioners (NARUC), National Association of Telecommunications Officers and Advisors (NATOA), National Cable and Telecommunications Association (NCTA), National Telecommunications Cooperative Association (NTCA), Satellite Industry Association (SIA), US Internet Industry Association (USIIA), United States Telecom Association (USTA), and Wireless Internet Service Providers Association (WISPA).

Officials from CTIA, NARUC, and NTCA did not provide comments. Officials from NATOA, NCTA, SIA, and USIIA provided technical comments that were incorporated, as appropriate.

USTA officials noted that our discussion of the effects of local franchising on deployment imply that franchise agreements have helped to ensure broad deployment of broadband, but that, in the view of USTA, franchise buildout requirements can deter entry and thus reduce deployment.

WISPA officials expressed concern about our findings regarding the taxation of Internet access and noted that it is important, in their view, that wireless Internet access provided by small providers not be taxed, and in fact, WISPA officials noted that small providers should be provided a tax incentive to encourage investment and expansion in underserved areas. Additionally, these officials expressed concern about the presentation of data on how households currently access the Internet from their homes. WISPA stated that these data understate the importance that wireless access will have toward the goal of universal broadband coverage both within and outside of users' homes. WISPA stated that the report accurately depicts that wireless Internet service providers (WISP) currently hold a minority market share, and WISPA officials note that without certain government policies to foster growth in the wireless industry, WISPs will be at a competitive disadvantage. WISPA officials also expressed concern that the report understates factors that are hindering the growth of the wireless Internet industry—most notably, the need for additional spectrum under 1 Ghz, such as the TV white spaces. Further WISPA noted that the data showing broadband penetration rates in urban, rural, and suburban areas should not be interpreted as indicating that access to broadband is lower in only rural areas. They suggested that differences in broadband penetration rates across these types of locations are not that great and that pockets of areas with no access exist in many areas. As such, WISPA suggests that policy response regarding spectrum availability, USF funding, and Rural Utilities Service be focused on engaging smaller providers that can bring broadband to areas not currently served by the larger incumbent providers.

End Notes

[1] Throughout this report, we refer to high speed Internet access over broadband technologies as broadband Internet access.

[2] See GAO, *Internet Access Tax Moratorium: Revenue Impacts Will Vary by State*, GAO-06-273 (Washington, D.C.: Jan. 23, 2006).

[3] While FCC states that its zip-code information is not meant to be a measure of broadband deployment, some parties have used it in this manner because there are no other official data on deployment of broadband across the country.

[4] The 3-mile limit applies to the path taken by the telephone wire, not necessarily a straight line between the central office and the customer's residence.

[5] With fiber feeders, DSL service can be extended beyond three miles from the central office.

[6] FCC defined "advanced service" as exceeding 200 Kbps both upstream and downstream and "high-speed" service as exceeding 200 Kbps in at least one direction, in order to distinguish these from existing data services based on widely available analog telephony and ISDN technology.

[7] Spectrum is a natural resource used to provide an array of wireless communication services. FCC regulates commercial entities' use of spectrum. With unlicensed spectrum, a number of users without licenses share a portion of the spectrum, adhering to certain technological specifications. In contrast, with licensed spectrum, FCC provides entities with a license to use a specific portion of the spectrum.

[8] In the past, companies with less than 250 broadband connections were not required to submit information to FCC through Form 477. FCC officials told us that many of the companies that are now required to report are very small and in rural areas. These officials stated that many of these companies are not reporting and that therefore the data may not fully represent broadband deployment.

[9] FCC requires providers to report on their broadband lines or wireless channels. While this may not exactly equate to subscribers, the number of lines and subscribers is related, and we use the word subscribers throughout this report as we refer to the 477 filings of companies.

[10] We used survey data from Knowledge Networks/SRI's *The Home Technology Monitor™: Spring 2005 Ownership and Trend Report*.

[11] A very small number of respondents to the survey accessed the Internet over a satellite connection, but none of the respondents reported any other means of wireless access.

[12] We refer to rural areas as areas outside metropolitan statistical areas (MSA); suburban areas as areas within an MSA but not a central city; and urban areas as a central city of an MSA.

[13] Our findings are not substantially different from those of other organizations. Based on 2003 data, the Census Bureau reported that 62 percent of American households had a computer—see U.S. Census Bureau, *Computer and Internet Use in the United States: 2003* (Washington, D.C., 2005). Additionally, the Department of Commerce reported that 20 percent of households—or 37 percent of online households—had broadband service, with DSL becoming increasingly popular. This study also found that broadband service was less commonly purchased in rural areas—see U.S. Department of Commerce, *A Nation Online: Entering the Broadband Age* (Washington, D.C., September 2004). Similarly, using survey data from 2005, the Pew Internet and American Life Project reported that 53 percent of Internet users subscribed to broadband service, that much of the growth in broadband service in recent years arose from DSL subscriptions, and that broadband service was less prevalent in rural areas when compared with broadband subscribership in suburban and urban areas.

[14] In a recent report, we also noted that the 477 data do not provide a full description of broadband services for certain segments of the population, such as Native Americans residing on tribal lands. See GAO, *Telecommunications: Challenges to Assessing and Improving Telecommunications for Native Americans on Tribal Lands*, GAO-06-189 (Washington, D.C.: Jan. 11, 2006).

[15] See FCC, *High-Speed Services for Internet Access: Status as of December 31, 2004* (Washington, D.C., July 2005).

[16] The problems related to tracking data on subscribership versus deployment/availability in Form 477 is not an issue with mobile wireless operators. Because mobile wireless broadband services are designed to be used while subscribers are mobile, those operators are directed to report the zip codes covered by their mobile wireless broadband networks, rather than the zip codes of the billing addresses of their subscribers.

[17] As noted earlier, some households might not be able to actually receive broadband over satellite if they do not have a clear view of the southern sky.

[18] UNEs are physical and functional elements of the telephone network, such as the telephone line, or loop, which, under the Telecommunications Act of 1996, incumbent telephone companies must make available to competitors for lease or purchase.

[19] In particular, we removed satellite providers, removed any companies we determined only provide service to business customers, removed a cable provider if we found that more than 1 of the largest 10 cable providers served the zip code, removed a cable provider if the respondent said that cable does not pass their residence, and removed telephone-based providers if the residence was further than 2.5 miles from the central office that served the respondent's home.

[20] See Tony H. Grubesic and Alan T. Murray, "Waiting for Broadband: Local Competition and the Spatial Distribution of Advanced Telecommunication Services in the United States," *Growth and Change* (2004), 139-165.

[21] We did not contact officials in Alaska and Hawaii, since the survey data from Knowledge Networks/SRI did not include households from these two states.

[22] This program also can provide loan guarantees, but to date, no loan guarantees have been requested.

[23] A recent GAO report, *Telecommunications: Challenges to Assessing and Improving Telecommunications for Native Americans on Tribal Lands*, GAO-06-189 (Washington, D.C.: Jan. 11, 2006) discusses how leadership in a community can help to improve telecommunications services on tribal lands. The report provides several examples of tribes addressing the barriers to deployment of telecommunications networks by partnering with private entities, providing technical training, and taking initiative to access federal grants.

[24] "High-income households" were defined as those having incomes in top 25 percent of all households, while "low-income households" were defined as those having incomes in the bottom 25 percent of all households.

[25] We define "young adults" as people between the ages of 18 and 33.

[26] See Debra J. Aron and David E. Burnstein, "Broadband Adoption in the United States: An Empirical Analysis" (paper presented at the 31st Annual Telecommunications Policy Research Conference, Arlington, Va., 2003).

[27] We recognize that our model does not fully include all the variables that would influence the adoption decision. As such, the parameter estimates will be biased. We are unable to assess the possible extent of this bias.

[28] U.S. Department of Agriculture, Office of Inspector General, *Audit Report: Rural Utilities Service Broadband Grant and Loan Programs*, Audit Report 0960 1-4-Te (Washington, D.C., Sept. 30, 2005).

[29] According to RUS officials, demand for the grant program exceeds available funding under the current program requirements.

[30] RUS officials noted that many different technologies were used in these 19 communities, so that the cost per household varied considerably across the grant recipients.

[31] For rural ILECs, the cost of service is based on "embedded costs"—or the historical costs of infrastructure, which is used to provide a variety of communications services. For nonrural ILECs, the cost of service is based on the forward-looking costs of providing only certain telecommunications services.

[32] See GAO, *Telecommunications: Strong Support for Extending FCC's Auction Authority Exists, but Little Agreement on Other Options to Improve Efficient Use of Spectrum,* GAO-06-236 (Washington, D.C.: Dec. 20, 2005).

[33] For example, see Tony H. Grubesic and Alan T. Murray, "Waiting for Broadband: Local Competition and the Spatial Distribution of Advanced Telecommunication Services in the United States," *Growth and Change*, vol. 35, no. 2 (2004): 139-165; and James E. Prieger, "The Supply Side of the Digital Divide: Is There Equal Availability in the Broadband Internet Access Market?" *Economic Inquiry*, vol. 41, no. 2 (2003): 346-363.

[34] For example, see Scott Wallsten, *Broadband Penetration: An Empirical Analysis of State and Federal Policies* (Washington, D.C.: AEI-Brookings Joint Center for Regulatory Studies, 2005); Scott J. Savage and Donald M. Waldman, "United States Demand for Internet Access," *Review of Network Economics*, vol. 3, no. 3 (2004): 228-247; and Debra J. Aron and David E. Burnstein, "Broadband Adoption in the United States: An Empirical Analysis" (paper presented at the 31st Annual Telecommunications Policy Research Conference, Arlington, Va., 2003).

[35] As such, we recognize that the parameter estimates will be biased. We are unable to assess the possible extent of this bias.

[36] We did not remove Verizon, since thought its acquisition of GTE, it serves a wide variety of locations as an incumbent exchange carrier.

[37] An alternative method to estimate these equations is the logit model. In a binary choice model, the differences between the logit and probit models are generally not significant. Differences can arise in the multinomial model, where there are three or more choices, because the logit model imposes independence conditions that sometimes do not reflect the conditions being modeled. Such was not the case in our models, since we are estimating binary choice equations.

[38] We did not find that the households that failed to answer the demographic questions were more or less likely to be online than were the households that answered these questions.

[39] BPL companies, such as Current Communications, offer the ability to improve the monitoring, detection, and management of the electrical distribution network through improved communication capabilities inherent in BPL equipment. These features increase the attractiveness of BPL to electrical companies, as the companies receive the benefits of improved network operation and potential revenues from broadband service.

[40] Discussions are underway for newer standards for Wi-Fi that would dramatically increase the transmission speeds.

In: Broadband and Rural America
Editor: Bethany J. Lenz

ISBN: 978-1-61668-637-6

Chapter 3

BROADBAND INTERNET'S VALUE FOR RURAL AMERICA

Peter Stenberg, Mitch Morehart, Stephen Vogel, John Cromartie, Vince Breneman and Dennis Brown

ABSTRACT

As broadband—or high-speed—Internet use has spread, Internet applications requiring high transmission speeds have become an integral part of the "Information Economy," raising concerns about those who lack broadband access. This report analyzes (1) rural broadband use by consumers, the community-at-large, and businesses; (2) rural broadband availability; and (3) broadband's social and economic effects on rural areas. It also summarizes results from an ERS-sponsored workshop on rural broadband use, and other ERS-commissioned studies. In general, rural communities have less broadband Internet use than metro communities, with differing degrees of broadband availability across communities. Rural communities that had greater broadband Internet access had greater economic growth, which conforms to supplemental research on the benefits that rural businesses, consumers, and communities ascribe to broadband Internet use.

Keywords: Internet, broadband, high-speed Internet, rural economies, rural economic growth, digital economy, telemedicine, rural, urban, Census data, June Agricultural Survey, Agricultural Resource Management Survey (ARMS), ERS, USDA

ACKNOWLEDGMENTS

The authors thank the many outstanding and practical comments from Mary Bohman, Patrick Sullivan, Jim Johnson, Robert Gibbs, Erik Dohlman, and the Peer Review Coordinating Council (USDA-Economic Research Service); John Allen (Utah State University); Mary Campanola (USDA-Rural Development); William Chambers (USDA-Farm Service Agency); and an anonymous university reviewer. Thank you to the workshop participants (listed in the appendix) whose research helped our research immeasurably. Thanks to USDA-National Agricultural Statistics Service and the Federal Communications Commission whose data were critical in our research. Special thanks to Chris Dicken, Ryan Williams, and David Nulph for their excellent GIS support. Also special thanks to Dale Simms and Cynthia A. Ray for their superlative editorial and design assistance.

SUMMARY

The Internet has become widely, but not universally, available. Two-thirds of U.S. adults had in-home Internet access by 2008. Rural businesses and consumers have become almost as likely as their urban counterparts to use the Internet, though broadband—or high-speed—access is less prevalent in rural areas than in more densely populated areas. The 2008 Farm Act reau-thorized USDA's telemedicine, distance learning, and rural broadband access grant and loan programs.

What Is the Issue?

Broadband access is viewed as necessary to fully utilize the Internet's potential. As the Internet economy has matured, more applications now require higher data transmission rates, even in the case of simple shopping websites. In a recessionary economy a number of Internet activities—including job searches and home businesses—may become more critical for households. Whereas an estimated 55 percent of U.S. adults had broadband access at home in 2008, only 41 percent of adults in rural households had broadband access. Evidence suggests that this shortfall in broadband use is involuntary, and may be due to the higher cost of broadband provision or lower returns to broadband investment in sparsely populated areas.

What Did the Study Find?

Analysis suggests that rural economies benefit generally from broadband availability. In comparing counties that had broadband access relatively early (by 2000) with similarly situated counties that had little or no broadband access as of 2000, employment growth was higher and nonfarm private earnings greater in counties with a longer history of broadband availability.

By 2007, most households (82 percent) with in-home Internet access had a broadband connection. A marked difference exists, however, between urban and rural broadband use—only 70 percent of rural households with in-home Internet access had a broadband connection in 2007, compared with 84 percent of urban households. The rural-urban difference in in-home broadband adoption among households with similar income levels reflects the more limited availability of broadband in rural settings.

Areas with low population size, locations that have experienced persistent population loss and an aging population, or places where population is widely dispersed over demanding terrain generally have difficulty attracting broadband service providers. These characteristics can make the fixed cost of providing broadband access too high, or limit potential demand, thus depressing the profitability of providing service. Clusters of lower service exist in sparsely populated areas, such as the Dakotas, eastern Montana, northern Minnesota, and eastern Oregon. Other low-service areas, such as the Missouri-Iowa border and Appalachia, have aging and declining numbers of residents. Nonetheless, rural areas in some States (such as Nebraska, Kansas, and Vermont) have higher-than-expected broadband service given their population characteristics, suggesting that policy, economic, and social factors can overcome common barriers to broadband expansion.

In general, rural America has shared in the growth of the Internet economy. Online course offerings for students in primary, secondary, post-secondary, and continuing education programs have improved educational opportunities, especially in small, isolated rural areas. And interaction among students, parents, teachers, and school administrators has been enhanced via online forums, which is especially significant given the importance of ongoing parental involvement in children's education.

Telemedicine and telehealth have been hailed as vital to health care provision in rural communities, whether simply improving the perception of locally provided health care quality or expanding the menu of medical services. More accessible health information, products, and services confer real economic benefits on rural communities: reducing transportation time and expenses, treating emergencies more effectively, reducing time missed at work, increasing local lab and pharmacy work, and generating savings for health facilities from outsourcing specialized medical procedures. One study of 24 rural hospitals placed the annual cost of not having telemedicine at $370,000 per hospital.

Most employment growth in the U.S. over the last several decades has been in the service sector, a sector especially conducive for broadband applications. Broadband allows rural areas to compete for low- and high-end service jobs, from call centers to software development, but does not guarantee that rural communities will get them.

Rural businesses have been adopting more e-commerce and Internet practices, improving efficiency and expanding market reach. Some rural retailers use the Internet to satisfy supplier requirements. The farm sector, a pioneer in rural Internet use, is increasingly comprised of farm businesses that purchase inputs and make sales online. Farm household characteristics such as age, education, presence of children, and household income are significant factors in adopting broadband Internet use, whereas distance from urban centers was not a factor. Larger farm businesses are more apt to use broadband in managing their operation; the more multifaceted the farm business, the more the farm used the Internet.

How Was the Study Conducted?

This report summarizes all available nationwide data on broadband use and availability and analyzes the data to isolate interactions between broadband use and economic activities. It also presents results from an ERS-sponsored workshop on rural broadband and ERS-commissioned research studies conducted by others. The aim is to assess the economic impact of not having broadband service on rural communities and their growth, community facilities, access to health care, and well-being, as requested by Congress on December 26, 2007.

We first analyze who uses broadband, what it is used for, and what differences exist between the average urban and rural user to better understand the perceived usefulness of the Internet, especially broadband. We then identify rural areas that have broadband (by ZIP Code), determine when they acquired broadband service, and develop measurements of broadband availability over time. We analyze the effect of broadband Internet access on the rural economy using quasi-experimental design (QED), which allows us to compare two sets of counties alike in most aspects other than broadband availability. The results are interpreted using further analysis of rural businesses and consumers, including farm business logistic regression analysis, rural-farm linkage logistic regression analysis, and workshop research exploring many complex interactions between the Internet and socioeconomic components of the rural community: rural community social interactions; telemedicine; distance education; and rural businesses, including retail, service, and farm.

INTRODUCTION

The Internet in its infancy was simply an alternative, and quite straightforward, communication device. Electronic mail was sent from one person to another. Intrinsically the Internet remains a communications device, but as it has grown more universally available, it has become more integrated into the rest of the economy. The simple e-mail system of sending a note has evolved to incorporate blogs, instant messaging (IM), text messaging, Facebook, and Twitter. Business, household, and government activities have moved and are moving onto Internet platforms (Greenstein and Prince, 2006; Leamer and Storper, 2001). Many Internet activities are self-sustaining as personal computers download upgrades and "patches" to their systems and automatic ordering, billing, and payment functions are conducted for households and businesses. The Internet is integral to the development and functioning of the digital or information economy.

Rural communities have not been left out, though from the outset equal access to the Internet has been a contentious issue. The farm sector of the rural economy helped to pioneer rural Internet use (Stenberg and Morehart, 2007), and rural businesses and households have become almost as likely as their urban counterparts to use the Internet (Stenberg and Morehart, 2008). Access to the Internet through broadband (i.e., high-speed) technologies, however, has been less prevalent in rural areas than in much more densely populated areas of the country (see box, "What Is Broadband?"). Broadband Internet access has become the crux of today's policy debate on equal access among urban and rural communities.

Broadband access is viewed as necessary to fully utilize the Internet's potential (Greenstein and Prince, 2006; Parker, 2000). As the Internet economy has matured, more applications now require higher data transmission rates, even in the case of simple shopping websites.

The broad scope of the research presented here complements earlier studies on rural telecommunication policy by Parker and Hudson (1992), Internet access in the Appalachian region by Oden and Strover (2002), and businesses in the digital economy by Malecki (2008).

This report examines (1) what role the Internet plays in the national economy; (2) how much, and for what purposes, consumers use the Internet, especially what differences might exist between broadband and non- broadband Internet use by rural and urban consumers, including such uses as telemedicine, distance learning, and community involvement; (3) what determines whether and how rural businesses use broadband Internet; and (4) how broadband Internet affects the broader rural economy. Specifically, we address a question posed to ERS by Congress on December 26, 2007, concerning broadband's impact on rural communities and their growth, community facilities, access to healthcare, and overall well-being.

While we attempt to measure the observable economic effects of broadband Internet access and use, this study makes no attempt to comprehensively examine all the issues surrounding the growth of the Internet. For one, there are inherent limitations in measuring the economic impact, or value, of a rapidly evolving technology. The telegraph age, for example, ushered in a number of unpredicted developments like the emergence of such companies as Sears Roebuck. This report neither forecasts new developments nor addresses social issues pertaining to Internet uses like privacy, chat rooms, or parental supervision.

WHAT IS BROADBAND?

The transmission capacity, or bandwidth, of Internet access has been a major impediment limiting the economic returns from online activity. The slower the Internet access speed, the less useful the Internet is. Dial-up, which was the primary access method before broadband access became more widely available, is the slowest way to connect to the Internet. The highest speed by which data can be transferred using dial-up is 56 kilobytes per second (kbps). In rural areas the speed often has been much less, with connection speeds of 14 kbps common. Effectively, this consigns rural dial-up users to using the Internet for text e-mail messages only. Anything requiring large graphics is simply not practical. High-speed Internet access is necessary to make use of much of what is now offered on the Internet.

Broadband is the term used to denote high-speed access to the Internet. Although the term has been used to refer to other services, such as digital television, the matter of most interest to consumers, providers, and policymakers is broadband Internet connectivity (Eisenberg, 2002). With the convergence of video, audio, text, graphics, and other analogous enduring and transient products and services into digital streams that can be transported across the Internet, broadband Internet connections have become a necessity for common Internet usages and applications.

The Federal Communications Commission (FCC), though altering the broadband definition recently, historically defined 200 kilobits per second in one transmission direction as the minimum speed for Internet service to be classified as broadband. Unfortunately, the definition includes a wide array of technologies ranging from the old ISDN and T- 1 lines to satellite service. Lumping very slow transmission and sometimes unreliable service in with superfast fiber-optic home service makes economic impact analysis and discussion of broadband Internet service from historical data challenging.

Most broadband Internet access in U.S. households is through DSL or cable modem technologies and is faster than the FCC standard. As of 2007, 55 percent of all households had broadband Internet access: 46 percent of these had DSL, 39 percent had cable modem, and 12 percent had wireless connections (PEW). DSL was the first technology to become widely deployed; cable and, more recently, fiber-optic lines are becoming the technologies of choice.

THE INTERNET ECONOMY

In 1995, there were roughly 16 million Internet users across the globe; by 2008 there were nearly 1.5 billion, about 22 percent of the world's population. Two-thirds of U.S. adults had in-home Internet access by 2008 (PEW). Domain names have grown from 30,000 in 1994 to 168 million in 2008 (Verisign). Hosts, also known as end-user computers, grew from 1,000 in 1984 to 570 million in 2008 (Internet Systems Consortium). The Bureau of Census reports online retail sales went from $31 billion in 2001 to $107 billion in 2007. Also according to U.S. census statistics, online wholesale trade in 2006 was an estimated $613 billion, or approximately 16 percent of sales. Online wholesale trade in farm products was an estimated $5 billion, or 4 percent of all wholesale farm product sales in 2006.

Online economic activities may be grouped into three broad categories: information sharing, purchase channels, and sales channels. Information sharing can range from the trivial to critical life or business issues—from chat rooms to medical or financial storehouses—and is the most common application for businesses and consumers (Hopkins and Morehart, 2001; Stenberg, 1999; Varian, 2003). Even when purchases are not consummated online, purchase decisions are facilitated through price discovery or consumer information gathering. Real estate and automobiles are just two of the markets that have been transformed by price discovery online (Borenstein and Saloner, 2001).

The Internet has led to new sources of supplemental income for some households. Crafts, for example, that used to be pitched only at annual State and county fairs are now marketed year-round to wider audiences, and the Internet has led to the rise of auction sites such as E-Bay where anyone can be a buyer and seller of new and used goods and services.

For businesses, the Internet has reduced geographic isolation, with information from collaborating businesses or customers instantly available. The effective market area for producers has increased, though many businesses have not taken advantage of this potential (Brynjolfsson and Smith, 2000; Malecki and Moriset, 2007).

The increased speed and quantity of information, however, cannot reduce the physical distance that passengers and goods must travel (Malecki and Moriset, 2007). It is not clear whether the number of business trips has lessened due to the Internet (i.e., substituting Internet communication for travel). Anecdotal evidence suggests that as the price of transportation soared in early 2008, businesses cut back on travel (Odlyzko, 2008). Although the moving of farm, agricultural, and other physical goods from point A to point B involves the same distances as always, the Internet has enhanced the ability to track shipments and increased the efficiency of shipping companies.

As the Internet expands the effective market area for businesses, it also increases competition. The financial system is a prime example. Banks were traditionally local in nature. Historically, each farm community had its own bank. The bank held deposits from local residents and, in turn, loaned funds out to local farmers and the business community. This has changed, however, over the last several decades. First, the regulatory limitations on the markets that banks could serve were relaxed or eliminated. With the new regulatory environment and the revolution in telecommunications, the number of locally owned banks has declined (DeYoung and Duffy, 2002; Keeton, 2001).

Depositors have increasingly turned to online banks or investment concerns that offer higher rates of return for their capital (DeYoung and Duffy). This has increased the financial returns of bank depositors, but has reduced the profit that local banks accrued from deposits. Many loan applicants have gone online to find the best rates and terms for home mortgages, home equity, and other loan instruments. This has reduced the cost for the loan applicant, but also eroded the profit from individual loans in the bank's loan portfolio. To address this shortfall, many banks have increased their volume of transactions, often by increasing their effective market area. This may have hastened the consolidation of the financial market over the last 10 years.

E-government is another major development in the digital economy affecting rural America. Four kinds of activities fall within this area: information dissemination, citizen/customer services, government business transac-tions, and governance (Gallegos, 2002; Lanvin, 2008). Information typically disseminated includes public holidays and events, regulatory actions, issue briefs, public schedules, and school lessons and lunch menus. Twelve percent of all farms and 22 percent of all farms with Internet access, for example, retrieved information from Federal websites, according to analysis of 2007 June Agricultural Survey (JAS) data.

Citizen or customer services include paying taxes and fees, lodging complaints, requesting additional information, scheduling of public facilities, and submitting applications for various programs. Four percent of all farms and 7 percent of all farms with Internet access conducted business with the U.S. Department of Agriculture over the Internet according to analysis of 2007 JAS data.

Government business transaction costs have fallen for governments able to conduct much of their own back-office operations online, including supply purchases, bill payments, travel arrangements, and grant/loan operations with other agencies (Lanvin, 2002; Crescia, 2006). Governance changes are evident in the virtual town hall meetings, online polling, campaigning, and voting that characterized the 2008 election cycle.

Both government and the citizenry have benefited from the development of e-government via decreased costs to deliver or obtain services, increased or enhanced provision of information and services, and improved feedback between the citizenry and government (Lanvin).

Broadband Internet Adoption and Use

Dial-Up versus Broadband Internet Use

That households and individuals greatly value the Internet, and especially broadband access to the Internet, is readily apparent from the data. Two major data sources directly address individual and household online proclivity and activity: the U.S. Bureau of the Census and PEW (PEW Internet & American Life Project). The Census Bureau has not collected thorough data on online activity since 2003, so we rely on the PEW surveys for our understanding of Internet users (see Appendix B for a description of this and other data used in this report). Aggregate e-retail, peer-to-peer, web- page access counts, and other such information are not used here because such measurements of volume give no information on individual behavior.

The PEW survey data suggest that rural and urban online behavior is alike if one controls for type of Internet access. In other words, rural and urban users with broadband Internet access have similar online behavior patterns vis-à- vis each other; rural and urban users with dial-up Internet access have similar online behavior patterns as well. Users with broadband Internet access, however, exhibit different online behavior than users with dial-up access.

Among the conclusions drawn from 2008 PEW survey data:

- Three-quarters of all adults used the Internet, with 69 percent having access at home.
- Fifty-five percent of all adults had broadband access at home.
- Only 41 percent of adults in rural households had broadband access at home in 2008.

Broadband access at home has increased dramatically for both urban and rural adults since 2001 (figure 1).

The number of online activities varies between dial-up users and broadband users, but not between rural and urban broadband users (Horrigan, 2008).[1] Most Internet users go online everyday and most of these send e-mail or get information off the Internet (table 1). Getting information, including visiting State and local government websites, is the most common activity. Many of the activities, such as hobbies and web-surfing, are of a personal nature. Still, 23 percent of adults used the Internet daily to conduct research for their job. Broadband Internet users were more likely than dial-up users to take part in any specific online activity, such as getting news online. The more data intensive the activity, the greater the difference is between dial-up and broadband user participation.

Table 1. Online activities, 2008

Activity that has ever been done by a user	All Internet users	Dial-up at home	Broadband at home
		Percent	
Use an online search engine	89	80	94
Check weather reports and forecasts	80	75	84
Get news online	73	61	80
Visit a State or local government website	66	55	72
Look online for information about the 2008 election	55	37	62
Watch a video on a videosharing site like YouTube or GoogleVideo	52	29	60
Look online for information about a job	47	36	50
Send instant messages	40	38	44
Read someone else's blog	33	15	40
Use a social networking site like MySpace or Facebook	29	21	33
Make a donation to charity online	20	9	23
Download a podcast	19	8	22
Download or share files using peer-to-peer networks such as BiTorrent or LiveWire	15	15	17
Create or work on your own blog	12	8	15

Source: Pew Internet & American Life Project Survey, April 2008.

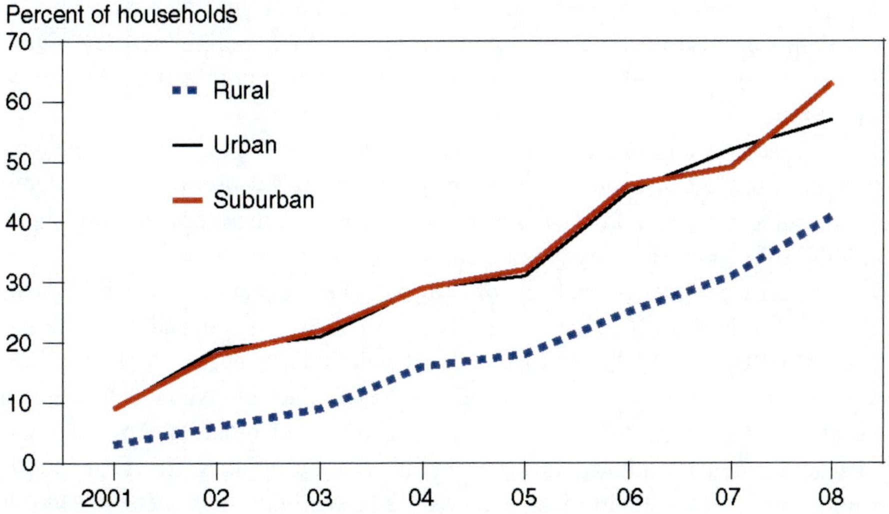

Source: Horrigan, 2008

Figure 1. Trends in home broadband adoption by region Percent of households

The Internet has reduced the economic involvement of the broker and other business middlemen in the economy. More bank transactions, for example, are taking place through ATMs or online instead of via tellers. According to a PEW survey in 2005, a quarter of all U.S. adults, or 44 percent of all adult Internet users, used the Internet for online banking. On any given day, 14 percent of all U.S. adults perform some online banking activity. Broadband users are especially heavy users of online banking services—over 60 percent of urban and nearly 50 percent of rural broadband users conducted some online banking activity in 2007.

As online data intensiveness increases, broadband access becomes more of a necessity. Nearly all online activities are becoming more sophisticated, using more data intensive processes; e-mail, for example, is becoming more data intensive as people are more inclined to attach photo and video files.

Broadband access is not uniform across rural and urban America, nor is the broadband access transmission rate identical across the country. Any shortfall in rural broadband availability is an implicit loss in economic opportunity for businesses, consumers, and governments.

Rural Internet Use

While the surveys conducted by PEW indicate popular online activities among adults, we turn to the Current Population Survey (CPS) to better understand who uses the Internet and where. The most recent CPS data are from a survey administered in October 2007. This large sample survey does not include any questions on what the Internet is used for or where it is accessed outside the home, but it provides a better understanding of who uses the Internet and rural-urban regional differences in household online proclivity.

Table 2 shows the share of households in which at least one person went online—no matter the technology—at home, school, work, or elsewhere in 2007. Over 71 percent of all households included one or more members that went online during the year. The CPS data suggest a variation across the country in the occurrence of going online (table 2). Households in the South were the least likely of the four Census regions to go online (see Appendix A for a description of the regions).

Nonmetro areas, in aggregate, had a lower percentage of individuals going online in 2007. While the variation in overall online use was insignificant between regions outside of the South, the same did not hold for nonmetro areas. Only in the Northeast was there not a significant dropoff in online activity going from metro to nonmetro areas.

Income differences have often been offered as a key explanation for the disparity in Internet use by households (Choudrie and Dwivedi, 2006; Flamm and Chaudhuri, 2007; Stenberg and Morehart, 2006). Lower income households clearly access the Internet less than higher income households (Figure 2). Income, of course, is not the whole story as income is highly correlated with or determined by education, age, and other factors, but household income by itself does raise an intriguing question: to what extent does use of the Internet lead to higher household income, and to what extent does higher household income lead to higher levels of Internet use?

Table 2. Households with at least one person going online at home or elsewhere, 2007

	Metro	Nonmetro	Total
		Percent	
Northeast	71.0	69.5	70.9
Midwest	74.0	65.7**	72.1
South	70.7	58.3**	68.3
West	75.5	68.6**	74.9
Total	72.6	63.3**	71.1

** Metro/nonmetro difference significant at 0.01.
Source: ERS using Bureau of the Census CPS data.

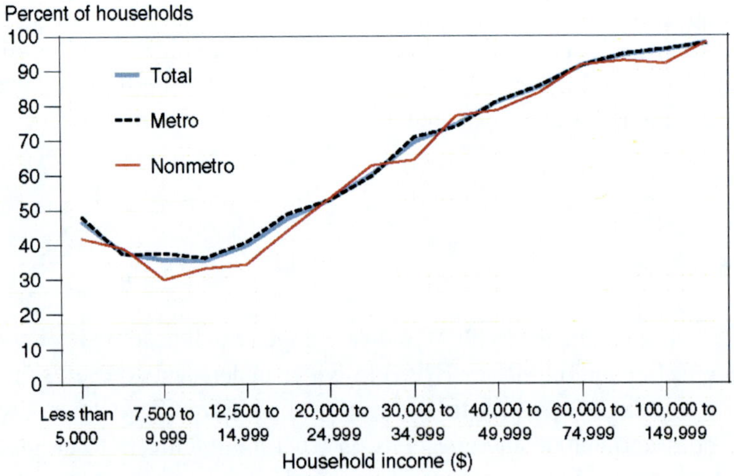

Source: ERS using Bureau of the Census CPS data.

Figure 2. Households accessing the Internet using any technology anywhere, 2007

Over 80 percent of households with annual incomes above $40,000 used the Internet during 2007. The rural-urban gap in accessing the Internet—either in-home or elsewhere—is not evident between rural and urban households of the same income (figure 2).

Sixty-two percent of all U.S. households had in-home Internet access in 2007 (table 3). The West had the highest share of households with in-home access, partially reflecting the more urbanized population distribution there. A significant dropoff in in-home Internet access is apparent between urban and rural households, especially outside the Northeast; fewer than half of rural households in the South had Internet access at home in 2007.

Income is a major factor in whether a household has in-home Internet access (figure 3). Over 70 percent of all households with incomes above $40,000 had in-home Internet access, and again rural-urban differences are largely nonexistent between households of the same income level. The steeper slope for in-home access rates versus going online anywhere suggests that affordability may be a factor in Internet access at home.

Table 3. Households with at least one person going online at home using any technology, 2007

	Metro	Nonmetro	Total
		Percent	
Northeast	64.1	61.0	63.7
Midwest	63.1	53.7**	60.9
South	61.7	46.4**	58.7
West	67.0	56.9**	66.1
Total	63.7	51 .9**	61.8

** Metro/nonmetro difference significant at 0.01.
Source: ERS using Bureau of the Census CPS data.

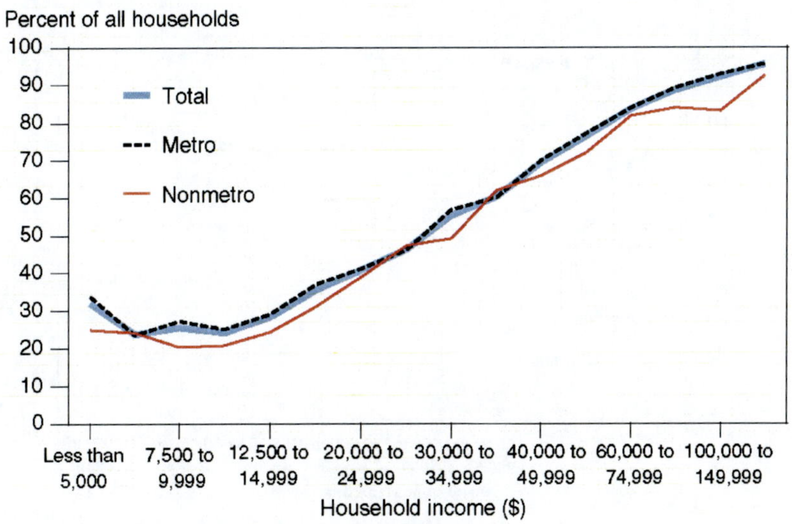

Source: ERS analysis of Bureau of the Census CPS data.

Figure 3. Home internet access by income, 2007 Percent of all households

Most households with in-home Internet access have broadband connections (table 4). This rate varies little across regions for urban households. The same cannot be said for rural households. A marked difference in broadband access exists between urban and rural residents, even in the Northeast. Only 70 percent of rural households with in-home Internet access had broadband access in 2007, versus 84 percent of urban households. The data suggest that broadband availability is an issue for rural areas across the country.

The rural-urban dichotomy in broadband access becomes even more apparent when household income is taken into account (figure 4). Income appears to be a minor factor in opting for broadband over dial-up for an in-house Internet connection. Generally, over 70 percent of Internet users, regardless of income, choose to pay for broadband (figure 4). Thus, the gap between rural and urban household use of broadband suggests that the availability of broadband services is more of a challenge for rural than urban households (unless there is some systemic difference between rural and urban house-holds that otherwise could explain the gap). Systemic household differences, if they exist, would have to explain why rural households are as likely as urban households to use the Internet but do not opt for broadband when they already use the Internet at home.

Table 4. Share of online households with broadband access, 2007

	Metro	Nonmetro	Total
	Percent of online households		
Northeast	87.3	68.8**	85.4
Midwest	82.9	70.6**	80.4
South	83.0	67.3**	80.5
West	85.3	75.2**	84.4
Total	84.4	69.7**	82.3

** Metro/nonmetro difference significant at 0.01.

Source: ERS using Bureau of the Census CPS data.

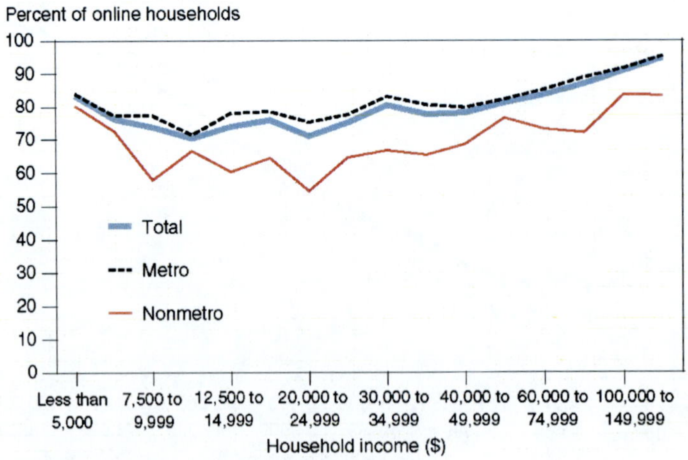

Source: ERS analysis of Bureau of the Census CPS data.

Figure 4. Households with broadband access by income, 2007

Further factors in rural broadband use

The presence of children in the household is a contributing factor in a household's having in-home Internet access (Stenberg and Morehart, 2006; Choudrie and Dwivedi, 2006). One way in which in-home Internet access may improve household well-being is through educational programs. The Internet has increased course offerings for students in primary, secondary, postsecondary, and continuing education programs, especially those attending small, isolated rural primary and secondary schools. The Internet has also improved interaction among students, parents, teachers, and school administrators in primary and secondary education. This is especially significant as studies have shown the importance of parental involvement in their children's education (Moore, 2007; Poley, 2008). As a result, education programs drive household demand for in-home Internet access. Analysis of the CPS data shows households with children have higher rates of in-home Internet access and households with teenage children are the most likely to have it (table 5).

Rural households, however, have uniformly less access to in-home Internet than urban households across all household composition types. Inasmuch as distance education is beneficial to economic well-being, continuation of this rural-urban dichotomy could put rural households at a disadvantage.

Once a household has in-home Internet access, the upgrade to broadband is seemingly not affected by household composition (table 6). The rural-urban gap, however, is more extreme and broadband's role in distance education would seem to put rural households at a further disadvantage.

Table 5. All types of Internet Access at Home by Household Composition, 2007

	Metro	Nonmetro	Total
	Percent of online households		
Not a parent	65.0	53.4**	63.3
No children under 18 years of age	69.7	59.4**	67.7
Only children less than 6	70.8	55.1**	68.7
At least one child 6-13 and none older than 13	72.8	65.4*	71.7
At least one child older than 13	81.4	76.5**	80.6
Total	**68.5**	**58.1****	**66.8**

** Metro/nonmetro difference significant at 0.01; * difference significant at 0.05.
Source: ERS using Bureau of the Census CPS data.

Table 6. Broadband in Homes with Internet Access by Household Composition, 2007

	Metro	Nonmetro	Total
	Percent of online households		
Not a parent	85.7	71 .4**	83.9
No children under 18 years of age	82.3	66.9**	79.7
Only children less than 6	90.5	78.4**	89.1
At least one child 6-13 and none older than 13	87.3	71 .8**	85.2
At least one child older than 13	85.4	71 .8**	83.3
Total	**85.2**	**70.3****	**83.1**

** Metro/nonmetro difference significant at 0.01.
Source: ERS using Bureau of the Census CPS data.

In a recessionary economy a number of Internet activities—including job searches and home businesses—may become more critical for households. The 2007 CPS data give some information on both activities.

Unemployed adults, while less frequent users of the Internet than employed persons, still had high "anywhere" access rates (table 7). People not in the labor force due to retirement or disability had the lowest rate of online activity. Unemployed individuals looking for work were more likely to use the Internet than other people not employed. Rural people in the labor force had a lower access rate than urban people.

The picture changes for home Internet access, where affordability likely becomes an issue for unemployed or disabled/retired persons (table 8). The dropoff in use for these groups holds for both rural and urban residents. These individuals likely go online at such locations as libraries and schools when in-home access becomes unaffordable.

A broadband connection is again the choice for most homes with in-home Internet access across all labor force categories (table 9). Regardless of labor force status, whether the household is in an urban or rural location, if a household has in-home Internet access, the household will most likely have a broadband connection.

Table 7. Online Activity Using Any Access Technology by Labor Force Status, 2007

	Metro	Nonmetro	Total
	Percent within labor force category		
Employed—At work	83.7	78.6**	82.9
Employed—Absent (on day of survey)	85.9	79.4**	84.8
Unemployed—On layoff	72.8	62.3**	70.6
Unemployed—Looking	77.7	72.8*	76.9
Retired—Not in labor force	52.6	43.7**	50.9
Disabled—Not in labor force	47.1	43.2*	46.1
Total of all adults	**77.2**	**70.6****	**76.1**

** Metro/nonmetro difference significant at 0.01; * difference significant at 0.05.

Source: ERS using Bureau of the Census CPS data.

Table 8. Home Internet access using any technology by labor force status, 2007.

	Metro	Nonmetro	Total
	Percent within labor force category		
Employed—At work	74.8	65.0**	73.3
Employed—Absent (on day of survey)	75.0	65.2**	73.5
Unemployed—On layoff	60.3	49.6**	58.1
Unemployed—Looking	63.8	51.4**	62.0
Retired—Not in labor force	48.8	41.3*	47.4
Disabled—Not in labor force	39.6	34.9*	38.4
Total of all adults	**68.5**	**58.1****	**66.9**

** Metro/nonmetro difference significant at 0.01; * difference significant at 0.05.

Source: ERS using Bureau of the Census CPS data.

Table 9. Broadband in Homes with Internet Access by Labor Force Status, 2007

	Metro	Nonmetro	Total
	Percent within labor force category		
Employed—At work	86.6	72.1**	84.6
Employed—Absent (on day of survey)	87.3	75.5**	85.6
Unemployed—On layoff	75.7	54.2**	72.0
Unemployed—Looking	83.7	72.0**	82.1
Retired—Not in labor force	75.0	59.0**	72.3
Disabled—Not in labor force	75.0	63.2**	72.4
Total of all adults	85.2	70.2**	83.1

** Metro/nonmetro difference significant at 0.01.
Source: ERS using Bureau of the Census CPS data.

Table 10. Internet users by home business status, 2007

	Metro	Nonmetro	Total
	Percent		
Households with home businesses	11.7	14.8*	12.2
Home businesses with any kind of home Internet access	83.3	70.1**	80.7
Proportion of home Internet access with broadband	87.6	71.3**	84.6

Source: ERS using Bureau of the Census CPS data.
Note: Difference between metro/nonmetro (*--significant at 0.05, ** -- significant at 0.01).

A small but significant number of households have home businesses covering a wide range of professions such as farmers, doctors, and artisans (table 10). Such households may become more commonplace in the current economic downturn as more households try to compensate for loss of jobs or reduced work hours by starting home businesses. Home businesses are more commonplace in rural areas than urban areas (table 10). In-home Internet access is much more common in households with home businesses (81 percent) than among all households in the aggregate (62 percent). This is true for both urban (83 versus 64 percent) and rural home businesses (70 versus 52 percent).

In summary, a broadband connection is almost the default for a great majority of online households. Analysis of the CPS data suggests that more rural households would have broadband connections if these connections were as readily available as in urban areas, implying lost economic opportunity for some rural households.

The data, however, also suggest that some of the shortfall in rural Internet activity may be due to other factors that precede the decision to get a broadband connection. These factors include the lower average income for rural households, higher average age of the rural population, and lesser educational attainment of rural residents as compared to their urban counterparts.

Geography of Rural Broadband Providers

Broadband provision follows a geographical pattern tied to population size and the urban-rural hierarchy. Limited provision is most strongly associated with low population size in a given area, but also exhibits regional patterns that reflect differences in urban concentration and challenges associated with mountainous terrain. The Federal Communications Commission (FCC) data on the number of broadband providers by ZIP Code area (see Appendix B) are the only source of geographically detailed information with national coverage. More robust findings would come from data measuring the actual number of broadband customers and variation in the price of service, rather than just the number of companies providing access. The FCC data, however, serve as the best available proxy of broadband accessibility.

Metro ZIP Code areas average 88 square miles apiece and include just over 16,700 people on average.[2] Nonmetro ZIP Code areas are generally much larger (131 square miles) with far fewer people (3,000) on average. Such population diversity drives geographic variation in the cost of broadband provision. While U.S. metro areas averaged close to nine providers per ZIP Code area in 2006, nonmetro ZIP Code areas averaged half that number (figure 5).

Despite significant expansion, the metro-nonmetro gap in number of broadband providers remains and has even widened by some measures since 2000, when broadband provision was much more limited in scope. It is not surprising that, as broadband access has expanded to encompass a large majority of Americans, the remaining areas of limited coverage increasingly reflect the higher costs associated with providing service to smaller populations.

THE ECONOMICS OF BROADBAND DELIVERY

The main economic principles underlying the diffusion and adoption of communication services across rural-urban space are twofold: companies invest where they earn the highest returns and households adopt if they can afford these services and either need or desire them (Davies, 1979; Rogers, 1995). The adoption and use of communication and information services, therefore, are not uniform across the country or among income groups.

Here we are interested in differences between rural and urban areas or, in Federal policy terminology, between high-cost and low-cost areas. Residents in rural areas have always faced higher costs for telecommunication services than those in urban areas and, at least for the foreseeable future, will continue to do so. Economies of scale for the current technology set are at the core of why they face higher costs (Stenberg, 2004).

Rural areas are characterized by low population density. With fewer people in any geographic space, the per capita costs of providing telecommunication services rise. Fewer customers share in the cost of the central office switches, loop maintenance, and other common components of the local telecommunication system.

Rural areas also have few large businesses or government operations. In the United States, private business and government use of telecommunication services has indirectly subsidized household use. In practice, this has meant that urban telecommunications service providers often charge higher rates to their business customers and lower rates to household customers than they would in a perfectly competitive market for telecommunication services (Egan, 1996). Rural telecommunications companies often do not have this luxury.

Rural telecommunications service providers must spend more per customer for maintenance and repair crews than urban providers. Rural maintenance and repair crews, especially those providing services in very remote regions, cover a larger territory than urban crews, resulting in more overtime, more travel expenditures, and all the other resultant expenditures that crews face when they are not near their home base (National Telephone Cooperative Association, 2000). Rural providers also need more resources per customer than urban telecommunications service providers, including duplicate facilities and backup equipment, to ensure network reliability (Egan, 1996).

Clusters of lower service provision in 2006 highlight clearly discernable regional patterns (Figure 6). The northern Great Plains, eastern Oregon, and northeastern New Mexico are sparsely populated. The Missouri-Iowa border area has experienced persistent population loss and an aging population. An extensive area of low service is evident in West Virginia and eastern Kentucky, but extends from Tennessee through upstate New York. This mountainous terrain, divided by innumerable ridges and narrow valleys, impedes broadband service provision to its widely dispersed, rural and small-town population. Though topography creates similar challenges in the West, people living there tend to concentrate more in towns and small cities, making broadband service less expensive to provide to each household.

Access to services of any kind—hospitals, grocery stores, public transportation, Internet—is strongly related to overall population size, the degree to which population is concentrated in urban centers, and the population of neighboring areas. In the case of broadband service, as measured by the FCC data, population size is the predominant explanatory feature. For all ZIP Code areas nationally, the very strong (0.6, or 60 percent) correlation between population size and service provision in 2000 increased to 0.75 by 2006 (figure 7). Thus, low-service areas that remain, though far fewer, are more likely to be in sparsely settled territory throughout nonmetro America.

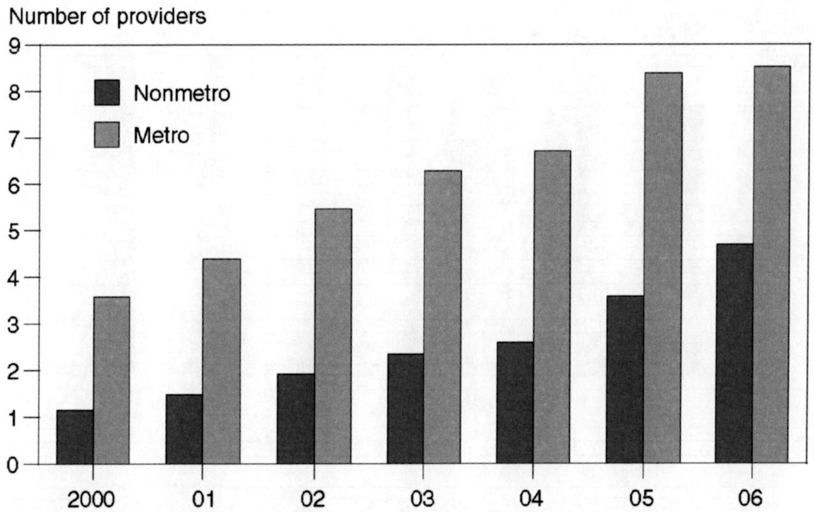

Source: USDA, Economic Research Service, using data from the FCC.

Figure 5. Average number of broadband providers per ZIP Code by metro versus nonmetro area, 2000-2006.

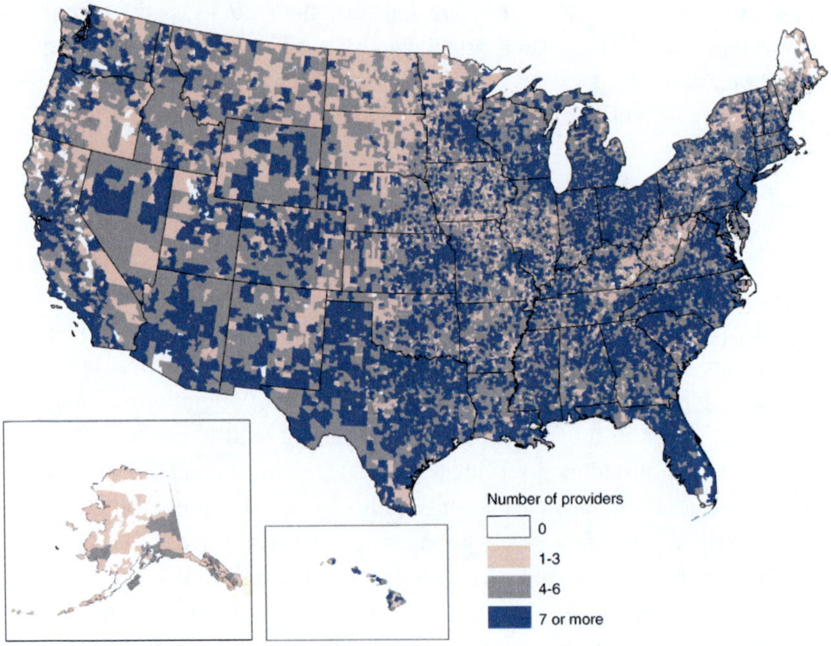

Source: USDA, Economic Reasearch Service, using data from the FCC.

Figure 6. Number of broadband provider, 2006

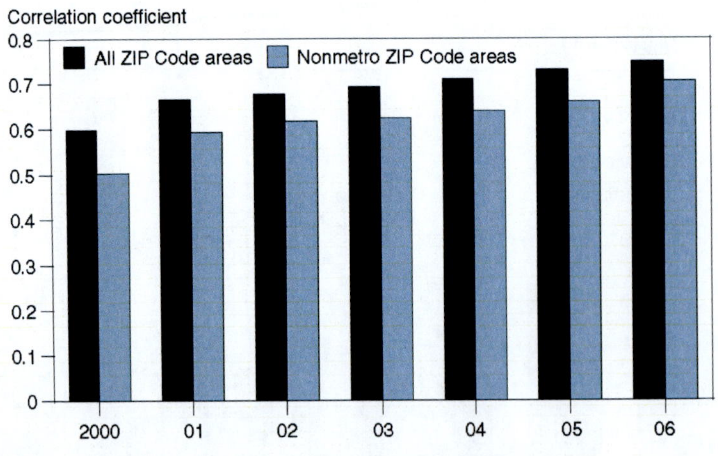

Source: USDA, Economic Research Service, using data from the FCC.

Figure 7. Strength of relationship between number of broadband providers and population size, 2000-2006.

Within nonmetro areas, population size is the strongest predictor of where broadband access is likely to be available. Other dimensions of nonmetro population distribution, however, strongly correlate as well, especially the degree of population concentration as measured by percent urban (figure 8). Correlation between broadband provision and the relative size of neighboring ZIP Code areas weakened from 2000 to 2006, as broadband provision increased.

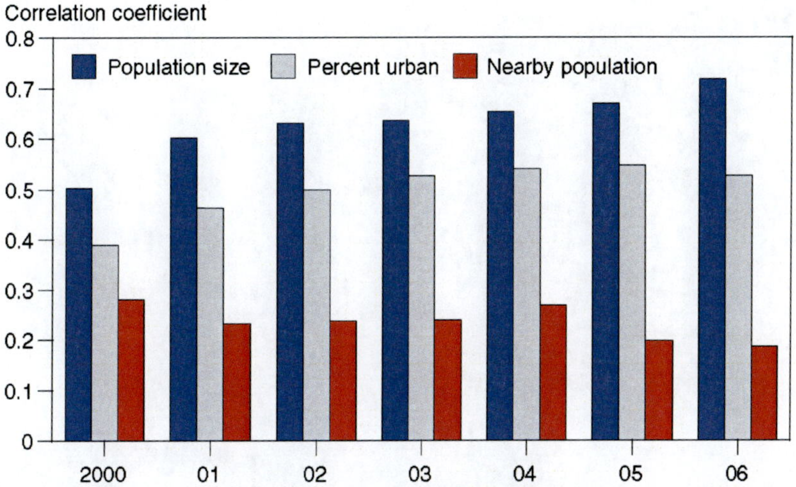

Source: USDA, Economic Research Service, using data from the FCC.

Figure 8. Strength of relationship between number of broadband providers in nonmetro ZIP Code areas and three population measures, 2000-2006.

Results of a regression analysis conform to expectations regarding the geography of broadband provision driven by relative costs per capita (for a description of the ordinary least squares model, see Appendix C). In combination, the three population-based measures explained 63 percent of variation in broadband availability. Population size contributed the largest effect, but percent urban and size of nearby populations also were significantly related to broadband provision in 2006.

Additional insights may be gained by examining where the number of broadband providers is higher or lower than expected, once these very strong and universal population effects are taken into account. This is accomplished by mapping the residuals from our regression model for 2006 (see Appendix C for an explanation of the model and the variables used). Residuals are measured for each ZIP Code area as the difference between the actual number of providers in that area and the number of providers predicted by the model, based on the area's population size, urban concentration, and proximity to nearby population centers. In the West, higher-than-expected service occurs in the most sparsely populated sections (Figure 9). In Nebraska and Kansas, including the more densely settled eastern parts, broadband provision is higher than expected, compared with most neighboring States. Vermont similarly stands out among New England States.

In Appalachia, broadband service provision is consistently lower than expected. The Ozarks in northeastern Arkansas show values similar to Appalachia, pointing to the role of topography in influencing broadband provision in the eastern half of the United States.

Higher-than-expected service in Vermont, however, suggests that economic and social factors may influence the level of broadband service in nonmetro ZIP Code areas. Higher levels of income and education may increase demand for broadband. Vermont's economy depends more heavily on tourism and recreation, and its population has high levels of college and technical training, especially in engineering, finance, and health. Both these features help explain differences in service levels compared with more southern Appalachian areas.

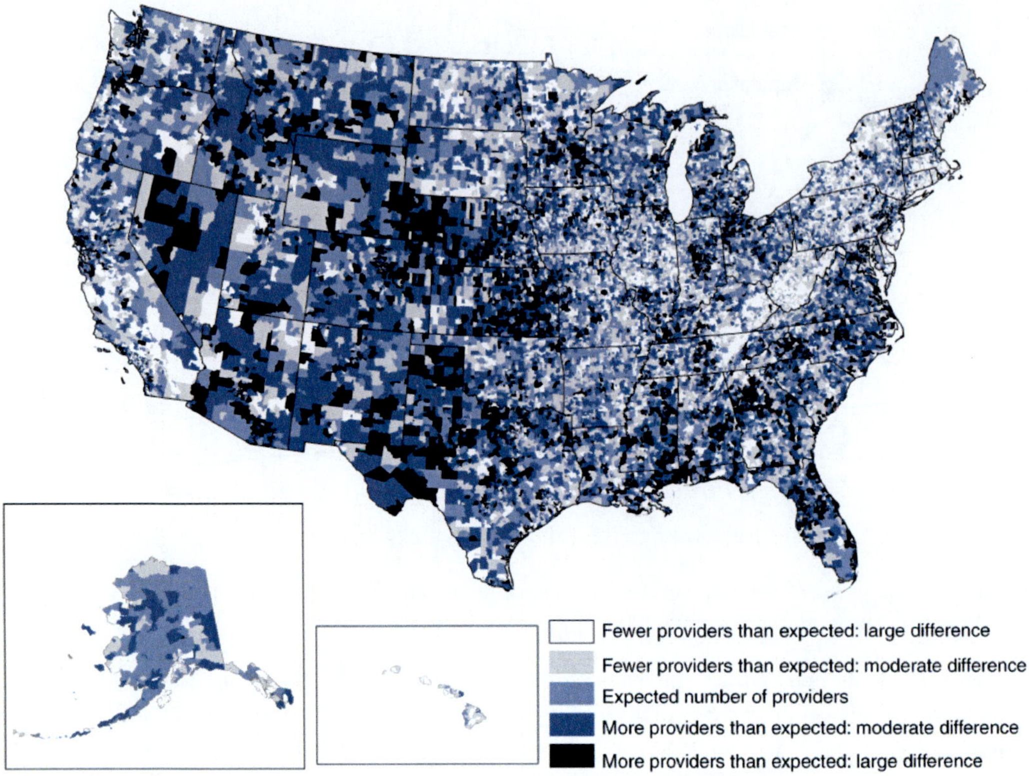

Note: The map shows residual regression values. The highest and lowest categories are greater than one standard deviation above and below the mean residual value of zero. Areas of moderate difference fall between one-quarter and one standard deviation above and below the mean. Areas with the expected number of providers are within one-quarter standard deviation above or below the mean.

Source: USDA, Economic Research Service, using data from the FCC.

Figure 9. Difference between the actual number of provider and the expected number based on population, 2006

The geography of broadband service provision in 2006—in particular, the contrast between higher- and lower-than-expected service areas—suggests several factors contributing to service gaps beyond basic population barriers. First, the variation in broadband coverage is less pronounced in the West, especially in the Intermountain West, compared with the Midwest and Appalachia. This may be due to a more concentrated population pattern (though this analysis accounts for some of this effect by including percent urban); a recreation-based economy, attracting tourists who increasingly demand broadband availability; or a rapidly growing population made up of younger, more educated individuals, including tech-savvy entrepreneurs whose businesses depend on being connected to urban-based clients.

Second, Appalachia's prominence as an underserved area suggests that topography significantly increases the cost of providing broadband service in this region. Education levels are below the national average, but no more than in many Coastal Plain States areas that are better served. The higher dependence on mining and other resource-based industries may play a role, but lower levels of broadband service exist even in areas of Appalachia where retirement and tourism have become important. And if a higher dependence on mining lowers Internet demand, the same is not true for agriculture. When population is taken into account, places with higher employment in agriculture exhibit higher levels of broadband support.

Finally, higher-than-expected service in States such as Nebraska, Kansas, and Vermont indicate that State-level policies and programs may be behind the widespread availability of broadband. Here, we face limitations of data that simply show the number of providers in a ZIP Code area and may not always reflect differences in costs or level of service. Still, conditions at the State level seemingly can transcend economic and social differences that tend to handicap some rural areas.

Broadband's Effect on the Rural Economy

Measuring the rural economic effects resulting from investment in broadband is challenging. Separating out the broadband effect from other causal factors in economic growth is difficult, especially given that broadband has not been available for long and its use has grown rapidly. The methodological approach that we take is called quasi-experimental design, and what is undertaken here may be considered an initial step toward ferreting out a causal relationship.

Quasi-experimental design (QED) is a statistical approach that simulates an ex-post laboratory experiment (Cook and Campbell, 1979). Like a laboratory or medical experiment, QED features both a treatment and control group. The treatment group is the group undergoing the "cure," which in this case includes areas with some minimum level of broadband availability.

The control group, or the untreated group, serves as the counterfactual to the treatment group. In theory, the counterfactual is what would have happened to the treatment group if they had not undergone the "cure." The control group provides the baseline forecast. Divergence in the post-treatment period is attributed to the effect resulting from the treatment.

Selection of control and treatment in QED (unlike a true laboratory experiment) is not perfectly random, hence the term "quasi." Treatment groups are self-selected. Control groups are selected based on their characteristic similarity with the initial, or pre-treatment, characteristics of the treatment group.

QED has been utilized in a large body of regional science research. It has been used in airport impact studies such as Farnsworth (1972) and Wheat (1970), fiscal policies such as Bender and Shwiff (1982), highway infrastructure studies such as Blum (1982) and Isserman (1987), and military base closure research such as Isserman and Stenberg (1994).

We use the year 2000 broadband density surface developed from the FCC broadband access data (see appendix C, specifically the section on enhancing the FCC data). In 2000, broadband was only starting to become widely available and it is the first year a broadband likelihood database could be constructed. Broadband access is based on the earliest reliable set of data from the FCC (according to our discussions with the FCC). Our 2000 likelihood data allow some effect resulting from broadband investment to start to appear in rural communities. Information technology takes time to be fully utilized after the technology's introduction (Greenstein, 2000; Bresnahan et al., 1999; Greenstein and Prince, 2006).

We selected 228 rural counties for our treatment group that had relatively high broadband availability in 2000. For each of these counties, we found a rural "twin," a county that most closely resembles the treatment county (outside of broadband availability) based on economic structure (farming, manufacturing, retail trade, Federal Government, and State/local government income as a percent of total income); spatial structure (population density, distance from various city sizes, and presence of interstate highway); and income (per capita, unearned, and transfer income) in 2000; as well as the growth in population and income from 1990 to 2000. Duplicate counties were not allowed in the control group (see appendix E for further discussion of the methodology used).

Our post-treatment period is 2002 through 2006. Year 2006 is the last year for which broadband data are available, and the 5-year period provides time for an economic effect from broadband service to manifest itself. Due to the rapid spread of broadband Internet access, the initial short period may be the only period when we are able to detect differences in economic outcomes resulting from the availability of broadband access.

We investigate changes in county employment and income in our QED analysis, and find that total employment grew faster in counties that had greater broadband Internet access sooner than in similarly situated rural counties without broadband access (table 11). Previous studies (Crandall et al., 2007) suggest that employment is not expected to be greatly influenced by broadband access. Simply put, the issue becomes whether the use of broadband Internet in business increases productivity, which subsequently either reduces actual employment (due to the productivity gain) or increases employment (as market share increases). At the county level, however, broadband availability may mean that the county's employers are more competitive with employers in other counties. This would attract both new jobs and potentially new employers.

Table 11. Difference in Employment Growth Rates between Early Broadband and Control Counties

	2002	2003	2004	2005	2006
Total number of jobs	0.003	0.0079*	0.0104*	0.0114*	0.0113
Total number of proprietors	-0.0068	0.0072*	0.0199*	0.0280*	0.0363*
Farm proprietors	-0.0001	0.0001	0.0009	0.00197	0.0058
Nonfarm proprietors	-0.0075	0.0048	0.0152*	0.0195*	0.0224*
Wage and salary jobs	0.0062*	0.0092*	0.0088*	0.0075*	0.0053*
Farm jobs	-0.0052	-0.0028	-0.004	-0.0050	-0.0010
Nonfarm jobs	0.00343	0.0076*	0.0096	0.0101	0.0087

Note: * significant at 10%.

Source: ERS using selected data from Bureau of the Census and Bureau of Economic Analysis data.

Table 12. Difference in Income and Population Growth Rates between Early Broadband and Control Counties

	2002	2003	2004	2005	2006
Population (number of persons)	0.0041*	0.0063*	0.0065*	0.0076*	0.0093*
Personal income	0.0141*	0.0064	0.0028	0.0037	-0.0012
Per capita personal income (dollars)	0.01 00*	-0.0002	-0.0047	-0.0049	-0.012
Private earnings	0.01 63*	0.0234*	0.0274*	0.0206*	0.01 92
Farm earnings	0.7545	0.0568	0.2863	0.4327	0.5483
Nonfarm earnings	0.0114*	0.01 14	0.0126	0.0068	0.0009

Source: ERS using selected data from Bureau of the Census and Bureau of Economic Analysis data.

Note: * significant at 10%.

Wage and salary jobs, as well as number of proprietors, grew faster in counties with early broadband Internet access. The farm sector seems largely to have been unaffected by broadband Internet access. The farm sector, however, seems more likely to embed broadband Internet access into productivity as its basic inputs are more fixed than other sectors of the economy. Subsectors of the counties' economies (not shown here), like wholesale trade, generally showed no significant effect from broadband access, though further analysis is warranted. The difference in nonfarm jobs starts to disappear as other counties get increased broadband access.

Income showed a mixed picture (table 12), though population showed greater growth in treatment counties than control counties. The normal year- to-year volatility of farm earnings due to weather and other causal factors not accounted for in the QED approach taken here may have been a factor in this outcome. Nonfarm earnings showed greater growth corresponding to broadband availability. The difference between control and treatment counties lessens over time as other counties get better broadband access.

Private earnings—all earnings, excluding farm earnings and Federal, State and local government earnings—were greater for the treatment counties than for the control counties. The results we obtained are consistent with the argument that broadband Internet access has a positive effect on rural communities.

Our analysis supports the hypothesis that investment in broadband Internet access leads to a more competitive economy. Further analysis, however, is needed to address the issue of causality more completely. Why and how broadband may lead to the results of the QED analysis was the subject of other ERS research at the ERS Broadband Workshop. It is the subject of the rest of the report.

RURAL COMMUNITIES AND BROADBAND INTERNET USE

Broadband Internet availability has direct implications for the well-being of communities. Limited or nonexistent broadband restricts Internet use in rural communities and subsequently the benefits derived from its use. Research suggests that broadband use fosters community involvement, enhances the provision of services such as health and education, and expands household income prospects through such activities as telework.

Community Interactions and the Internet

In the Internet's infancy some researchers warned that its use would weaken community ties, splinter common interests, and erode levels of voluntary or community participation, according to Stern et al. (2008). Local communities would be destroyed as people went online and found their own virtual communities of shared interest.

Sociological research over the past decade, however, has found this concern largely unjustified. In fact, Internet use has been shown to bolster community vitality through civic engagement and community participation (Stern and Dillman, 2006; Stern et al., 2008; Wellman et al., 1996). The concern has now evolved into fear that those with antiquated or no connections to the Internet are systematically being left out of community activities (Stern et al., 2008).

By use of a logistic regression model, Stern et al. showed that the use of the Internet is associated with higher degrees of community participation across a variety of groups and organizations. People use the Internet to receive information via e-mail from organizations or to search out information groups. Using broadband technologies corresponds positively with higher levels of passive and active community participation. Stern et al. concluded that the **quality** of Internet access is also important to a community's sociological well-being, regardless of factors such as income, age, education, and race.

Stern and others' data suggest that rural communities that rely most on volunteerism to function might be at a disadvantage as a consequence of either a lack of Internet access or an unwillingness to adopt broadband technologies. Furthermore, the lack of broadband service may in itself discourage the development of proficiencies in using the Internet. In other words, sluggish dial-up service may preclude individuals from growing efficient in daily Internet use, and thereby attenuate their contribution to their local communities.

Broadband Internet services, however, offer more definitive, or at least economically measurable, benefits to rural communities. These benefits covered during the 2008 ERS broadband workshop span medical, educational, and job services. One particular benefit is telemedicine.

Telemedicine and Telehealth

Rural communities have long faced challenges in getting adequate local health care. Telemedicine and telehealth have been hailed as vital to health care provision in rural communities, whether simply improving the perception of locally provided health care quality or expanding the menu of medical services (Goetz and Debertin, 1996). More accessible health information, products, and services confer real economic benefits on rural communities and their residents: reducing transportation time and expenses, treating emergencies more effectively, reducing time missed at work, increasing local lab and pharmacy work, and providing savings to health facilities from outsourcing specialized medical procedures (Capalbo and Heggem, 1999; Whitacre, 2008).

These benefits have been recognized by Federal policy. The USDA administers a number of programs aimed at improving in-clinic medical technology and broadband Internet access to fully utilize medical technologies in rural clinics. The 2008 Farm Act (the Food, Conservation, and Energy Act of 2008) expands these programs.

Telemedicine studies have primarily been case studies on how hospitals have adapted to telemedicine or cost-benefit studies for hospitals adapting particular telemedicine applications. Whitacre (2008) examined the economic benefits from a community perspective. Rural hospital services affect a community directly in the health of its citizenry and financially since rural hospitals in the States studied are most often partially funded by local sales taxes. As a consequence, understanding the full economic potential of telemedicine is important in understanding the economic benefit of such programs for rural communities.

Whitacre visited hospitals in 24 rural communities across Arkansas, Kansas, Oklahoma, and Texas. Four telemedicine benefits were catalogued: (1) hospital savings resulting from outsourcing specific procedures, (2) transportation savings accruing to patients, (3) income savings resulting from reduction in missed work, and (4) increases in local lab and pharmacy work.

Whitacre found wide variation in the way rural hospitals use teleradiology, teleoncology, and telepsychiatry, the specific telemedicine services analyzed in the study. The biggest benefit noted by hospital staff was in improved turnaround for patients. Cost savings for the hospital were typically not great, though annual cost savings varied significantly. Transportation savings to patients also varied considerably. The estimated transportation cost savings for patients not having to go to a more distant hospital after the initial visit was estimated to range from $2,000 to $110,000 per year per hospital across the 24 hospitals.

Estimated savings to patients who would have missed additional work time had they gone beyond their local hospital ranged from $3,000 to $70,000 for the 24 rural hospitals. Additional pharmacy and lab work that was gained by rural hospitals or their local affiliates was estimated to range from $31,000 to $1.5 million per annum. The cost of not having telemedicine thus was estimated to average $370,000 per annum for the 24 rural hospitals. Communities with larger hospitals (2,000 or more patient encounters per month) would be forgoing over $500,000 per year if telemedicine were not offered.

Distance Education

Education has long been shown to contribute positively to individuals' economic well-being as well as to national economic growth (through labor productivity gains). Education, also a major factor in the well-being of rural America, has been undergoing a transformation because of broadband Internet, with promising economic consequences. The institutional changes taking place between students, parents, faculty, and education administrators are already evident.

School-to-school distance education systems have opened to other provider groups that reach beyond schools or learning centers (Poley, 2008). Learners include students in K-12 and higher education as well as new immigrants, continuing education students, and individuals taking courses for personal development. Providers of distance education include universities, community colleges, private companies, communities, professional organizations, primary and secondary schools, and individuals.

Learners "attend" anytime during the 24-hour day, at work sites, learning centers, and in the home (Poley). Elementary school learners can visit educational sites that were introduced in the classroom, complete research projects (often with the assistance of parents), and check up on assignments while home sick. Their parents also can interact with teachers more freely. Secondary school students can enroll in online Advance Placement courses and other college-prep and college-level courses.

Distance education for post-secondary institutions also saves resources—for schools by reducing overhead and for rural households by reducing travel costs. It enables motivated learners to complete a course of study that might otherwise be incompatible with their work schedules and parental responsibilities (Poley). The promise for rural residents is increased access to educational resources, at lower costs than without distance learning.

The Service Sector in the Internet Economy

Most employment growth in the United States over the last several decades has been in the service sector, a sector especially conducive for broadband applications. Analysis of Bureau of Economic Analysis data shows services making up 50 percent of real private gross domestic product (GDP), 60 percent of personal consumption expenditures, 16 percent of private investment, and 25 percent of total trade in 2007.

As the sector has grown, information technology has increased the sector's productivity and globalized its marketplace. Information technology and the service sector are primary drivers of economic growth (Mann, 2008). As the global economy has grown, so have markets for information technology and services that process or trade through the Internet (Mann, 2006). Information technologies have allowed services to be fragmented between nontraded and tradable segments, such as back-office operations of financial firms, airline reservation offices, and software development for private business.

During the ERS-sponsored workshop on rural broadband in 2008, Mann identified four overarching effects of the Internet on the service sector:

- Greater international division of labor in services or standardization of services,
- Greater supply of intermediate services or fragmentation of services,
- Greater demand for intermediate services or fragmenting of services, and
- Globalization of both the supply and demand for services.

This has led to increasing private service sector exports and imports (table 13), with the United States exporting more services than it imports.

Not all service sector trade is uniformly distributed. Mann (2008) concluded that the trade in services has harmed low-wage service work like call centers and may continue to do so. Codification of work processes, such as in the software industry, also puts some high-wage work at risk of going overseas. Broadband allows rural areas to compete for low- and high-end service jobs, such as software development, but does not guarantee that rural communities will get them.

Table 13. Trade in private services, 1997-2006

	Exports			Imports		
	1997	**2002**	**2006**	**1997**	**2002**	**2006**
Private services ($ million)	**83,929**	**122,207**	**187,771**	**43,154**	**72,604**	**116,524**
	Percent of private services					
Education services	10	10	8	3	4	4
Financial services	15	18	23	14	13	12
Insurance services	3	4	5	14	30	29
Telecommunications	5	3	3	19	6	4

Source: Mann, 2008.

Telework

Telework is one part of the service sector that could be greatly enhanced by broadband Internet. Might rural communities benefit? Do rural communities have people that would be interested in and able to work online? Are there firms that would be interested in rural sourcing rather than global sourcing some of their services? Morris and Goodridge (2008) addressed this during the ERS workshop. They surveyed businesses across the country and found that approximately 12 percent were engaged in global offshoring of some service support activity, mostly (52 percent) to India.

The major advantages with offshoring cited by businesses were reducing costs, freeing up management time, and offering "24/7" customer access. Disadvantages were increased incidences of customer dissatisfaction and morale problems among domestic employees. The businesses indicated that 57 percent of customers were dissatisfied with outsourced services and only 15 percent were satisfied. Around three-fourths of these businesses would be interested in bringing back some of the jobs if rural employees could be recruited. The most common reasons cited for bringing back service jobs were increased customer satisfaction (63 percent), favorable government tax incentives (53 percent), and ease of identifying skilled labor in rural America (51 percent).

Morris and Goodridge conducted three surveys in 2006 and 2007, with one drawing on individuals across the entire country. Of rural residents, 37 percent were very interested in working from home to earn additional income, 39 percent were moderately or somewhat interested, and 21 percent were not interested. Homemakers were least likely to be interested in working at home (40 percent), though 30 percent would be interested in working anywhere from 11 to 20 hours a week at home.

Fifty percent of retirees, however, were interested in returning to the workforce, citing flexible schedule, supplemental income, social interaction, and intellectual stimulation as inducements. Fifty-four percent of retirees nationwide stated that they had broadband Internet access; the rate for rural retirees is uncertain, but is likely much lower given our analysis of Current Population Survey data. The relative lack of broadband access in rural communities may inhibit rural telework opportunities for retirees and others.

Broadband Availability Leads to Use

Several questions arise regarding policy intervention in broadband markets. Is there pent-up demand for broadband in areas that are unserved or under- served? Does greater availability beget correspondingly greater use of broadband Internet? When terrestrial broadband (DSL or cable) first becomes available, pent-up demand would be evident if conversion rates to broadband access are higher than in communities that have had terrestrial broadband for some time or are served only by satellite broadband. If the adoption rate is not higher, then broadband may be oversupplied or satellite is sufficient. A higher adoption rate would be another indicator that households do indeed value the Internet and broadband's advantages over the dial-up alternative.

Broadband Adoption on the Farm

The paucity of national geographically specific data presents a challenge in trying to analyze whether availability leads to broadband adoption. Data from USDA's June Agricultural Surveys provide a unique opportunity to examine geographically specific rural changes in Internet access methods (see Appendix B for a discussion of the June Agricultural Surveys). Figure 10 shows the conversion to broadband Internet access by farms across the country between 2005 and 2007. Unfortunately, a change in area identifiers did not allow us to match data for Illinois and Arkansas; hence, these States are omitted from the map and our analysis. By conversion we mean farms that did not already have broadband Internet access converted to broadband Internet access; farms may or may not have had dial-up Internet access.

The data presented in figure 10 show sharp differences in conversion rates across the country. When cross-referenced with the FCC broadband availability data (as estimated for likelihood for specific farm locations in December 2004 and December 2006, see Appendix C), our analysis of farm use of broadband supports the hypothesis that people embrace terrestrial broadband when given the option. Roughly 24 percent of all farms using the Internet in 2005 already had broadband Internet access of some type, and so could not convert. Conversions were nearly nonexistent in areas where broadband was available mostly via satellite.

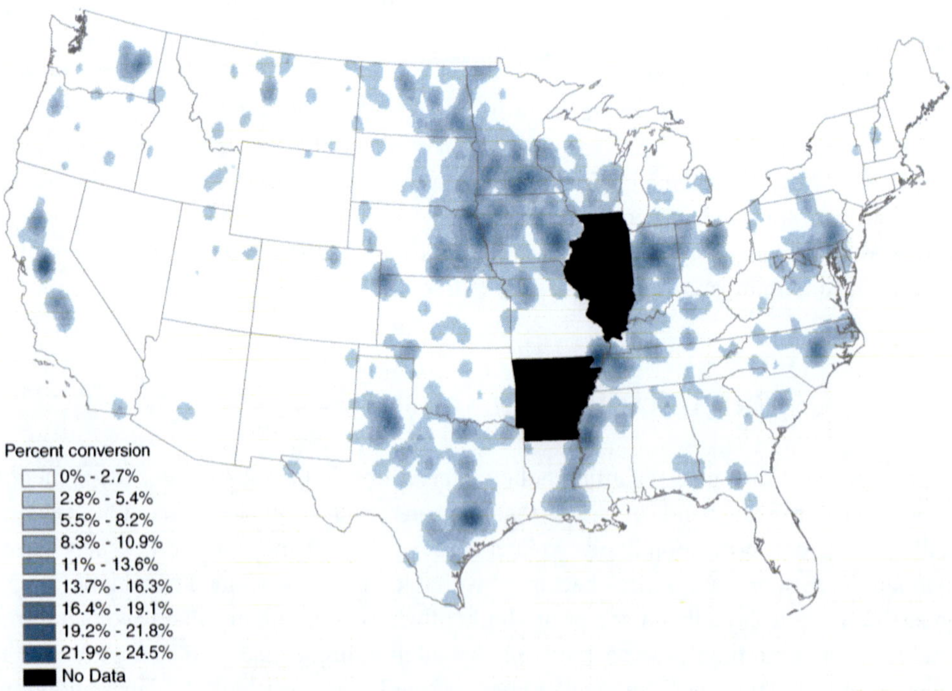

Source: ERS, using June Agricultural Survey data from USDA's National Agricultural Statistics Service.

Figure 10. Conversion to broadband use among U.S. farms, 2005-07

Farms were unlikely to make the direct jump from no Internet use to broadband Internet access; farms that already had dial-up Internet access were more likely to acquire broadband Internet access. DSL service was the most common broadband Internet access option among farms, whereas cable and fiber optics have shown the largest gains in highly urbanized areas over the last few years. The preponderance of DSL service for farms indicates both the rural location of most farms and Internet users finding satellite a less desirable option.

Rural Broadband Availability and Adoption

Some States collect data on broadband availability, enabling a more refined analysis of broadband deployment and adoption. Renkow (2008) examined broadband availability and adoption in two such States—Kentucky and North Carolina. Kentucky had both broadband availability and adoption data. North Carolina had broadband adoption data only. Broadband availability and adoption increased substantially during the last several years, with the largest proportional gains occurring in counties that had been the least well-served at the beginning of the period, typically rural counties. North Carolina had higher rates of adoption by all households than Kentucky (figure 11).

Renkow found that population density was more important than income in driving broadband deployment. The relative insensitivity of local income to patterns of broadband deployment may indicate that broadband providers perceive demand as being highly income inelastic. If so, cost of physical infrastructure would be the primary consideration in extending capacity into unserved or underserved areas.

The growth of broadband availability even in the most sparsely populated counties, however, is striking. More than three-quarters of households in all but 8 of Kentucky's 120 counties had broadband available to them by 2007. Seven of these eight counties were the least densely populated counties in the State. Interestingly, county adjacency to major urban areas was not related to broadband provision, as within-region clustering seems common. In addition, Renkow found evidence to suggest that the broadband loan program administered by USDA's Rural Development Utilities Programs stimulated broadband deployment in rural areas.

Broadband Internet Use and Rural Businesses

One of the salient features of the Internet is its capacity to provide information quickly and cheaply compared to other dissemination methods (Henderson et al., 2000). Wider and more convenient access may reduce the costs of communicating, transacting, and sourcing information. With improved information and knowledge, individuals' perception of products and services provided would be more accurate, thereby improving the adoption of worthwhile technologies and discarding those that have little value (Hooker et al., 2001; Just and Just, 2001). As a result, Internet use may lead to greater efficiency in the agricultural and other rural business sectors (Borenstein and Saloner, 2001; Gloy and Akridge, 2000; Greenstein and Prince, 2006).

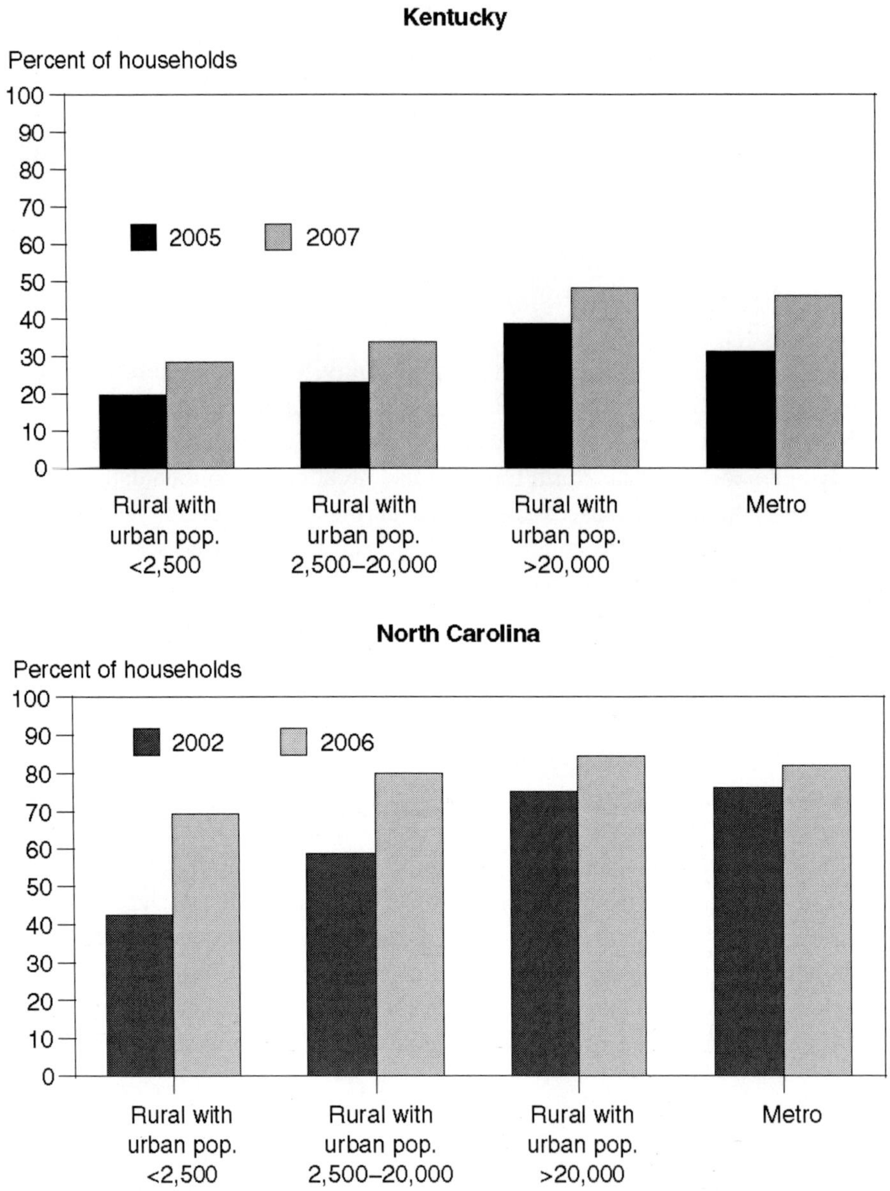

Source: Renkow, 2008.

Figure 11. Residential broadband adoption in Kentucky and North Carolina

Crandall (2008) pointed out during the ERS workshop that the effect of information and communications technology (ICT) on productivity growth is clear. Overall U.S. labor productivity growth from 1995 to 2000 was 2.5 percent per year, with an estimated 30 percent of it ascribed to ICT-producing and 56 percent due to ICT-consuming sectors of the economy (Fuss and Waverman, 2005). Measuring broadband Internet's contribution to this, however, is challenging due to data limitations and the problems of separating out overlapping causal effects (Crandall). As a consequence, empirical studies directly linking broadband to regional productivity growth are largely nonexistent.

Crandall and some of his colleagues, however, studied the effect of broadband on output and job growth. In their attempt to establish a relationship between State gross domestic product (GDP) or job growth and broadband deployment, they conducted a cross-section regression analysis on variables capturing local economic characteristics (unionization, business tax, education, wage rates), quality-of-life characteristics (climate, mean temperature), and broadband lines per capita. They found that total nonfarm employment growth was significantly related to broadband lines per capita. The results for GDP were not statistically significant. The strongest effects of broadband Internet on employment growth were in finance and insurance, real estate, and education services. The results largely supported an earlier study by Gillett et al. (2006).

Rural Businesses and Broadband

Pociask (2005) found evidence that rural small businesses did not use broadband as much as their urban counterparts. He attributed the lower usage to fewer employees, on average, in rural businesses and higher prices for rural broadband service. Socioeconomic characteristics—such as rural-urban differences in age, education, and affluence—may also play a role (Pociask, 2005; Stenberg, 2000). But what does this gap in broadband adoption mean for rural businesses?

Lamie et al. (2008) in their ERS workshop paper examined rural small business adoption of e-commerce practices. The rural businesses in their study were primarily manufacturing and retailing firms that fell into one of a number of e-commerce classifications: traditional local businesses that increased their market ranges and sales through e-commerce, virtual businesses (all marketing and sales conducted through e-commerce), businesses that used e-commerce primarily to reduce marketing inputs and costs, and businesses that used e-commerce primarily for business-to-business (B2B) or business-to-consumer (B2C) transactions.

Most businesses in their study used e-commerce because it provided an opportunity for increased profits and enhanced sustainability. E-commerce may benefit a firm in product development, inventory management, manufacturing, marketing and sales, and customer service. The 28 rural businesses in the case study had varied experiences in the application of e-commerce. Most felt that e-commerce activities benefited their operations. Economic returns from their e-commerce activities were enhanced if the business served a niche market, took advantage of public and private IT service providers in the maintenance of their e-commerce operation, and integrated e-commerce into multiple aspects of the business operation.

Rural Retailers and Broadband

Retailers are a particularly important type of rural business. They are present in nearly every community, are often major local employers, and often serve as a social hub. In their workshop presentation, Stoel and Ernst (2008) examined the attitudes and beliefs of rural retail business owners (specifically apparel, hardware, and grocery) that may act as impediments to accepting the Internet in their businesses. Included were owners' attitudes toward use of the Internet in their business, the perceived ease of use, and the Internet's usefulness in operational efficiency, strategic positioning, and other applications.

In their survey of 181 retail business owners, Stoel and Ernst found that retail store owners that used the Internet were less enamored of the Internet than nonusers, perhaps due to a fuller realization of some of its shortcomings or to a begrudging compliance with suppliers' demand that they use it. Broadband Internet access, however, did appear to facilitate using the Internet for operational effectiveness and business strategic positioning. Rural broadband users seemed to capitalize on the Internet's capacity to increase operational effectiveness and exploit market niches. Broadband users perceive the Internet to be easier to use than non-broadband users (which may say something about slow-speed toleration in business Internet operations).

Ernst and Stoel (2008), in a more thorough survey of rural grocers, found that these retailers felt that they had to be more price competitive because consumers received more information and explored more options via the Internet. As a result, businesses were expanding their markets and commercial business was moving faster. Rural grocers did not believe that e-commerce reduced their profits or threatened their existence. Rather, customers were more familiar with their business because of the Internet.

Farm Businesses and Broadband

Agriculture is another rural business sector that benefits from the Internet. For farm operators with Internet access in 2000, 98 percent used it to gather information. Price tracking (82 percent) was the next most common application (Hopkins and Morehart, 2001).

With growth in e-commerce, horticulture and other specialty farm products are increasingly sold direct to households. E-commerce has increased efficiencies in existing relationships along the food marketing chain, reduced the cost of expanding market area, and brought about new services such as supermarket home delivery and direct-to-consumer sales (Kinsey and Buhr, 2003).

Not all types of agricultural production lend themselves readily toward direct sales from producer to consumer. Still, the wholesale and retail food industry has enhanced its productivity with Internet adoption (Akridge, 2003; Beurskens, 2003; Henderson et al., 2000; Stricker et al., 2003; Zilberman et al., 2002).

Respondents to the 2007 Agricultural Resources Management Survey (ARMS) were asked if they had Internet access and if it was "high-speed." A majority of farms (63 percent) reported using the Internet in their farm business (figure 12). Among those using the Internet, the predominant access method was broadband and this group of users accounted for over 60 percent of U.S. farm production. This is consistent with other estimates of farm broadband use. USDA's National Agricultural Statistics Service (NASS) reported that, for the first time in 2007, the majority of farm Internet users were connecting with broadband Internet technologies (USDA/NASS, 2007).

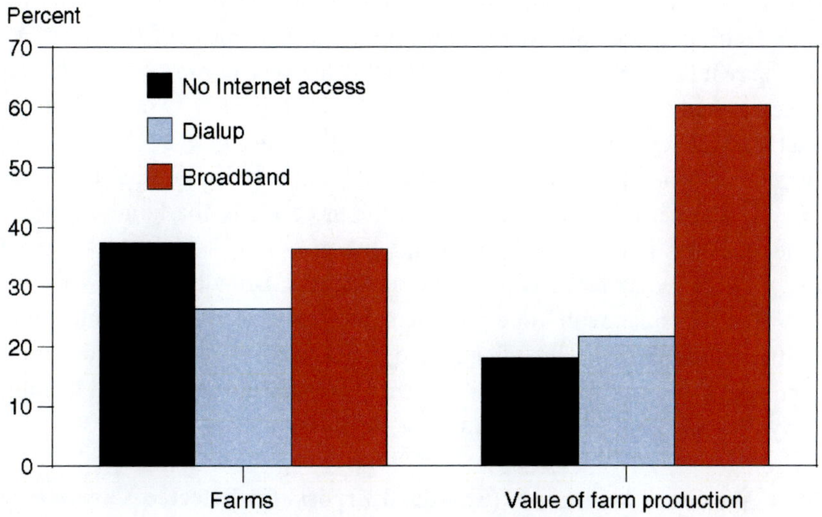

Source: ERS analysis of 2007 ARMS (USDA, NASS and ERS).

Figure 12. Distribution of farms and value of farm production by Internet use, 2007

Prior research has identified several demographic and socioeconomic attributes that have consistently distinguished those who use the Internet from those who do not (Forman, 2005; Stenberg, 2006; Stenberg and Morehart, 2007). These include income, education, age, and number of children. In our analysis of the determinants of broadband Internet use among farm households, we include household income, education level attained by the farm operator, age of the farm operator, off-farm work by spouse, presence of school-age children, number of hired farmworkers, rural-urban farm location, county net migration, and number of local broadband providers.

Income has consistently been cited as a primary factor determining Internet use (Stenberg, 2006). The greater the income level, the more likely that work is highly skilled. In addition, the more highly skilled the work, the more likely that computer technologies and the Internet are part of the work environment. As household income increases, regardless of location, the likelihood of Internet use increases (Stenberg, 2000).

Educational attainment has long been recognized as a determinant in income level (Becker, 1964). The prevalence of the Internet and computer technologies in educational institutions provides additional exposure and experience as years of formal education increase. Consequently, the greater an indi-**vidual' s education, the greater the likelihood** of Internet use at home or in the workplace (Stenberg, 2006).

Many (Oden and Strover, 2002; Grant and Meadows, 2002; Stenberg and Morehart, 2007) have cited age as a factor in determining the likelihood of Internet use. The literature suggests that older individuals are reticent about adopting the Internet, while the young readily adopt. The average age of farm operators claiming no Internet use in 2007 was 62, compared with 54 for those who accessed the Internet using broadband (table 14).

Only about a third of spouses on farms with no Internet use reported working off-farm, compared to more than 50 percent on farms that used the Internet. On the one hand, off-farm employment may provide more income and exposure to Internet technologies, instigating home or farm adoption. On the other hand, a spouse who works off the farm may indicate financial stress and lesser wherewithal to invest in farm-specific Internet use. Households with school-age children are expected to have a higher awareness of the Internet and more demand for bandwidth-intensive applications (Grant and Meadows, 2002). In keeping with this, the percentage of farms with school-age children was nearly two times higher in 2007 when Internet use was reported than when it was not (table 14).

We hypothesize that the greater the size and complexity of the farm business, as evidenced by the number of hired farmworkers, the more likely the farm is using broadband to access the Internet. Farms with broadband Internet access had twice the number of farmworkers, on average, as farms with no Internet access. Broadband use is also expected to be higher in ZIP Code areas with more providers, as competition for customers likely lowers the price differential between broadband and dial-up. The mean number of providers, however, showed little variation across Internet use categories (table 14) reflecting the predominantly rural location of farm operations.

Table 14. Weighted Means and (Standard Errors) for Selected Variables, 2007

		Internet use			All family farms
Variable	**Name**	**No Internet**	**Dial-up**	**Broadband**	**1997**
Continuous variables:					
Operator age	OP_AGE	61.90	55.15	53.85	57.20
		(0.45)	(0.34)	(0.38)	(0.20)
No. broadband providers	NOPROVIDERS06	6.80	6.71	6.83	6.79
		(0.09)	(0.09)	(0.09(	(0.05)
Household income	TOTHHI	61,614	77,831	121,141	87,523
		(3,912.25)	(2,396.71)	(5,286.60)	(2,548.60)
No. farmworkers	NOWORKERS	0.42	0.63	0.98	0.68
		(0.04)	(0.03)	(0.03)	(0.01)
Dummy variables:					
Population change	MIGCLS	0.58	0.60	0.56	0.58
		(0.01)	(0.01)	(0.01)	(0.01)
Children	CHILD	0.24	0.40	0.42	0.35
		(0.01)	(0.02)	(0.02)	(0.01)
Spouse working off-farm	SPOFF	0.34	0.52	0.51	0.45
		(0.01)	(0.02)	(0.02)	(0.01)
College education	COLLEGE	0.12	0.22	0.36	0.23
		(0.01)	(0.01)	(0.02)	(0.01)
Urban	RURAL	0.21	0.21	0.25	0.22
		(0.01)	(0.01)	(0.02)	(0.01)
Not urban or rural		0.61	0.64	0.60	0.61
		(0.01)	(0.01)	(0.02)	(0.01)
Rural		0.18	0.14	0.15	0.17
		(0.01)	(0.01)	(0.01)	(0.01)

Source: ERS using FCC and 2007 ARMS (USDA, NASS and ERS).

Table 15. Multinomial logistic regression results

Variable	Dial-up	Broadband	Dial-up	Broadband	Dial-up	Broadband
	Estimate	*Estimate*	*Std.Err*	*Std.Err*	*T value*	*T value*
(Intercept)	-0.77	-0.57	0.07	0.09	-11 .26**	-6.27**
OP_AGE	-50.75	-65.10	9.08	10.49	-5.59**	-6.21**
$(OP\_AGE)^2$	-27.76	-32.38	6.75	6.99	-4.13**	-4.64**
NOWORKERS	0.08	0.15	0.03	0.03	3.32**	5.18**
TOTHHI above 50k	0.27	0.44	0.07	0.06	4.01**	8.03**
NOPROVIDERS06	-11.45	-12.88	5.69	6.52	-2.01*	-1.97*
$(NOPROVIDERS06)^2$	-12.22	-5.32	4.98	6.69	-2.45**	-0.08
MIGCLS	0.11	-0.02	0.09	0.08	1.20	-0.26
CHILD	0.17	0.29	0.09	0.12	2.10*	2.52*
SPOFF	0.29	0.12	0.10	0.12	2.93**	1.03
COLLEGE	0.70	1.34	0.11	0.11	5.89**	11.96**
Not urban or rural	-0.24	-0.38	0.12	0.10	-1.96*	-3.60**
Rural	-0.09	0.10	0.06	0.06	-1.39	1.61

LR = 380954.54; AIC = 3905755.62; McFadden R^2 = 0.089; McFadden Adj R^2 = 0.089.
Note: ** - significant at 0.01, *- significant at 0.05. Equations simultaneously estimated.
Source: ERS using FCC and 2007 ARMS (USDA, NASS and ERS).

Maximum-likelihood methods were used to estimate a multinomial logit model that estimates the relationship between farm household socioeconomic characteristics and the type of Internet connection used. For the most part, coefficient signs and variable significance are consistent with expectations (table 15). The model fit—as indicated by the McFadden r-squared value of 0.089—is somewhat poor, even for a cross-sectional analysis. The coefficients represent the log-odds of a farm household using dial-up or broadband Internet access, relative to the base class (no Internet use by the farm). That is, what is the chance that, instead of not using the Internet at home, the household has an in-home dial-up or broadband Internet connection? These are not the odds of using dial-up or broadband, only the odds relative to not being connected to the Internet at all. The sign of the coefficient gives the direction of the relationship: increase or decrease in probability due to the predictor. For example, as the age of the farm operator increases, the probability of having dial-up or broadband Internet access relative to no Internet access declines, as indicated by the negative and significant coefficients. (More discussion on the underlying methodology of this analysis is presented in Appendix D.)

Other significant model results include:

- Larger farm businesses, as indicated by more hired workers, have a higher probability of broadband Internet access.
- Farm households with income above $50,000 have a higher probability of broadband Internet access.
- The relative probability of broadband Internet use does not increase as the number of providers in an area increases.
- Having school-age children in the household is associated with higher probability of broadband Internet use.

- Operators with at least a college degree are more likely to use broadband.
- Farms located in mixed urban/rural areas are less likely to use broadband than those in urban areas.

The model's results suggest that household characteristics such as age, education, presence of children, and household income are significant factors in adopting broadband Internet use. Farm business complexity, as measured by the number of farmworkers, was also related to the use of the Internet and broadband Internet access. Distance from urban centers was not a factor in Internet use. Our proxy for county economic well-being—population migration—was not significant and may be indicative of cross currents that are present; counties under economic distress may invest in broadband to help mitigate the distress, or may not have the economic wherewithal for broadband investment.

The relationship between Internet/broadband use and farm location is less clear. Farms in mixed urban-rural areas were less likely to use dial-up or broadband Internet. This may be a result of cost or availability of service. More isolated farms, as measured by the rurality of the county, had mixed, though not significant, results. These results warrant further analysis.

Table 16. Logistic regression results[1]

Choice variable: Bypass the nearest town?	**Type of purchase**		
	Farm inputs	**Farm equipment**	**Farm credit**
Farm level variables:	*Odds ratios*		
Log (gross farm sales)	1.1 258***	1.1 223***	1.1 938***
Operator's years experience	1.0063	1.01 92***	.9964
Years of education	1.071 0**	1.0675*	1.0244
Internet farm purchase	2.0649***	2.01 48***	1.5553*
County-level variables:			
No. of farm inputs merchants	.9360***		
No of farm equipment dealers		.9303**	
Remote county	.7456*	.7417*	1.0081
Log (highway miles)	1 .3623***	1.0956	1 .2733**
Log (population density)	1.1 720**	1.1713*	1.0674
Log (per capita income)	1.2686	1.0424	1.4158
	Model statistics		
No. of observations	2,793	2,793	2,793
Log pseudolikelihood	-1,861.19	-1,620.98	-1,817.38
Wald 2	44.33	41.48	45.88
Prob > 2	0.0000	0.0000	0.0000
Pseudo R2	.0386	.0388	.0343

[1] *Odds ratio*$_i$ = exp $(\hat{\beta}_i)$. Significance level of the coefficient estimates $(\hat{\beta}_i)$:* p<.1; ** p<.05; *** p<.01.

Source: ERS analysis of 2004 ARMS data (USDA's NASS and ERS) and other data.

Farm-Rural Linkages in the Internet Economy

The Internet may change the economic relationship between farms and their local economies. Using data drawn from the 2004 ARMS, we investigate how Internet use affects the geography of farm input purchases.

Conceptually, farmers may choose to (1) purchase inputs in the nearest local town, (2) bypass the nearest local town but purchase inputs within the market reach of the nearest farm service center, or (3) bypass the farm service center altogether. Purchasing patterns are examined for three broad categories of resource inputs: farm inputs (feed, seed, and fertilizer), farm machinery and equipment, and farm credit. Comparing input purchases in each of these three mutually exclusive categories allows us to observe the changing nature of farm/local area interrelationships and the use of the Internet. We present results using logistic regression from one of these models in table 16.

Our results suggest that the market reach of the nearest town may no longer define what the farm operator perceives as local. Making farm purchases over the Internet is the strongest factor increasing the likelihood of the operator bypassing the nearest town and even the more distant farm service center (table 16).

As farm operators increase their participation in e-commerce, their relationships with local suppliers are likely to weaken. Farm operators may increasingly opt for distant suppliers to secure lower prices or better access to niche inputs. Suppliers with an established Internet presence, including local ones, would appear better positioned to retain customers within the local economy.

CONCLUSION

The Internet, and more specifically broadband Internet, has become an integral part of the broader economy. Since the digital or information economy incorporates computer processes, telephony, information storage and use, hardware, and software, however, it is challenging to separate out the Internet's contribution to economic growth and well-being. In fact, the Internet's economic contribution has more than occasionally been estimated using the residual in economic growth not otherwise explained.

Rural communities are invested in the digital economy, though equal access across the rural-urban landscape is questionable. Rural and farm households are almost as likely as urban households to use the Internet, but are less likely to use broadband. Rural businesses are less likely than urban businesses to use the Internet. Broadband access is less prevalent in rural areas than in more densely populated areas, and analysis of CPS data suggests that lower broadband use in rural areas may be involuntary.

Broadband provision follows a geographical pattern strongly tied to population size and the urban-rural hierarchy. Lack of broadband is most strongly associated with low population size in the area. Low broadband provision also exhibits very strong regional patterns that reflect differences in urban concentration and topography.

Broadband users use the Internet more intensively than dial-up users and now outnumber dial-up users. The high adoption rate of broadband technologies by urban Internet users indicates that people value what the Internet has to offer. Rural Internet users have less in-home broadband access, and this is likely due to its higher cost and limited availability in rural settings. Government policies to encourage deployment of broadband services in rural areas have increased availability and in some cases encouraged competitive pricing. Unfortunately, there are little national data that link Internet access choice with the cost of service. As a result, it is difficult to distinguish between financial and other motives when examining broadband adoption. Nonetheless, by using the Internet more intensively, on more activities, using newer (and presumably better) technologies, and bringing the Internet into their homes, schools, and workplaces, society has clearly indicated that it values Internet access.

More activities are shifting to the Internet. Some of these activities have great potential value for the rural economy. Education programs and offerings—primary, secondary, higher education, and continuing education—have become richer on the Internet. Telework is becoming a more practical option for workers and businesses. Some medical services may lend themselves readily to the Internet environment, with potential cost savings for rural residents and medical clinics that offer in-situ services not otherwise readily available in rural settings. Rural businesses are adopting more e-commerce and Internet practices, enhancing economic vitality and expanding market reach. Individuals are using the Internet to get involved with their communities.

Analysis of farm businesses indicates that household characteristics such as age, education, presence of children, and household income are significant factors in adopting broadband Internet use. Distance from urban centers was not a factor in Internet access. Larger farm businesses are more apt to use broadband in managing their operation; the more multifaceted the farm business, the more the farm uses the Internet.

Analysis suggests that rural economies benefit generally from broadband Internet availability. In comparing counties that had broadband access relatively early (by 2000) with similarly situated counties that had little or no broadband access as of 2000, employment growth was higher and nonfarm private earnings greater in counties with a longer history of broadband availability.

Government policies that encourage deployment of broadband services have broadened their availability in rural America. The 2008 Farm Act (Food, Conservation, and Energy Act of 2008) reauthorized USDA's telemedicine, distance learning, and rural broadband access grant and loan programs. The American Recovery and Reinvestment Act of 2009 provided $2.5 billion to USDA for loans and grants to increase broadband provision in rural areas. As much as these funds address the needs of unserved and underserved communities, rural broadband availability will increase.

More research is needed in many areas to better understand broadband and rural economies. Better information on broadband availability, use, cost, and technical characteristics is needed to gain a better understanding of broadband Internet's potential effect on rural economies. Broadband availability data are only now being collected below the ZIP Code area; the more geographically granular the data the better our understanding will become on unserved and underserved areas. Price data are largely unavailable, hindering economic analysis of supply and demand of the regional broadband market.

Detailed broadband Internet use data also have not been collected since 2003. Our understanding of broadband users' online behavior is limited. For example, we know certain Internet-based education activities are taking place, but the extent to which Internet practices are taking place in rural education systems (in either the school or at home) is still unclear.

REFERENCES

Akridge, Jay T. (2003). "E-Business in the Agricultural Input Industries," *Review of Agricultural Economics*, *25*, 1, pp. 3-13.

Becker, Gary (1964). *Human Capital*, New York: Columbia University Press.

Bender, Bruce & Steven Shwiff (1982). "The Appropriation of Rents by Boomtown Governments," *Economic Inquiry*, vol. 20, pp. 84-103.

Beurskens, Frank (2003). "The Economics of Dot.coms and E-commerce in the Agrifood Sector," *Review of Agricultural Economics*, *25*, 1, pp. 22-28.

Blum, U. (1982). "Effects of Transportation Investments on Regional Growth: A Theoretical and Empirical Investigation," *Papers of the Regional Science Association, vol. 49*, pp. 169-84.

Borenstein, Severin & Garth Saloner (2001). "Economics and Electronic Commerce," *Journal of Economic Perspectives*, *15*, 1, pp. 3-12.

Bresnahan, Timothy F., Brynjolfsson, Erik & Hitt, Lorin M. (1999). *Information Technology, Workplace Organization, and the Demand for Skilled Labor: Firm-Level Evidence*, National Bureau of Economic Research, NBER Working Paper Series, Working Paper 7136.

Brynjolfsson, Erik & Smith, Michael D. (2000). "Frictionless Commerce? A Comparison of Internet and Conventional Retailers," *Management Science*, *46*, 4(April), pp. 563-85.

Capalbo, Susan M. & Heggem, Christine N. (1999). "Valuing Rural Health Care: Issues of Access and Quality," *American Journal of Agricultural Economics*, *81*, 3, pp. 674-679.

Choudrie, Jyoti & Dwivedi, Yogesh K. (2006). "Investigating Factors Influencing Adoption of Broadband in the Household," *The Journal of Computer Information Systems*, *46(4)*, pp. 25-34.

Cook, Thomas D. & Campbell, Donald T. (1979). *Quasi-Experimentation: Design and Analysis Issues for Field Settings*, Chicago: Rand McNally College Publications.

Crandall, Robert W. (2008). "*Broadband, Jobs, and Regional Growth: The Limited Evidence.*" Paper presented at the Economic Research Service's Broadband in the Rural Economy workshop, Washington, DC, Sept.

Crandall Robert, Lehr William & Litan Robert (2007). "The Effects of Broadband Deployment on Output and Employment: A Cross-sectional Analysis of U.S. Data," *Issues in Economic Policy*, The Brookings Institution, No. 6, July.

Crescia, Elena (2006). "Measuring e-gov Impact: The Experience of São Paulo, Brazil," *e-estratégia pública*, Feb. 9.

Davies, Stephen (1979). *The Diffusion of Process Innovations*, Cambridge University Press.

DeYoung, Robert & Duffy Denise (2002). "The Challenges Facing Community Banks: In Their Own Words," *Economic Perspectives*, Chicago Federal Reserve Bank.

Egan, Bruce L. (1996). *Information Superhighways Revisited: The Economics of Multimedia*, Boston: Artech House.

Eisenberg, Jon (2002). "*Broadband: Bringing Home the Bits.*" Paper presented at the 2002 TPRC meetings, Arlington, VA, Sept.

Ernst, Stan & Stoel Leslie (2008). "*Rural Grocers and Technology Adoption: Attitude Matters, Size Matters More.*" Paper presented at the Economic Research Service's Broadband in the Rural Economy workshop, Washington, DC, Sept.

Farnsworth, Don (1972). *The Ohio County Airport Study 1964-1972*, Ohio Department of Commerce, Division of Aviation, Columbus, Ohio.

Federal Communications Commission (2008). *Trends in Telephone Service*, http://hraunfoss.fcc.gov/edocs_public/attachmatch/DOC-284932A1.pdf .

Flamm, K. & Chaudhuri, A. (2007). "An Analysis of the Determinants of Broadband Access," *Telecommunications Policy*, *31*, 312-326.

Forman, Chris (2005). "The Corporate Digital Divide: Determinants of Internet Adoption," *Management Science*, *51*, 4, pp. 641-654.

Fuss, Melvyn & Waverman, Leonard (2005). "*Canada's Productivity Dilemma: The Role of Computers and Telecom*," Bell Canada's Submission to the Telecommunications Review Panel, Appendix E-1.

Gallegos, Hugo C. (2002). "*E Government: Definitions, Benefits, Development, Examples*," La Sociedad Digital.

Gillett, Sharon E., Lehr, William H., Osorio, Carlos A. & Sirbu, Marvin A. (2006). *Measuring the Economic Impact of Broadband Deployment*, Final Report, National Technical Assistance, Training, Research, and Evaluaton Project, U.S. Department of Commerce, Economic Development Administration.

Gloy, B. A. & Akridge, J. T. (2000). "Computer and Internet adoption on large U.S. farms," *International Food and Agribusiness Management Review*, *3*, 323-338.

Goetz, Stephan & Debertin David (1996). "Rural-urban Locational Choices of Medical Doctors: A County-Level Analysis," *Review of Agricultural Economics*, *vol.18*, pp. 547-63.

Grant, August E. & Meadows, Jennifer H. editors (2002). *Communication Technology Update*, eighth edition, Woburn, MA: Elsevier Science.

Greene, W. H. (1993). *Econometric Analysis*, third ed. Prentice-Hall, Upper Saddle River, NJ.

Greenstein, Shane & Prince, Jeff (2006). *The Diffusion of the Internet and the Geography of the Digital Divide in the United States*, National Bureau of Economic Research, NBER Working Paper Series, Working Paper 123182.

Greenstein, Shane (2000). *Building and Delivering the Virtual World: Commercializing Services for Internet Access*, National Bureau of Economic Research, Working Paper 7690, May.

Henderson, J., Dooley, F. & Akridge, J. (2000). "*Adoption of E-Commerce strategies for agribusiness firms*," http://www.ebscohost.com.

Hooker, N. H., Helig, J. & Ernst, S. (2001). "*What is unique about EAgribusiness?*" http://agecon.lib.umn.edu/.

Hopkins, Jeff & Morehart Mitch (2001). "Farms, the Internet, & E-Commerce: Adoption and Implications," *Agriculture Outlook*, Nov., pp. 17-20.

Horrigan, John B. (2008). "*Broadband Adoption, Rural America, and Economic Decisions*." Paper presented at the Economic Research Service's Broadband in the Rural Economy workshop, Washington, DC, Sept.

Internet Systems Consortium, https://www.isc.org/.

Isserman, Andrew & Rephan, Terance (1995). "The Economic Effects of the Appalachian Regional Commission: An Empirical Assessment of 26 Years of Regional Development Planning," *Journal of the American Planning Association*, *61*, 3(Summer), pp. 345-64.

Isserman, Andrew M. & Stenberg, Peter L. (1994). *The Recovery of Rural Economies from Military Base Closures: Control Group Analysis of Two Decades of Experience*, Research Paper 9408, Regional Research Institute, West Virginia University.

Isserman, Andrew M. (1987). *Highways and Rural Economic Development: Results from Quasi-Experimental Approaches*, Research Paper 8907, Regional Research Institute, West Virginia University.

Just, D. R. & Just, R. E. (2001). "Harnessing the Internet for farmers," *Choices: The Magazine of Food & Resource Issues, 16(2)*, http://www. ebscohost.com.

Keeton, William R. (2001). "The Transformation of Banking and Its Impact on Consumers and Small Businesses," *Economic Review*, Kansas City Federal Reserve Bank.

Kinsey, Jean & Buhr, Brian (2003). "*E-Commerce: A New Business Model for the Food Supply/Demand Chain,*" Working Paper 03-0 1, The Food Industry Center, University of Minnesota, February.

Lamie, R. David, Barkley, David L. & Markley, Deborah M. (2008). "*Positive Examples and Lessons Learned from Rural Small Business Adoption of E-Commerce Strategies*." Paper presented at the Economic Research Service's Broadband in the Rural Economy workshop, Washington, DC, Sept.

Lanvin, Bruno (2008). "*Where is e-government going in 2020?*" Paper presented at the e-leaders conference, The Hague, March 6-7.

Leamer, Edward E. & Storper, Michael (2001). *The Economic Geography of the Internet Age*, National Bureau of Economic Research, Working Paper 8450, Aug.

Maddala, G. S. (1983). *Limited-Dependent and Qualitative Variables in Econometrics*, Cambridge University Press, New York.

Malecki, Edward J. (2008). "*The Rural Digital Economy: Competition and the Geography of Telecommunications.*" Paper presented at the Economic Research Service's Broadband in the Rural Economy workshop, Washington, DC, Sept.

Malecki, Edward J. & Bruno Moriset (2007). *The Digital Economy: Business Organization, Production Processes and Regional Developments*, London: Routledge.

Mann, Catherine L. (2008) "*Information Technology, Services, and the Global Economy,*" Paper presented at the Economic Research Service's Broadband in the Rural Economy workshop, Washington, DC, Sept.

Mann, Catherine L. (2006). *Accelerating the Globalization of America: The Role for Information Technology*, Institute for International Economics: Washington DC.

Massachusetts Broadband Institute, http://www.masstech.org/broadband resources.html.

McFadden, Daniel (1974). "Conditional Logit Analysis of Qualitative Choice Behavior," in Zarembka, P. ed., *Frontiers in Econometrics*, New York: Academic Press.

Moore, Michael Grahame, editor (2007). Handbook of Distance Education, NY: Routledge.

Morris, Douglas E. & Goodridge, Lyndon E. (2008). "*What skills are at the end of broadband cables in rural America? Do they match up with firms wishing to engage rural sourcing?*" Paper presented at the Economic Research Service's Broadband in the Rural Economy workshop, Washington, DC, Sept.

National Telephone Cooperative Association, "NTCA Members Internet/ Broadband Survey Report," Nov. 2000.

Oden, Michael & Strover, Sharon (2002). *Links to the Future: The Role of Information and Telecommunications Technology in Appalachian Economic Development*, Appalachian Regional Commission, Washington, DC.

Odlyzko, Andrew (2008). "The Internet's role in aggravating and alleviating the energy crises," *Internet Evolution*, April 8.

Parker, Edwin B. (2000). "Closing the Digital Divide in Rural America," *Telecommunications Policy*, *24(4)*, pp. 281-90.

Parker, B. Edwin & Hudson, Heather E. (1992). *Electronic Byways: State Policies for Rural Development Through Telecommunications*, Boulder: West View Press.

PEW Internet & American Life Project, http://www.pewinternet.org/index. asp.

Pociask, Stephen B. (2005). "*Broadband Use by Rural Small Businesses*," Washington, DC: Small Business Administration, Office of Advocacy.

Poley, Janet (2008). "*Rural Distance Education*." Paper presented at the Economic Research Service's Broadband in the Rural Economy workshop, Washington, DC, Sept.

Renkow, Mitchell (2008). "*Residential Broadband Availability and Adoption: Recent Evidence from Kentucky and North Carolina*." Paper presented at the Economic Research Service's Broadband in the Rural Economy workshop, Washington, DC, Sept.

Rogers, Everett M. (1995). *Diffusion of Innovations*, New York: The Free Press.

Stenberg, Peter L. & Morehart, Mitchell (2008). "Characteristics of Farm and Rural Internet Use in the United States," *Handbook of Research on Global Diffusion*, Yogesh Dwivedi, et al. (eds.), Idea Group, Inc..

Stenberg, Peter L. & Morehart, Mitchell (2007). "The Internet in U.S. Farm and Rural Businesses," in Irene Johansson (eds.), *Entrepreneurship and Development-Local Processes and Global Patterns*, University West, Sweden.

Stenberg, Peter L. & Morehart, Mitchell (2006). "*Determinants of Internet Use for Rural and Farm Economic Sectors*", Telecommunication Policy Research Conference, Arlington, VA, Sept. 29.

Stenberg, Peter L. (2006). "Investment and Household Adoption of Communication and Information Services Across the United States" in *The Emerging Digital Economy: Entrepreneurship, Clusters and Policy*, Borge Johansson, Charlie Karlsson, and Roger Stough (eds.), Springer: Berlin, Heidelberg, New York, pp. 263-76.

Stenberg, Peter L. (2004). "United States Policy for Spatial Investment in CIT", *Regulation and Capital Market Evaluation in Telecommunications*, H.J. Piepenbrock,F. Schuster, and E.O. Ruhle (eds.), Springer-Verlag Publishing, New York.

Stenberg, Peter L. (2000) "*Rural Telecommunication Policy in the U.S.: Issues and Economic Consequences.*" Electronic proceedings for the conference, European Rural Policy at the Crossroads, The Arkleton Centre for Rural Development Research, University of Aberdeen, Scotland, UK, 29 June-1 July, http:\\www.abdn.ac.uk/arkleton/conf2000/ papers/stenberg.doc, Sept.

Stenberg, Peter L. (editor) (1999). *Telecommunications in Rural Economic Development.* Proceedings from a workshop, Western Rural Development Center, Oregon State University, Corvallis, OR, Dec.

Stern, Michael J., Adams, Alison E. & Boase, Jeffery (2008). "*The Role of the Internet in Rural Community Participation: Examples from Survey Data*." Paper presented at the Economic Research Service's Broadband in the Rural Economy workshop, Washington, DC, Sept.

Stern, Michael & Dillman, Don A. (2006). "Community Participation, Social Ties, and Use of the Internet," *City & Community*, *vol. 5*, pp.409-24.

Stoel, Leslie & Ernst, Stan (2008). "*Comparing Rural Retailer Internet Users and Non-users: Access Speed, Demographics, Attitudes, and Beliefs.*" Paper presented at the Economic Research Service's Broadband in the Rural Economy workshop, Washington, DC, Sept.

Stricker, S., Sumner, D. A. & Mueller, R. A. E. (2003). "*Wine on the Web in a Global Market: A Comparison of E-Commerce Readiness and Use in Australia, California, and Germany.*" Paper presented at the EFITA 2003 Conference, July 5-9.

USDA, National Agricultural Statistics Service (2007). *Farm Computer Usage and Ownership*, Washington, DC: U.S. Department of Agriculture.

Varian, Hal R. (2003) "*Economics of Information Technology.*" Raffaele Mattioli Lecture, Bocconi University, Milano, Italy, revised March 23.

Verisign, http://www.verisign.com/.

Wellman, Barry, Janet Salaff, Dimitrina Dimitrova, Laura Garton, Milena Gulia, and Caroline Haythornthwaite (1996). "Computer Networks as Social Networks: Collaborative Work, Telework, and Virtual Community," *Annual Review of Sociology, vol. 22*, pp. 213- 238.

Wheat, Leonard (1970). *The Effect of Airline Service on Manufacturing Growth in Cities Below 40,000 Population*, U.S. Department of Commerce, Economic Development Administration, Washington, DC.

Whitacre, Brian E. (2008). "*Estimating the Economic Impact of Telemedicine in a Rural Community.*" Paper presented at the Economic Research Service's Broadband in the Rural Economy workshop, Washington, DC, Sept.

Zilberman, David, Metcalfe, Mark & Heiman, Amir (2002). "*Economics and Adoption of Information Technology with Evidence from California.*" Paper for the Card E-Commerce research mini-conference, Feb. 9.

APPENDIX A. BUREAU OF THE CENSUS REGIONS

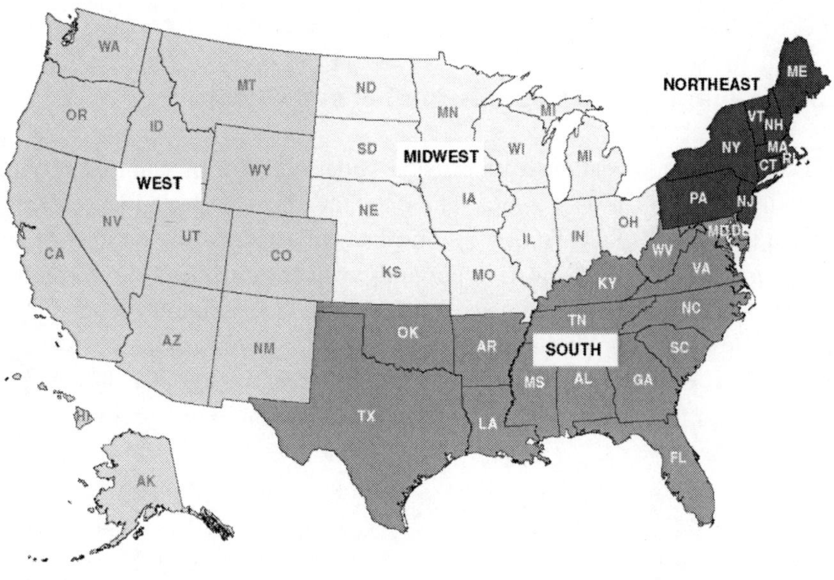

Figure A-1. Bureau of the Census Regions

Appendix B. Data Used in the Study

Agricultural Resources Management Survey

Most of the farm-level data used in the analysis are from the 2007 Agricultural Resources Management Survey (ARMS), conducted annually by USDA's Economic Research Service and National Agricultural Statistics Service. The survey collects information needed to measure the financial condition (farm income, expenses, assets, and debts) and operating characteristics of farm businesses, the cost of producing agricultural commodities, and the well-being of farm operator households. The target population of the survey is operators of farm businesses representing agricultural production in the 48 contiguous States. A farm is defined as an establishment that sold or normally would have sold at least $1,000 of agricultural products during the year. Farms can be organized as proprietorships, partnerships, family corporations, nonfamily corporations, or cooperatives. Data are collected from one operator per farm, the senior farm operator. A senior farm operator is the operator who makes most of the day-to-day management decisions. For this study, operator households organized as nonfamily corporations or cooperatives and farms run by hired managers are excluded.

June Agricultural Survey Data

The June Agricultural Survey is conducted by USDA's National Agricultural Statistics Service every year to provide estimates of farm numbers and land in farms, crop acres planted, grains and oilseeds in storage, livestock inventories, and land values. In 1997, 1999, 2001, 2003, 2005, and 2007, the surveys included questions about Internet access. In 2005 and 2007, the questions addressed broadband Internet access. The 2007 computer usage estimates are based on responses from over 31,400 agricultural operations and represent all sizes and types of farms.

Federal Communications Commission Form-477 Data

The Federal Communications Commission collects data from Internet service providers twice annually: in June and December. The data have been collected since December 1999 through what is called the FCC Form-477. The data collected include number of lines, various company characteristics, and most importantly where broadband is currently provided. Initially, very small providers were not required to file the form, but eventually all providers were required to return the form to the FCC every 6 months. On the form, a company specifies each ZIP Code in which it has customers. Recently, the FCC has started to collect more detailed information on type of broadband service and location of broadband service provision.

PEW Surveys

PEW data are collected by Princeton Survey Research Associates. Sample sizes for each survey are around 2,500, with total sampling error of +/-2 percent and sampling error of +/-3 percent for Internet users. Data were collected from various surveys and reports available from PEW Internet & American Life Project's website (http://www.pewinternet.org/index.asp). Urban, as defined by the U.S. Census Bureau, includes any population concentration of 2,500 or more people.

Current Population Survey

The U.S. Census Bureau conducts monthly Current Population Surveys. The surveys constitute 57,000 households containing 134,000 persons. On an irregular basis, the Bureau's monthly survey includes questions on Internet and related technologies. These special surveys provide broad- based and statistically reliable information on the ways that information technologies, in general, and broadband more specifically are transforming the way we live, work, and learn. As of this writing, the last survey on information technology took place in October 2007, but only asked four Internet-related questions, whereas an October 2003 survey asked detailed questions on what, where, and how the Internet was used. Estimation conducted in this report had a sampling error no greater than +/-2 percent. Urban, as defined by the U.S. Census Bureau, includes any population concentration of 2,500 or more people.

Appendix C. Using the FCC Data

Broadband Availability Over Time

The growth in broadband availability has been rapid. This can be seen in the following maps (Figure C-1 - C-4) developed from the FCC data that show broadband availability from December 2000 through December 2006. Each dot in the maps represents a ZIP Code area that has at least one provider. The dots are located at the population center for their corresponding ZIP Codes. Dots are used to represent broadband availability as rural broadband service most commonly radiates out from towns into the surrounding countryside. These maps give a contrasting view of broadband availability from the isopleth map (Figure 6, p. 16) shown in the main body of this report. Some persistent wilderness areas and the development of broadband service along arterial roadways become apparent in the dot maps over time (from 2000 to 2006).

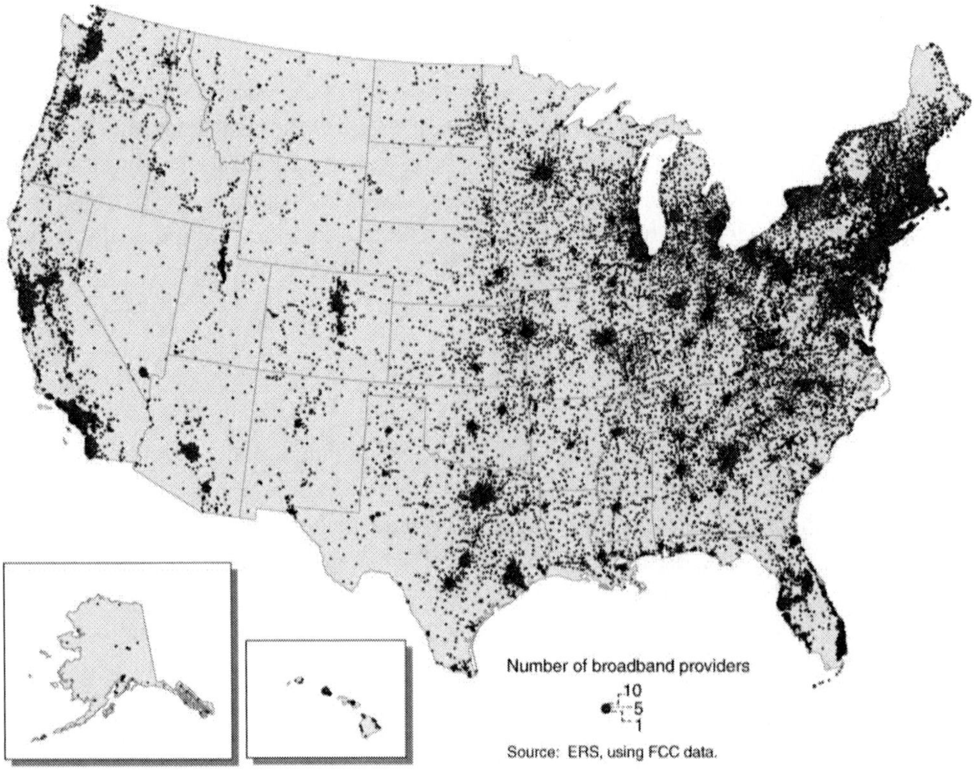

Figure C-1. Number of high-speed service providers by ZIP code, 2000

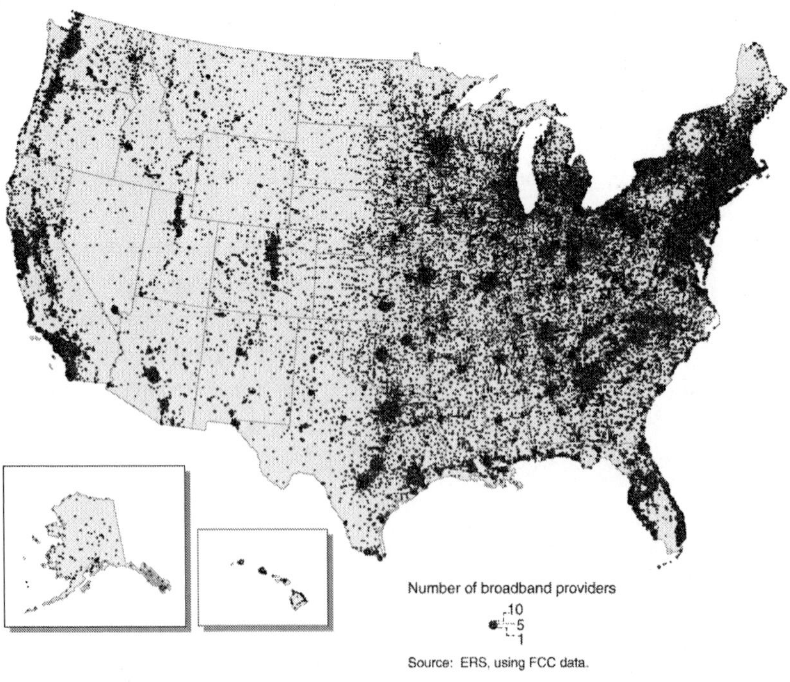

Figure C-2. Number of high-speed service providers by ZIP code, 2002

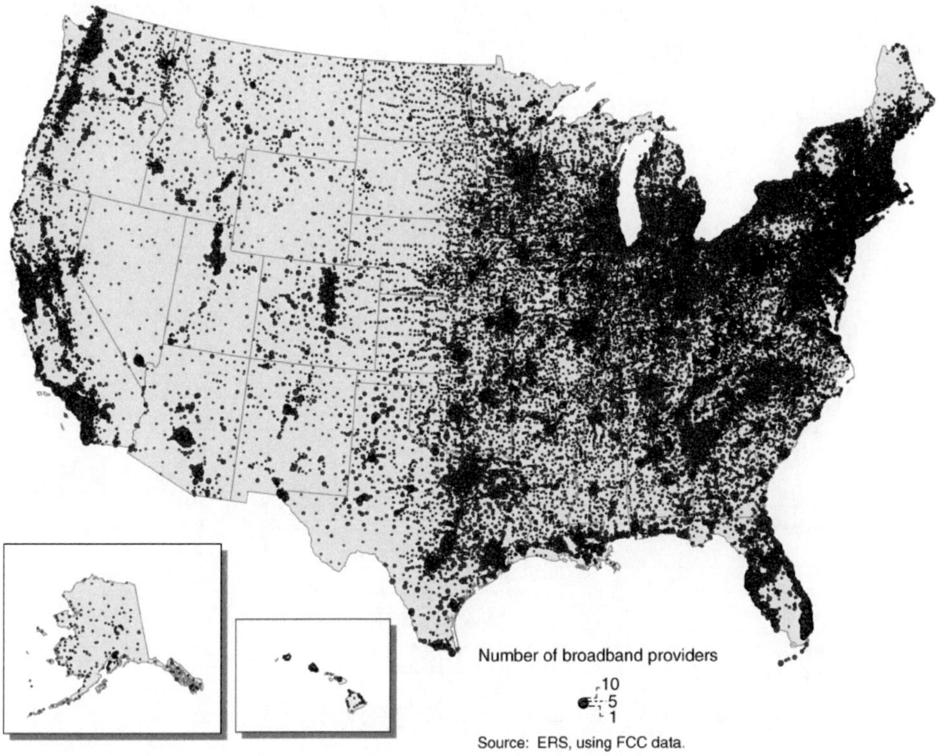

Figure C-3. Number of high-speed service providers by ZIP code, 2004

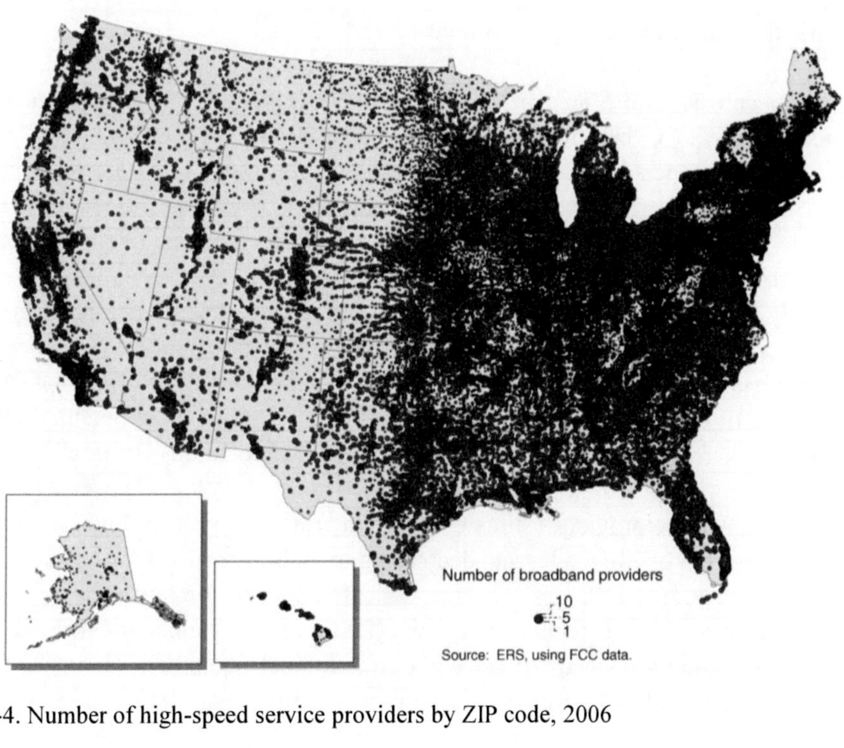

Figure C-4. Number of high-speed service providers by ZIP code, 2006

Broadband deployment has generally followed population density. This can be seen from the population density map (figure C-5).

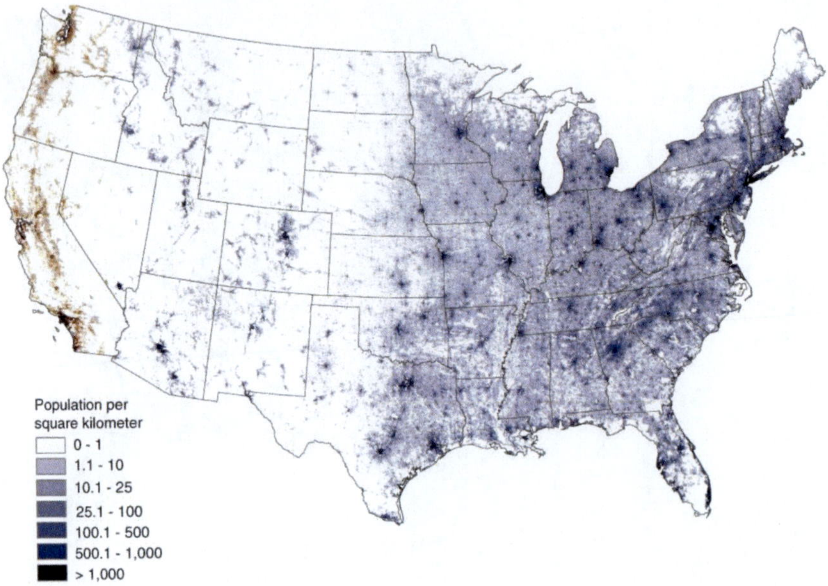

Source: ERS using Bureau of the Census, Census of the Population, 2002

Figure 5. Population density across the unified states, 200

Enhancing the FCC Data

Population and adjoining area affect broadband availability. These facts underlie our enhancement of the FCC broadband availability data. From the FCC data we developed broadband availability density maps that constitute our most basic measure for a number of our research applications. This basic broadband database is composed of equally sized 2-km sub-ZIP Code zonal building blocks. The databases are further refined and adapted to each line of research in our broader rural broadband Internet study.

Essentially, these 2-km grid cells show the likelihood of having broadband available at any location within the lower 48 States at different points of time. To generate these broadband probability surfaces, we first locate the number of providers at the population centroid of the ZIP Code, where providers are most likely to maximize potential customers. Next, we pass a kernel density function over the provider locations using an 8-km search radius.

The resulting surface provides estimates of provider access with the highest likelihood at the population centroid of ZIP Codes and decreasing probability toward the edge of the search radius. This search radius is larger than the typical limitation of DSL Internet service of 15,000 feet (4.5 km); due to technical reasons, DSL service cannot go beyond a certain distance from its signal's point of origin without additional equipment along the telephone line. The 8-km search radius helps to balance our assumption that all providers within a ZIP

Code are centrally within the ZIP Code. Likelihood of service increases with more providers within a ZIP Code. Overlapping provision areas increase the likelihood of service to any location within the overlap, so high provision in adjoining zonal areas further increases the likelihood of broadband availability.

Our density map was tested against June Agricultural Survey data of farm broadband use. The June Agricultural Survey (JAS) data are a geographic- based survey of farms in the lower 48 States. Internet use data have been collected since 1997. The JAS Internet data give geographic- and time- specific use and non-use of broadband Internet. (See Appendix B for further discussion of the JAS data.)

The density map matched very well with the JAS data in all areas except what is essentially the Great Plains region. The challenge here is the large geographic size of some ZIP Code areas, suggesting that the population centroid indicates less well the broadband Internet service area. The location of schools is used to further define the likelihood of broadband Internet service in an area; schools are useful because of their widespread use of broadband Internet. With the additional data, the surface map was adjusted to include additional provision areas. The resulting broadband density is essentially a likelihood measure—the probability of broadband Internet access for any given point in geographic space. Likelihood of broadband Internet access is centered in urban areas and radiates out from these urban centers (Figure C-6). The FCC data and the various selected indices that we developed here, one of which is shown in figure C-7, form the basis for much of the analysis in this report.

Our data were left as geo-specific or reconstituted into either county or ZIP Code areas for our analysis. Reconstitution into ZIP Code or county-level estimates required using population weighting of the provider likelihood estimate. A county aggregation of the data can be seen in figure C-7.

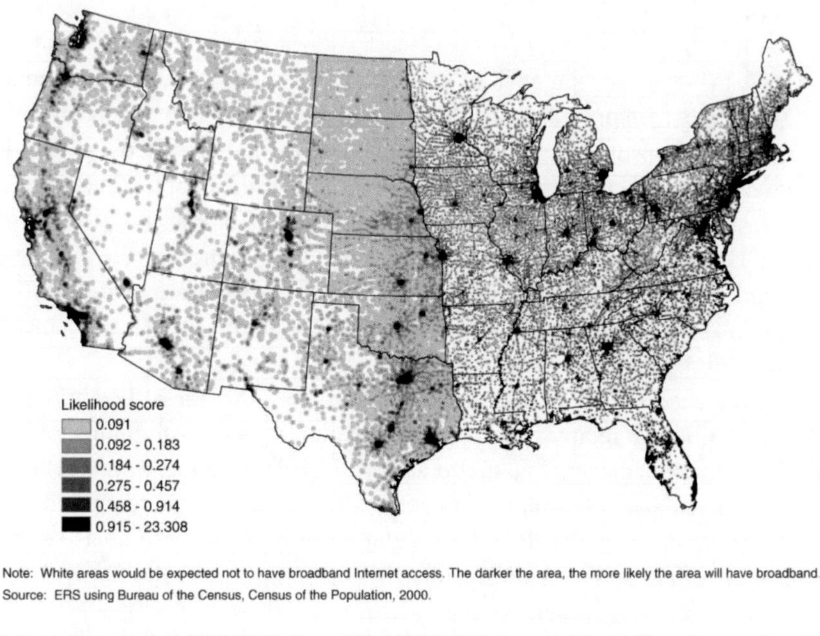

Note: White areas would be expected not to have broadband Internet access. The darker the area, the more likely the area will have broadband.
Source: ERS using Bureau of the Census, Census of the Population, 2000.

Figure C-6. Broadband availability based on 2005 FCC ZIP reporting and distance to school in the west.

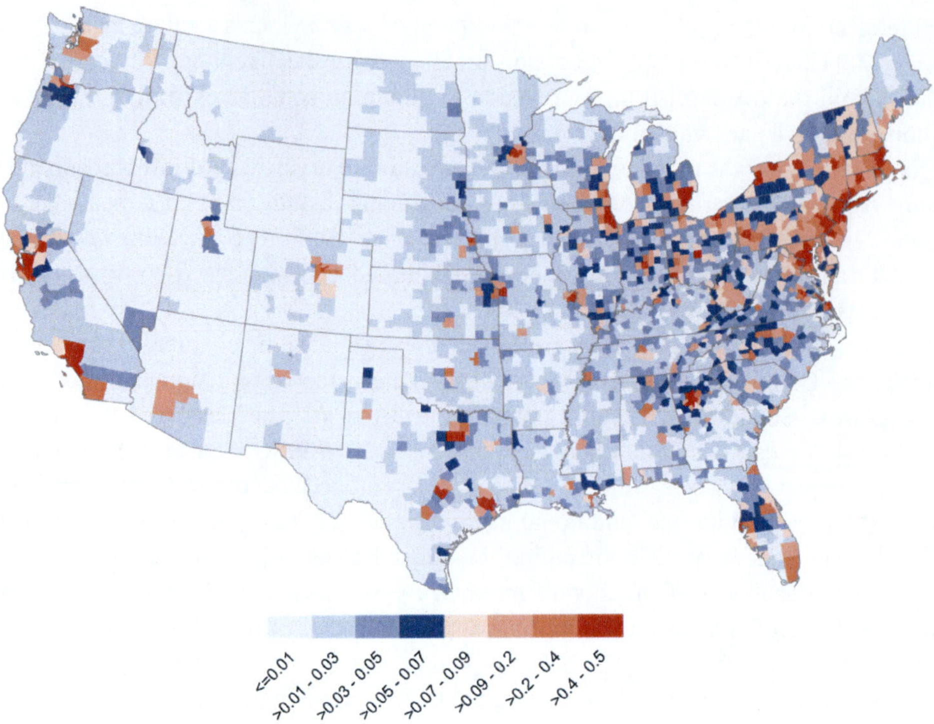

Source: ERS using Bureau of the Census Census of the population, 2000

Figure C-7. Country representation of average broadband provision per square kilometer, 2000

Estimation of Expected Number of Broadband Providers Using Population Measurements

To estimate the expected number of broadband providers, we began with the FCC's measure of the number of providers in a ZIP Code area in an unmodified form. To avoid disclosure of information on individual companies, areas that recorded 1-3 providers were combined into one category and assigned a value of 2. We used a population-based regression analysis to measure the relationship between the number of broadband providers and three components of U.S. population distribution: the overall population in the ZIP Code area in 2005, the percent of the ZIP Code's population living in urban areas as defined by the U.S. Census Bureau in 2000 (the most recent data on urban-rural population), and a measure of the ZIP Code's accessibility to nearby populations.

ZIP Code area accessibility is used to distinguish areas that might be similarly sized but are differently positioned relative to large population centers. All things being equal, a nonmetro ZIP Code area adjacent to a metro area will likely have more providers than a more isolated area. To calculate accessibility for each ZIP Code area *a*, we identified all other ZIP Code areas within 200 miles, divided the population of each of these areas by the distance (squared) to area *a*, and summed the results. Values range from 0.0002 in Prudhoe Bay, Alaska to 150 million in Los Angeles (table C-1).

The independent variables were logged for the regression analysis to meet the assumption of linearity with the dependent variable. Parameter estimates represent the change in the number of broadband providers that occurs with a 1-percent increase in the independent variable. For example, a 1-percent increase in population size will increase the number of providers by 1 .07. The standardized parameter estimates indicate the relative strength of the influence of each independent variable. Population size has twice the effect on number of broadband providers compared with percent urban, which in turn has a stronger effect than accessibility.

Table C1. Descriptive statistics and linear regression results measuring the effect of population size, percent urban, and population accessibility on number of broadband providers, 2006

	Mean	Standard deviation	Minimum	Maximum	Parameter estimate	Standardized estimate
Dependent variable:						
Number of broadband providers, 2006	6.59	3.58	0	21	n/a	n/a
Independent variables:						
Population size, 2005	9,922	13,878	0	114,726	1.08	0.55
Percent urban, 2000	39.93	43.90	0	100	2.54	0.22
Population accessibility, 2005	20,944.09	1,045,331.36	0	150,709,763	0.21	0.13

Note: Descriptive statistics for the independent variables are shown for unlogged values; logged values were used in the regression analysis. All parameter estimates are significant at the .01 level. The proportion of variation in broadband provision explained by the regression (adjusted R-square) equals .63.

Source: USDA, Economic Research Service, using data from the FCC and U.S. Census Bureau.

Parameter estimates may be used to calculate the expected number of broadband providers in a ZIP Code area. Residuals from the regression show the difference between the actual and expected number of providers (Figure 9, p. 18). For example, Grafton, West Virginia's ZIP Code area had 10,316 people, was 59 percent urban, and had an accessibility measure of 385. Taking the natural logs of these values (4, 0.2, and 2.6, respectively), multiplying them by their parameter estimates, and summing the results show a predicted value for Grafton of 5.4 providers. The number of broadband providers in this ZIP Code area, as reported by the FCC in 2006, was 2, so the model indicates that Grafton has roughly 3 fewer providers than predicted.

APPENDIX D. MODELING BROADBAND USE ON THE FARM

Discrete choice models are interpreted in terms of an underlying behavioral model, the so-called random utility maximization (RUM) model. The decisionmaker chooses the alternative with the highest utility. Let x_{ij} be an attribute vector of alternatives j that individual i faces; let fl be the impacts of the changes of the attributes; let e_{ij} be a random component. The random utility function of alternative j for individual i can then be written as $U_{ij} = \beta' x_{ij} + \varepsilon_{ij}$. Suppose that alternative j is chosen and that alternative k is not chosen. Individual i will choose j to maximize the random utility function, if and only if $U_{ij} > U_{ik}$ for any $k \sim i$. Since e_{ij} is a random component of the individual utility function, the probability that individual *i* actually chooses alternative *j* is written as $P(U_{ij}>U_{ik})$ for any $k \sim i$. The true utilities of the alternatives are considered random variables, so the probability that the alternative is chosen is defined as the probability that it has the greatest utility among the available alternatives.

Using the conceptual framework put forth by Flamm and Chaudhuri (2007), we assume that the underlying utility of Internet use is related to economic and demographic attributes (Flamm and Chaudhuri, 2007). The purchase- decision outcome (*j*) consists of one of three choices: no purchase, dial-up, or broadband. The x_{ij} are represented by the explanatory variables described in the farm businesses and broadband section. Under certain restrictive assumptions, the probability that individual *i* chooses alternative *j* can be expressed as

$$P_{ij} = P(U_{ij}>U_{ik}) = \exp(\beta' x_{ij})/\sum_{k=1}^{K} \exp(\beta' x_{ik}).$$

The model that results under certain distribution assumptions about *e* is usually called the conditional logit model. Apart from the assumptions underlying the RUM model, the conditional logit model implies that "the choice probabilities have the property which is called independence of irrelevant alternatives" (McFadden, 1974). This means that the ratio of the probabilities of choosing two alternatives is independent of the characteristics of all other choice possibilities. The conditional logit model is sometimes called the multinomial logit model. Following Greene (1993) and Maddala (1983), this latter term is reserved for models where the probabilities of the individual making a certain choice are functions of the characteristics of the individual, while the term conditional logit model is used when the choice probabilities are functions of characteristics of the choice alternatives. The way the problems are set up is different in these two models. However, the likelihood functions will be the same.

APPENDIX E. QUASI-EXPERIMENTAL DESIGN

Quasi-experimental design (QED) is a statistical approach that simulates an ex-post laboratory experiment featuring both a treatment and control group. Selection of control and treatment in QED, unlike a true laboratory experiment, is not perfectly random, hence the term "quasi." Treatment groups are self-selected. Control groups are selected based on their characteristic similarity with the initial, or pre-treatment, characteristics of the treatment groups. The QED approach taken here follows those of Isserman and Rephan (1995). SAS and other software were used in the analysis. A couple of the SAS routines used here were initially developed by Isserman.

The closeness between counties that is used to select the control counties is derived using a discrete measure called Mahalanobis distance. Mahalanobis distance measures the similarity between the treatment county and each county that could potentially be part of a control group. The measure is derived from the differences between the treatment county's and another county's characteristics' measures. The Mahalanobis distance is

$$MAHAL_{bj}=(X_b-X_j)^T \Sigma^{-1}(X_b-X_j),$$

where b is the treatment county, j is the potential control group county, X is the vector of variables that measure a county's characteristics, and Ó is the variance-covariance matrix of the variables calculated over all possible control counties. There are a number of ways to compare treatment versus control groups in QED. In the application here, there is one control county for each treatment county. No control county is allowed to appear more than once in the control group. The pairwise counties are the basic unit of analysis. The difference in growth is computed for each pair. The mean and standard deviations of these differences are computed, as are t-statistics between the treatment and control groups.

Robustness checks were made by analyzing prior-period growth rates. A tautology did not exist between the selection of control counties and their post-economic growth measures as the selection of control counties employs a large array of spatial and socioeconomic factors. Control and treatment county growth rates were more similar in the prior period, 1997-2000, than in the treatment period, 2002-2006. Selection criteria for treatment groups were relaxed and strengthened (i.e., cutoff points in broadband likelihood were increased and decreased). No appreciable change in outcomes was found as a result of these changes; the model was not sensitive to minor changes in treatment group inclusiveness.

More analysis on the robustness, however, needs to be completed to further substantiate the results and address more completely the issue of causality. Treatment group selection and control group characteristic variables will be further varied to test sensitivity in selection process.

APPENDIX F. ECONOMIC RESEARCH SERVICE BROADBAND WORKSHOP, SEPTEMBER 29-30, 2008

Broadband in the Rural Economy

Keynote: Rural Digital Economy
Edward Malecki, The Ohio State University

Internet and Rural Business Activity

Broadband Deployment and Economic Development in Kentucky and North Carolina
Mitch Renkow, North Carolina State University

Rural Broadband Internet Use and Rural Economy
Peter Stenberg, Economic Research Service, RRED

Comparing Rural Retailer Internet Users and Non-Users: Access Speed, Demographics, Attitudes and Beliefs.
Leslie Stoel and Stan Ernst, Ohio State University

Rural Grocers and Technology Adoption: Attitude Matters. Size Matters More.
Stan Ernst and Leslie Stoel, The Ohio State University

Food and Nonfarm Rural Business

Internet Marketing of Nursery and Greenhouse Products
Enefiok Ekanem and Fisseha Tegegne, Tennessee State University

Positive Examples and Lessons Learned from Rural Small Business Adoption of E Commerce Strategies
David Lamie, David Barkley, Clemson University, and Deborah Markley, University of Missouri

IT and E-Commerce Companies
John Leatherman, Kansas State University, and Hanas Cader, South Carolina State University

Farm and Rural Households

Farm Businesses and Broadband Internet Use
Mitchell Morehart and Peter Stenberg, Economic Research Service
Farming and the Internet: Reasons for Nonuse
Brian Briggeman and Brian Whitacre, Oklahoma State University

What Skills Are at the End of Broadband Cables in Rural America? Do They Match Up with Firms Wishing to Engage Rural Sourcing?
Doug Morris and Lyndon Goodridge, University of New Hampshire

Digital Economy

IT in the Global Economy
Catherine Mann, Peterson Institute for International Economics and Brandeis University

Effects of Broadband Deployment on Output and Employment
Robert Crandall, Brookings Institution

Home Broadband Adoption in the United States: Patterns, Barriers, and Consequences
John Horrigan, Pew Internet & American Life Project

Community Internet Use

The Role of the Internet in Rural Community Participation—Examples from Recent Survey Data
Michael Stern and Alison E. Adams, Oklahoma State University

Rural Distance Education
Janet Poley, University of Nebraska-Lincoln and President of the American Distance Education Consortium

Economic Impact of Rural Telemedicine
Brian Whitacre, Oklahoma State University

End Notes

[1] John Horrigan and a number of other researchers whose works are discussed at length in this report presented their research at the Economic Research Service's broadband workshop, September 2008. For a complete listing of participants and papers, see the workshop agenda in appendix F.

[2] Metro and nonmetro categories used here are based on the ERS Rural-Urban Commuting Areas and are defined using criteria similar to OMB's county- based metro and nonmetro areas. Nonmetro ZIP Code areas are defined as those outside urban centers of 50,000 or more and their surrounding commuting zones

In: Broadband and Rural America
Editor: Bethany J. Lenz
ISBN: 978-1-61668-637-6

Chapter 4

BROADBAND LOAN AND GRANT PROGRAMS IN THE USDA'S RURAL UTILITIES SERVICE

Lennard G. Kruger

SUMMARY

Given the large potential impact broadband access may have on the economic development of rural America, concern has been raised over a "digital divide" between rural and urban or suburban areas with respect to broadband deployment. While there are many examples of rural communities with state of the art telecommunications facilities, recent surveys and studies have indicated that, in general, rural areas tend to lag behind urban and suburban areas in broadband deployment.

Citing the lagging deployment of broadband in many rural areas, Congress and the Administration acted in 2001 and 2002 to initiate pilot broadband loan and grant programs within the Rural Utilities Service (RUS) at the U.S. Department of Agriculture (USDA). Subsequently, Section 6103 of the Farm Security and Rural Investment Act of 2002 (P.L. 107-171) amended the Rural Electrification Act of 1936 to authorize a loan and loan guarantee program to provide funds for the costs of the construction, improvement, and acquisition of facilities and equipment for broadband service in eligible rural communities. The RUS/USDA houses two assistance programs exclusively dedicated to financing broadband deployment: the Rural Broadband Access Loan and Loan Guarantee Program and the Community Connect Grant Program.

The 110th Congress considered reauthorization and modification of the loan and loan guarantee program as part of the farm bill. The Food, Conservation, and Energy Act of 2008 became law on June 18, 2008 (P.L. 110-246). Title VI (Rural Development) contains authorizing language for the broadband loan program.

Meanwhile, on May 11, 2007, RUS released a Proposed Rule seeking to revise the broadband loan program rules and regulations. Some key issues pertinent to a consideration of the RUS broadband programs include restrictions on applicant eligibility, how "rural" is defined with respect to eligible rural communities, how to address assistance to areas with preexisting broadband service, technological neutrality, funding levels and mechanisms, and the appropriateness of federal assistance. The final rule will reflect language in the enacted farm bill statute (P.L. 110-246). Ultimately, modification of rules, regulations, or criteria associated with the RUS broadband program will likely result in "winners and losers" in terms of which companies, communities, regions of the country, and technologies are eligible or more likely to receive broadband loans and grants.

On February 17, 2009, President Obama signed P.L. 111-5, the American Recovery and Reinvestment Act (ARRA). Broadband provisions of the ARRA provide a total of $7.2 billion, primarily for broadband grants. The total consists of $2.5 billion for RUS broadband loans, grants, and loan/grant combinations, and $4.7 billion to the National Telecommunications and Information Administration (NTIA) at the Department of Commerce (DOC) for a newly established Broadband Technology Opportunities Program. On July 1, 2009, RUS and NTIA jointly released the first Notice of Funds Availability (NOFA) detailing requirements, rules, and procedures for applying for ARRA grants, loans, and loan grant combinations

This report will be updated as events warrant.

Background: Broadband and Rural America

The broadband loan and grant programs at RUS are intended to accelerate the deployment of broadband services in rural America. "Broadband" refers to high-speed Internet access and advanced telecommunications services for private homes, commercial establishments, schools, and public institutions. Currently in the United States, broadband is primarily provided via cable modem (from the local provider of cable television service) or over the telephone line (digital subscriber line or "DSL"). Other broadband technologies include fiber optic cable, mobile wireless, fixed wireless, satellite, and broadband over power lines (BPL).

Broadband access enables a number of beneficial applications to individual users and to communities. These include e-commerce, telecommuting, voice service (voice over the Internet protocol or "VOIP"), distance learning, telemedicine, public safety, and others. It is becoming generally accepted that broadband access in a community can play an important role in economic development. A February 2006 study by the Massachusetts Institute of Technology for the Department of Commerce's Economic Development Administration marked the first attempt to measure the impact of broadband on economic growth. The study found that "between 1998 and 2002, communities in which mass-market broadband was available by December 1999 experienced more rapid growth in employment, the number of businesses overall, and businesses in IT-intensive sectors, relative to comparable communities without broadband at that time."[1]

Subsequently, a June 2007 report from the Brookings Institution found that for every one percentage point increase in broadband penetration in a state, employment is projected to increase by 0.2 to 0.3% per year. For the entire U.S. private non-farm economy, the study projected an increase of about 300,000 jobs, assuming the economy is not already at full employment.[2] Similarly, an August 2009 report from the USDA Economic Research Service found that counties with a longer history of broadband availability had higher employment growth and higher nonfarm private earnings than similarly situated counties with little or no broadband access since 2000.[3]

Access to affordable broadband is viewed as particularly important for the economic development of rural areas because it enables individuals and businesses to participate fully in the online economy regardless of geographical location. For example, aside from enabling existing businesses to remain in their rural locations, broadband access could attract new business enterprises drawn by lower costs and a more desirable lifestyle. Essentially, broadband potentially allows businesses and individuals in rural America to live locally while competing globally in an online environment.

Given the large potential impact broadband may have on the economic development of rural America, concern has been raised over a "digital divide" between rural and urban or suburban areas with respect to broadband deployment. While there are many examples of rural communities with state of the art telecommunications facilities,[4] recent surveys and studies have indicated that, in general, rural areas tend to lag behind urban and suburban areas in broadband deployment. For example:

- An August 2009 report from the USDA Economic Research Service found that 70% of rural households with in-home Internet access had a broadband connection in 2007, compared with 84% of urban households.[5]
- April 2009 data from the Pew Internet & American Life Project indicate that while broadband adoption throughout the nation is growing – with home broadband adoption at 63% of adult Americans—broadband users make up larger percentages of non-rural users (67%) than rural users (46%).[6]

The comparatively lower population density of rural areas is likely the major reason why broadband is less deployed than in more highly populated suburban and urban areas. Particularly for wireline broadband technologies—such as cable modem and DSL—the greater the geographical distances among customers, the larger the cost to serve those customers. Thus, there is often less incentive for companies to invest in broadband in rural areas than, for example, in an urban area where there is more demand (more customers with perhaps higher incomes) and less cost to wire the market area.

The terrain of rural areas can also be a hindrance, in that it is more expensive to deploy broadband technologies in a mountainous or heavily forested area. An additional added cost factor for remote areas can be the expense of "backhaul" (e.g., the "middle mile") which refers to the installation of a dedicated line which transmits a signal to and from an Internet backbone which is typically located in or near an urban area.

Cable modem and DSL currently comprise about 80% of residential broadband deployment nationwide.[7] However, because of the challenges of deploying these technologies in low population density areas, other broadband technologies have been identified as perhaps offering potential in rural areas. These include mobile wireless (cellular), fixed wireless (WIMAX, wi-fi), satellite, and broadband over powerlines (BPL).

PILOT BROADBAND LOAN AND GRANT PROGRAMS

Given the lagging deployment of broadband in rural areas, Congress and the Administration acted to initiate pilot broadband loan and grant programs within the Rural Utilities Service of the U.S. Department of Agriculture. While RUS had long maintained telecommunications loan and grant programs (Rural Telephone Loans and Loan Guarantees, Rural Telephone Bank, and more recently, the Distance Learning and Telemedicine Loans and Grants) none were exclusively dedicated to financing rural broadband deployment. Title III of the FY200 1 agriculture appropriations bill (P.L. 106-387) directed USDA/RUS to conduct a "pilot program to finance broadband transmission and local dial-up Internet service in areas that meet the definition of 'rural area' used for the Distance Learning and Telemedicine Program."

Subsequently, on December 5, 2000, RUS announced the availability of $100 million in loan funding through a one-year pilot program "to finance the construction and installation of broadband telecommunications services in rural America."[8] The broadband pilot loan program was authorized under the authority of the Distance Learning and Telemedicine Program (7 U.S.C. 950aaa), and was available to "legally organized entities" not located within the boundaries of a city or town having a population in excess of 20,000.

The FY2001 pilot broadband loan program received applications requesting a total of $350 million. RUS approved funding for 12 applications totaling $100 million. The FY2002 agriculture appropriations bill (P.L. 107-76) designated a loan level of $80 million for broadband loans, and on January 23, 2002, RUS announced that the pilot program would be extended into FY2002, with $80 million in loans made available to fund many of the applications that did not receive funding during the previous year.[9]

Meanwhile, the FY2002 agriculture appropriations bill (P.L. 107-76) allocated $20 million for a pilot broadband grant program, also authorized under the Distance Learning and Telemedicine Program. On July 8, 2002, RUS announced the availability of $20 million for a pilot grant program for the provision of broadband service in rural America. The program was specifically targeted to economically challenged rural communities with no existing broadband service. Grants were made available to entities providing "community-oriented connectivity" which the RUS defined as those entities "who will connect the critical community facilities including the local schools, libraries, hospitals, police, fire and rescue services and who will operate a community center that provides free and open access to residents."[10]

In response to the July 8, 2002, Notice of Funds Availability, RUS received more than 300 applications totaling more than $185 million in requested grant funding. RUS approved 40 grants totaling $20 million. The pilot program was extended into FY2003, as the Consolidated Appropriations Resolution of 2003 (P.L. 108-7) allocated $10 million for broadband grants. On September 24, 2003, 34 grants were awarded to eligible applicants who did not receive funding during the previous year.

RURAL BROADBAND ACCESS LOAN AND LOAN GUARANTEE PROGRAM

Building on the pilot broadband loan program at RUS, Section 6103 of the Farm Security and Rural Investment Act of 2002 (P.L. 107-171) amended the Rural Electrification Act of 1936 to authorize a loan and loan guarantee program to provide funds for the costs of the construction, improvement, and acquisition of facilities and equipment for broadband service in eligible rural communities.[11] Section 6103 made available, from the funds of the Commodity Credit Corporation (CCC), a total of $100 million through FY2007. P.L. 107-171 also authorized any other funds appropriated for the broadband loan program.

Beginning in FY2004, Congress annually blocked mandatory funding from the CCC. Thus— starting in FY2004—the program was funded as part of annual appropriations in the Distance Learning and Telemedicine account within the Department of Agriculture appropriations bill. Every fiscal year, Congress approves an appropriation (loan subsidy) and a specific loan level (lending authority) for the Rural Broadband Access Loan and Loan Guarantee Program. **Table 1** shows—for the life of the program to date—loan subsidies and loan levels (lending authority).

The Rural Broadband Access Loan and Loan Guarantee Program is codified as 7 U.S.C. 950bb. Specifically, Treasury rate loans, 4% loans, and loan guarantees are authorized for entities providing broadband service for "eligible rural communities," defined as any area of the United States that is not contained in an incorporated city or town with a population in excess of 20,000 inhabitants.[12] RUS is required to be technologically neutral in determining whether or not to make a loan, and is instructed to give priority to rural communities with no existing residential broadband service. Loans are used for financing new or improved existing broadband provider facilities. Loans cannot be used to finance installations or equipment at customers' premises.

Table 1. Appropriations Funding for the Rural Broadband Access Loan and Loan Guarantee Program

	Direct Appropriations (subsidy level)	Loan Levels Specified in Annual Appropriations
FY2001 (pilot)	—	$100 million
FY2002 (pilot)	—	$80 million
FY2003	a	$80 million
FY2004	$13.1 million	$602 million
FY2005	$1 1.7 15 million	$550 million
FY2006	$10.75 million	$500 million
FY2007	$10.75 million	$500 million
FY2008	$6.45 million	$300 million
FY2009	$15.619 million	$400 million
FY2010 (request)	$38.495 million	$532 million
FY2010 (H.Rept. 111-279)	$28.96 million	$400 million

a. Program received $40 million composed of $20 million from FY2002 plus $20 million from FY2003 of mandatory funding from the Commodity Credit Corporation, as directed by P.L. 107-171. In the FY2004, FY2005, and FY2006 appropriations bills, mandatory funding from the CCC was canceled.

On January 30, 2003, the RUS published in the *Federal Register* the regulation (7 C.F.R. part 1738) establishing the Rural Broadband Access Loan and Loan Guarantee Program, as authorized by P.L. 107-171.[13] According to the regulation, entities eligible to receive loans include corporations, limited liability companies, cooperative or mutual organizations, Indian tribes, and public bodies. Specifically **not** eligible are individuals, partnerships, and any entity serving 2% or more of the telephone subscriber lines in the United States. All applicants are required to demonstrate adequate credit support—a minimum of 20% of requested loan amount, including cash on hand equivalent to one full year of operating expense.[14] Of loans approved, approximately 53% have been made to corporations, 37% to LLCs, 6% to cooperatives, 3% to municipalities, and 1% to a tribal authority.[15]

To be eligible for 4% loans, applicants must be proposing to serve a community with no existing broadband service, a population of 2,500 or less, and a service area with population density of no more than 20 persons per square mile. Additionally, the community must be located in a county with a per capita income of less than or equal to 65% of the national per capita income.

On March 25, 2008, RUS announced its largest loan ever, $267 million to Open Range Communications. The $267 million loan, accompanied by an over $100 million investment from the private sector, will enable Open Range to provide wireless broadband (Wi-Max technology) and satellite connectivity to 518 rural communities in 17 states.[16]

Applications for the Rural Broadband Access Loan and Loan Guarantee program are accepted at any time. The maximum loan amount for 4% loans is $7.5 million. There is no maximum for treasury rate loans, and the minimum level for all loans is $100 thousand. In 2003, the average loan was $11.2 million, while in 2006, the average loan was $44 million.[17] Loans are made for the term equal to the expected service life of financed facilities. Further information, including application materials and guidelines, is available at http://www.usda.gov/rus/telecom/ broadband.htm.

Community Connect Broadband Grants

The Consolidated Appropriations Act of 2004 (P.L. 108-199) appropriated $9 million "for a grant program to finance broadband transmission in rural areas eligible for Distance Learning and Telemedicine Program benefits authorized by 7 U.S.C. 950aaa." On July 28, 2004, RUS published its final rule on the broadband grant program, called the Community Connect Grant Program (7 C.F.R. part 1739, subpart A).[18] Essentially operating the same as the pilot broadband grants, the program provides grant money to applicants proposing to provide broadband on a "community-oriented connectivity" basis to currently unserved rural areas for the purpose of fostering economic growth and delivering enhanced health care, education, and public safety services. Funding for the broadband grant program is provided through annual appropriations in the Distance Learning and Telemedicine account within the Department of Agriculture appropriations bill. **Table 2** shows a history of appropriations for the Community Connect Broadband Grants.

Table 2. Appropriations for the Community Connect Broadband Grants

Fiscal Year	Appropriation
FY2002	$20 million
FY2003	$10 million
FY2004	$9 million
FY2005	$9 million
FY2006	$9 million
FY2007	$9 million
FY2008	$13.4 million
FY2009	$13.4 million
FY2010 (req)	$13.4 million
FY2010 (H.Rept. 111-279)	$17.9 million

Source: Compiled by CRS from appropriations bills.

Eligible applicants for broadband grants include incorporated organizations, Indian tribes or tribal organizations, state or local units of government, cooperatives, private corporations, and limited liability companies organized on a for profit or not-for-profit basis. Individuals or partnerships are not eligible.

Funded projects must: serve a rural area of 20,000 population or less[19] where broadband service does not exist, serve one and only one single community, deploy free basic broadband service (defined as 200 kbps in both directions) for at least two years to all community facilities, offer basic broadband to residential and business customers, and provide a community center with at least ten computer access points within the proposed service area while making broadband available for two years at no charge to users within that community center.

Since the inception of the RUS broadband grant program, $83.7 million in grant money has been awarded to 173 communities. Awardees must contribute a matching contribution equal to 15% of the requested grant amount.

RUS typically publishes an annual Notice of Funding Availability (NOFA) in the *Federal Register*, which specifies the deadline for applications, the total amount of funding available, and the maximum and minimum amount of funding available for each grant. Further information, including application materials and guidelines, is available at http://www.usda.gov/rus/telecom/ commconnect.htm.

OTHER BROADBAND PROGRAMS

Prior to enactment of the American Recovery and Reinvestment Act (P.L. 111-5), which established the Broadband Technology Opportunities Program at the National Telecommunications and Information Administration/Department of Commerce, the Rural Broadband Access Loan and Loan Guarantee Program and the Community Connect Broadband Grants were the only federal programs *exclusively* dedicated to deploying broadband infrastructure.

Table 3. Number of Customers Receiving New or Improved Telecommunication Services (Broadband) through USDA Financing of Telecommunications Facilities (millions)

FY2001	FY2002	FY2003	FY2004	FY2005	FY2006	FY2007	FY2008	FY2009	FY2010
0.31	0.31	0.38	0.37	0.23	0.30	0.36	0.78	0.37	0.35

Sources: U.S. Dept. of Agriculture, *2006 Performance and Accountability Report*, November 2006, p. 82; U.S. Dept. of Agriculture, *FY2010 Budget Summary and Performance Plan*, p. 51.

Note: Customers are defined as access lines financed by the programs.

There also exist other federal programs that provide financial assistance for various aspects of telecommunications development.[20] Though not explicitly or exclusively devoted to broadband, many of those programs are used to help deploy broadband technologies in rural areas. For example, since 1995, the RUS Rural Telephone Loan and Loan Guarantee program—which has traditionally financed telephone voice service in rural areas under 5,000 inhabitants—has required that all telephone facilities receiving financing must be capable of providing DSL broadband service at a rate of at least 1 megabyte per second.[21] An October 2006 survey of RUS traditional telephone loan program borrowers found that 92% of those borrowers were providing broadband to all of the telephone exchanges in their service territories.[22] According to FY2010 budget documents, 85% of loans funded in FY2008 were used for fiber-to-the home (FTTH) broadband infrastructure, and more than 105 telephone exchanges will be upgraded to FTTH as a result of financing in FY2008.[23]

Another RUS telecommunications program, Distance Learning and Telemedicine grants, is used to support deployment of broadband technologies specifically for telemedicine and distance learning applications. **Table 3** shows the number of customers receiving broadband due to USDA financing of telecommunications facilities.

The other major vehicle for funding telecommunications development in rural areas is the Universal Service Fund (USF).[24] Subsidies provided by USF's Schools and Libraries Program and Rural Health Care Program are used for a variety of telecommunications services, including broadband access. While the USF's High Cost Program does not *explicitly* fund broadband infrastructure, subsidies are used, in many cases, to upgrade existing telephone networks. Regarding the USF High Cost Program, the Congressional Budget Office has found that "current policy implicitly provides funds for broadband in rural areas," adding that:

> Whether such upgrades are motivated by the intention to provide broadband or better conventional telephone service is not immediately clear. However, the fact that wireline carriers as a whole have been losing subscribers and long-distance revenue over the past half decade suggests that at least part of the new investment in local loops has been made with the expectation of generating revenue from broadband subscriptions.[25]

In the 110th Congress, legislation to reform universal service—which could have a significant impact on the amount of financial assistance available for broadband deployment in rural and underserved areas—was introduced. With none of this legislation enacted in the 110th Congress, further legislation regarding universal service could be considered by the 111th Congress. For more information on universal service, see CRS Report RL3 3979, *Universal Service Fund: Background and Options for Reform*, by Angele A. Gilroy.

In addition to federal support for broadband deployment, there are programs and activities ongoing at the state and local level. Surveys, assessments, and reports from the Alliance for Public Technology and the Communications Workers of America,[26] the California Public Utilities Commission,[27] the National Governors Association,[28] and the National Conference of State Legislatures[29] have explored state and local broadband programs. A related issue is the emergence of municipal broadband networks (primarily wireless and fiber based) and the debate over whether such networks constitute unfair competition with the private sector.

CRITICISMS OF RUS BROADBAND PROGRAMS

Broadband loan and grant programs have been awarding funds to entities serving rural communities since FY2001. Since their inception, a number of criticisms have emerged.

Loan Approval and Application Process

Perhaps the major criticism of the broadband loan program is that not enough loans are approved, thereby making it difficult for rural communities to take full advantage of the program. As of June 22, 2009, the broadband loan program received 225 applications, requesting a total of $4.7 billion in loans. Of these, 97 applications were approved (totaling $1.8 billion), 120 were returned (totaling $2.7 billion), and 8 are pending (totaling $170 million).[30] According to RUS officials, 28% of available loan money was awarded in 2004, and only 5% of available loan money was awarded in 2005.[31]

The loan application process has been criticized as being overly complex and burdensome, requiring applicants to spend months preparing costly market research and engineering assessments. Many applications are rejected because the applicant's business plan is deemed insufficient to support a commercially viable business. The biggest reason for applications being returned has been insufficient credit support, whereby applicants do not have sufficient cash-onhand (one year's worth is required in most cases). The requirement for cash-on-hand is viewed as particularly onerous for small start up companies, many of whom lack sufficient capital to qualify for the loan. Such companies, critics assert, may be those entities most in need of financial assistance.

In report language to the FY2006 Department of Agriculture Appropriations Act (P.L. 109-97), the Senate Appropriations Committee (S.Rept. 109-92) directed the RUS "to reduce the burdensome application process and make the program requirements more reasonable, particularly in regard to cash-on-hand requirements." The Committee also directed USDA to hire more full-time employees to remedy delays in application processing times.

At a May 17, 2006 hearing held by the Senate Committee on Agriculture, Nutrition, and Forestry, the Administrator of the RUS stated that RUS is working to make the program more user friendly, while at the same time protecting taxpayer investment:

> As good stewards of the taxpayers' money, we must make loans that are likely to be repaid. One of the challenges in determining whether a proposed project has a reasonable

> chance of success is validating the market analysis of the proposed service territory and ensuring that sufficient resources are available to cover operating expenses throughout the construction period until such a time that cash flow from operations become sufficient. The loan application process that we have developed ensures that the applicant addresses these areas and that appropriate resources are available for maintaining a viable operation.[32]

According to RUS, the loan program was initially overwhelmed by applications (particularly during a two week period in August 2003), and as the program matured, application review times have dropped.[33] On May 11, 2007, RUS released a Proposed Rule which sought to revise regulations for the broadband loan program. In the background material accompanying the Proposed Rule, RUS stated that the average application processing time in 2006 was almost half of what it was in 2003.[34]

Eligibility Criteria

Since the inception of the broadband grant and loan programs, the criteria for applicant eligibility has been criticized both for being too broad and for being too narrow. An audit report released by USDA's Office of Inspector General (IG) found that the "programs' focus has shifted away from those rural communities that would not, without Government assistance, have access to broadband technologies."[35] Specifically the IG report found that the RUS definition of rural area has been "too broad to distinguish usefully between suburban and rural communities,"[36] with the result that, as of March 10, 2005, $103.4 million in loans and grants (nearly 12% of total funding awarded) had been awarded to 64 communities located near large cities. The report cited examples of affluent suburban subdivisions qualifying as rural areas under the program guidelines and receiving broadband loans.[37]

On the other hand, eligibility requirements have also been criticized as too narrow. For example, the limitation of assistance only to communities of 20,000 or less in population excludes small rural towns that may exceed this limit, and also excludes many municipalities seeking to deploy their own networks.[38] Similarly, per capita income requirements can preclude higher income communities with higher costs of living (e.g. rural Alaska), and the limitation of grant programs only to underserved areas excludes rural communities with existing but very limited broadband access.[39]

Loans to Communities with Existing Providers

The IG report found that RUS too often has given loans to communities with existing broadband service. The IG report found that "RUS has not ensured that communities without broadband service receive first priority for loans," and that although RUS has a system in place to prioritize loans to unserved communities, the system "lacks a cutoff date and functions as a rolling selection process—priorities are decided based on the applicants who happen to be in the pool at any given moment."[40] The result is that a significant number of communities with some level of preexisting broadband service have received loans. According to the IG report, of 11 loans awarded in 2004, 66% of the associated communities served by those loans had existing service. According to RUS, 31% of communities served by all loans (during the period 2003 through early 2005) had preexisting competitive service (not including loans used to upgrade or expand existing service).[41] In some cases, according to the IG report, "loans were issued to companies in highly competitive business environments where multiple providers competed for relatively few customers."[42] At the May 1, 2007 hearing before the House Subcommittee on Specialty Crops, Rural Development, and Foreign Agriculture, then-RUS Administrator James Andrews testified that of the 69 broadband loans awarded since the program's inception, 40% of the communities approved for funding were unserved at the time of loan approval, and an additional 15% had only one broadband provider.[43]

Awarding loans to entities in communities with preexisting competitive service raised criticism from competitors who already offer broadband to those communities. According to the National Cable and Telecommunications Association (NCTA), "RUS loans are being used to unfairly subsidize second and third broadband providers in communities where private risk capital already has been invested to provide broadband service."[44] Critics argued that providing loans in areas with preexisting competitive broadband service creates an uneven playing field and discourages further private investment in rural broadband.[45] In response, RUS stated in the IG report that its policies are in accordance with the statute, and that they address "the need for competition to increase the quality of services and reduce the cost of those services to the consumer."[46] RUS argued that the presence of a competitor does not necessarily mean that an area is adequately served, and additionally, that in order for some borrowers to maintain a viable business in an unserved area, it may be necessary for that company to also be serving more densely populated rural areas where some level of competition already exists.[47]

Follow-Up Audit by USDA Office of Inspector General

In 2008, as directed by the House Appropriations Committee (H.Rept. 110-258, FY2008 Agriculture appropriations bill), the IG reexamined the RUS broadband loan and loan guarantee program to determine whether RUS had taken sufficient corrective actions in response to the issues raised in the 2005 IG report. The IG concluded "the key problems identified in our 2005 report—loans being issued to suburban and exurban communities and loans being issued where other providers already provide access—have not been resolved."[48]

Specifically, the follow-up IG report found that between 2005 and 2008, RUS broadband borrowers providing services in 148 communities were within 30 miles of cities with 200,000 inhabitants, including communities near very large urban areas such as Chicago and Las Vegas.

The IG report also found that since 2005 "RUS has continued providing loans to providers in markets where there is already competing service."[49] Of the 37 applications approved since September 2005, 34 loans were granted to applicants in areas where one or more private broadband providers already offered service. These 34 borrowers received $873 million to service 1,448 communities. The IG report found that since 2005, 77% of communities which were expected to receive service from a project financed by an approved RUS broadband loan had at least one existing broadband provider present, 59% had 2 or more existing providers, and 27% had 3 or more existing providers.[50]

In an official response to the follow-up IG report, RUS fundamentally disagreed with the IG criticisms, stating that the loans awarded between 2005 and 2008 were provided "in a way entirely consistent with the statutory requirements of the underlying legislation governing administration of the program, the regulations and guidance issued by the Department to implement the statute, and the intent of Congress."[51] Specifically RUS argued that its May 11, 2007 Proposed Rule, and the subsequent changes to the broadband loan and loan guarantee statute made by the 2008 farm bill, both addressed concerns over loans to non-rural areas and to communities with preexisting broadband providers. However, the Final Rule based on the Proposed Rule and the 2008 farm bill had not yet been released and implemented during the 2005-2008 period examined by the IG, and RUS was compelled by law to continue awarding broadband loans under the existing law and rules. RUS pointed out that they have not approved any requests for loan assistance received since enactment of the 2008 Farm Bill on June 18, 2008.

Issues During Reauthorization

The previous authorization for the Rural Broadband Access Loan and Loan Guarantee program expired on September 30, 2007. The 110th Congress considered reauthorization of the program as part of the farm bill. Modification of rules, regulations, or criteria associated with the RUS broadband program will likely result in "winners and losers" in terms of which companies, communities, regions of the country, and technologies are eligible or more likely to receive broadband loans and grants. The following are some key issues which were pertinent to the debate over reauthorization of the RUS broadband loan and loan guarantee program.

Restricting Applicant Eligibility

The RUS broadband program was criticized for excluding too many applicants due to stringent financial requirements (e.g. the requirement that an applicant have a year's worth of cash-onhand) and an application process—requiring detailed business plans and market surveys—that some viewed as overly expensive and burdensome to complete. During the reauthorization process, Congress considered whether the criteria for loan eligibility should be modified, and whether a more appropriate balance could be found between the need to make the program more accessible to unserved and often lower-income rural areas, and the need to protect taxpayers against bad loans.

Definition of "Rural Community"

The definition of which communities qualify as "rural" had been changed twice by statute since the broadband loan program was initiated. Under the pilot program, funds were authorized under the Distance Learning and Telemedicine Program, which defines "exceptionally rural areas" (under 5,000 inhabitants), "rural areas" (between 5,000 and

10,000) and "mid-rural areas" (between 10,000 and 20,000). RUS determined that communities of 20,000 or less would be eligible for broadband loans in cases where broadband services did not already exist.

In 2002, this definition was made narrower by the Farm Security and Rural Investment Act (P.L. 107-171), which designated eligible communities as any incorporated or unincorporated place with fewer than 20,000 inhabitants, and which was outside any standard metropolitan statistical area (MSA). The requirement that communities not be located within MSA's effectively prohibited suburban communities from receiving broadband loans. However, in 2004, the definition was again changed by the FY2004 Consolidated Appropriations Act (P.L. 108-199). The act broadened the definition, keeping the population limit at 20,000, but eliminating the MSA prohibition, thereby permitting rural communities near large cities to receive loans. Thus the current definition used for rural communities is the same as what was used for the broadband pilot program, except that loans can now be issued to communities with preexisting service.

The definition of what constitutes a "rural" community is always a difficult issue for Congressional policymakers in determining how to target rural communities for broadband assistance. On the one hand, the narrower the definition the greater the possibility that deserving communities may be excluded. On the other hand, the broader the definition used, the greater the possibility that communities not traditionally considered "rural" or "underserved" may be eligible for financial assistance.

A related issue is the scope of coverage proposed by individual applications. While many of the loan applications propose broadband projects offering service to multiple rural communities, RUS identified a trend towards larger regional and national proposals, covering hundreds or even more than a thousand communities.[52] The larger the scope of coverage, the greater the complexity of the loan application and the larger the possible benefits and risks to taxpayers.

Preexisting Broadband Service

Loans to areas with competitive preexisting service—that is, areas where existing companies already provide some level of broadband—sparked controversy because loan recipients are likely to compete with other companies already providing broadband service.

During reauthorization, Congress was asked to more sharply define whether and/or how loans should be given to companies serving rural areas with preexisting competitive service.[53] On the one hand, some argued that the federal government should not be subsidizing competitors for broadband service, particularly in sparsely populated rural markets which may be able only to support one provider. Furthermore, keeping communities with preexisting broadband service eligible may divert assistance from unserved areas that are most in need. On other hand, many suburban and urban areas currently receive the benefits of competition between broadband providers—competition which can potentially drive down prices while improving service and performance. It is therefore appropriate, others argued, that rural areas also receive the benefits of competition, which in some areas may not be possible without federal financial assistance. It was also argued that it may not be economically feasible for borrowers to serve sparsely populated unserved communities unless they are permitted to also serve more lucrative areas which may already have existing providers.

Technological Neutrality

The 2002 farm bill (P.L. 107-171) directed RUS to use criteria that are "technologically neutral" in determining which projects to approve for loans. In other words, RUS is prohibited from typically valuing one broadband technology over another when assessing loan applications. As of November 10, 2008, 37% of approved and funded projects employed fiber-to-the-home technology, 17% employed DSL, 25% fixed wireless, 19% hybrid fiber-coaxial (cable), and 2% broadband over powerlines (BPL).[54] No funding has been provided for projects utilizing satellite broadband.[55]

While decisions on funded projects were required to be technologically neutral, RUS (through the Secretary of Agriculture) had the latitude to determine minimum required data transmission rates for broadband projects eligible for funding. According to the statute, "the Secretary shall, from time to time as advances in technology warrant, review and recommend modifications of rate-ofdata transmission criteria for purposes of the identification of broadband service technologies."

Some argued that the minimum speed thresholds should be raised to ensure that rural areas receive "next-generation" broadband technologies with faster data rates capable of more varied and sophisticated applications. On the other hand, significantly raising minimum data rates could exclude certain technologies—for example typical data transmission rates for fiber and some wireless technologies exceed what is offered by "current generation" technologies such as DSL and cable. Proponents of keeping the minimum threshold at a low level argued that underserved rural areas are best served by any broadband technology that is economically feasible to deploy, regardless of whether it is "next" or "current" generation.

Broadband Loan Program Reauthorization (P.L. 110-246)

The Food, Conservation, and Energy Act of 2008 became law on June 18, 2008 (P.L. 110-246). Section 6110, "Access to Broadband Telecommunications Services in Rural Areas," reauthorized the RUS broadband loan and loan guarantee program and addressed many of the criticisms and issues raised during the reauthorization process. The following summarizes broadband-related provisions that change previous law.

Eligibility and Selection Criteria

- Defines rural area as any area other than (1) a city or town that has a population of greater than 20,000 and (2) an urbanized area contiguous and adjacent to a city or town with a population greater than 50,000. The Secretary may, by regulation only, consider not to be rural an area that consists of any collection of census blocks contiguous to each other with a housing density of more than 200 housing units per square mile and that is contiguous with or adjacent to an existing boundary of a rural area.

- Provides that the highest priority is to be given to applicants that offer to provide broadband service to the greatest proportion of households currently without broadband service. Eligible entities are required to submit a proposal to the Secretary that meets the requirements for a project to offer to provide service to a rural area and agree to complete build out of the broadband service within three years.
- Prohibits any eligible entity that provides telecommunications or broadband service to at least 20% of the households in the United States from receiving an amount of funds under this section for a fiscal year in excess of 15% of the funds authorized and appropriated for the broadband loan program.
- Directs the Secretary of Agriculture "from time to time as advances in technology warrant," to review and recommend modifications in rate-of-data transmission criteria for the purpose of identifying eligible broadband service technologies. At the same time, the Secretary is prohibited from establishing requirements for bandwidth or speed that have the effect of precluding the use of evolving technologies appropriate for use in rural areas.

Loans to Communities with Existing Providers

- Prohibits the Secretary from making a loan in any area where there are more than 3 incumbent service providers unless the loan meets all of the following requirements: (1) the loan is to an incumbent service provider that is upgrading service in that provider's existing territory; (2) the loan proposes to serve an area where not less than 25% of the households are offered service by not more than 1 provider; and (3) the applicant is not eligible for funding under another provision of the Rural Electrification Act. Incumbent service provider is defined as an entity providing broadband service to not less than 5% of the households in the service territory proposed in the application. Also prohibits the Secretary from making a loan in any area where not less than 25% of the households are offered broadband service by not more than 1 provider unless a prior loan has been made in the same area.

Financial Requirements

- Directs the Secretary to consider existing recurring revenues at the time of application in determining an adequate level of credit support. Requires the Secretary to ensure that the type, amount, and method of security used to secure a loan or loan guarantee is commensurate to the risk involved with the loan or loan guarantee, particularly when the loan or loan guarantee is issued to a financially healthy, strong, and stable entity. The Secretary is also required, in determining the amount and method of security, to consider reducing the security in areas that do not have broadband service.

- Allows the Secretary to require an entity to provide a cost-share in an amount not to exceed 10% of the amount of the loan or loan guarantee.
- Retains the current law rate of interest for direct loans—which is the rate equivalent to the cost of borrowing to the Department of Treasury for obligations of comparable maturity or 4%.
- Directs that loan or loan guarantee may have a term not to exceed 35 years if the Secretary determines that the loan security is sufficient.
- In case of substantially underserved trust areas (for example, Indian lands), where the Secretary determines a high need exists for the benefits of the program, the Secretary has the authority to provide loans with interest rates as low as 2% and may waive nonduplication restrictions, matching fund requirements, credit support requirements, or other regulations.

Loan Application Requirements

- Allows the Secretary to require an entity that proposes to have a subscriber projection of more than 20% of the broadband service market in a rural area to submit a market survey. However, the Secretary is prohibited from requiring a market survey from an entity that projects to have less than 20% of the broadband market.
- Requires public notice of each application submitted, including the identity of the applicant, the proposed area to be served, and the estimated number of households in the application without terrestrial-based broadband. Authorizes the Secretary to take steps to reduce the costs and paperwork associated with applying for a loan or loan guarantee under this section by first-time applicants, particularly those who are smaller and start-up Internet providers.
- Allows the Secretary to establish a pre-application process under which a prospective applicant may seek a determination of area eligibility. Provides that an application, or a petition for reconsideration of a decision on such an application, that was pending on the date 45 days before enactment of this act and that remains pending on the date of enactment of this act is to be considered under eligibility and feasibility criteria in effect on the original date of submission of the application.

Other Provisions

- Authorizes the Rural Broadband Access Loan and Loan Guarantee program at $25,000,000 to be appropriated for each of fiscal years 2008 through 2012.
- Requires that the Secretary annually report to Congress on the rural broadband loan and loan guarantee program. The annual report is to include information pertaining to the loans made, communities served and proposed to be served, speed of broadband service offered, types of services offered by the applicants and recipients, length of time to approve applications submitted, and outreach efforts undertaken by USDA.

- Section 6111 provides for a National Center for Rural Telecommunications Assessment. The Center is to assess the effectiveness of broadband loan programs, work with existing rural development centers to identify appropriate policy initiatives, and provide an annual report that describes the activities of the Center, the results of research carried out by the Center, and any additional information that the Secretary may request. An appropriation of $1,000,000 is authorized for each of the fiscal years 2008 through 2012.
- Section 6112 directs the Chairman of the Federal Communications Commission (FCC), in coordination with the Secretary, to submit to Congress a report describing a comprehensive rural broadband strategy. Requires the report to be updated during the third year after enactment.

Implementation of P.L. 110-246

The final RUS/USDA rule that will implement the Rural Broadband Access Loan and Loan Guarantee program, as modified by P.L. 110-246, is still pending and has not been released.

Meanwhile, pursuant to section 6112 of P.L. 110-246, the FCC released on May 22, 2009 its report on rural broadband strategy, entitled *Bringing Broadband to Rural America*.[56] The report made a series of recommendations including: improved coordination of rural broadband efforts among federal agencies, states, and communities; better assessment of broadband needs, including technological considerations and broadband mapping and data; and overcoming challenges to rural broadband deployment. According to Acting FCC Chairman Michael Copps, the report is "a prelude to, and a building block for" the national broadband plan which the FCC is developing and will deliver to Congress by February 17, 2010, as mandated by the American Recovery and Reinvestment Act (P.L. 111-5).

APPROPRIATIONS

FY2008

The President's FY2008 budget proposal requested a $6.45 million (subsidy) to support a loan level of $300 million. Noting that this is a $200 million reduction from the FY2007 level, the budget documents stated that the "funding is sufficient to meet expected demand," and that

> Regulations are being changed to correct certain weaknesses that have become apparent since the program was established a few years ago. The new regulations will ensure that program funds are focused on rural areas that are lacking existing providers, and that applicants meet high enough standards to ensure long term success.[57]

The FY2008 budget proposal requested no funding for the Community Connect Broadband Grant program.

On July 19, 2007, the House Appropriations Committee approved the FY2008 Agriculture, Rural Development, Food and Drug Administration, and Related Agencies appropriations bill (H.R. 3161; H.Rept. 110-258). The Committee approved $6.45 million to support a loan level of $300 million for the broadband loan program, and $17.82 million for broadband community connect grants (twice the FY2007 level). In report language, the Committee expressed concern over broadband loans to areas with existing providers, and directed the USDA Office of the Inspector General to conduct a comprehensive follow-up study reexamining the RUS broadband loan program. Specifically, the report is directed to detail: how many unserved households were included in approved RUS Broadband Loan Program applications; how many applications were granted to applicants proposing to serve areas where one or more private broadband providers already offered service; how many approved loans (and their total amount) have defaulted since the program's inception; and how many applicants who have been approved for loans have subsequently withdrawn from the program due to the eventually discovered infeasibility of the approved project.

In report language, the House Appropriations Committee also expressed concern over the administration of the broadband loan program, and noted USDA's failure to obligate available resources to fund broadband projects. According to the Committee, USDA will carry over $10.643 million from FY2007 to FY2008, which will support an additional program (loan) level of $495 million. Regarding the proposed RUS broadband loan program rule, the Committee "expects the Department to prioritize deployment of broadband service to households with no or limited broadband access." Finally, the Committee recommended $250,000 to the USDA Economic Research Service to research deployment of broadband service to households with no or limited access to broadband, and to study the economic impact of not having broadband on rural communities and their growth, community facilities, access to healthcare, and well being.

Also on July 19, 2007, the Senate Appropriations Committee approved its version of the FY2008 agriculture appropriations bill (S. 1859; S.Rept. 110-134). The Committee approved $1 0.643 million to support a loan level of $495 million for the broadband loan program, and $8.9 million for broadband grants.

On December 26, 2007, the President signed the Consolidated Appropriations Act, 2008 (P.L. 110-161). The bill provided $6.45 million to support a loan level of $300 million for the broadband loan program, and $13.4 million for broadband community connect grants. The Joint Explanatory Statement accompanying P.L. 110-161 directed USDA to evaluate and report on the potential of a combination loan/grant broadband program to expand the reach and more effectively utilize broadband resources.

FY2009

The Bush Administration's FY2009 budget proposal requested a $11.619 million (subsidy) to support a loan level of $297.923 million. The budget justification noted that available funding in recent years has significantly outpaced demand and that $495 million was carried forward to FY2008, in addition to the FY2008 funding. The Administration proposed a rescission of the FY2008 subsidy of $6.45 million, an unobligated balance that would otherwise be carried forward into FY2009. According to the budget justification, the program "routinely carries over the entire year's appropriations, so the funds are not necessary to support the program's demand."

As in past budget requests, the Administration proposed no funding for the Community Connect Broadband Grants (down from $13.4 million in FY2008). According to the budget justification, "building broadband infrastructure in rural America is expensive and requires extensive amounts of capital. A loan program is more suited to supporting this kind of activity."

On July 21, 2008, the Senate Appropriations Committee reported the FY2009 agriculture appropriations bill (S. 3289; S.Rept. 110-426). The Committee approved $11.618 million to support a loan level of $297.923 million for the broadband loan program, the same level as requested by the Administration. Unlike the Administration, the Committee provided $13 .406 million for broadband grants.

The Omnibus Appropriations Act, 2009 (P.L. 111-8) appropriated $15.619 million to support a loan level of $400.487 million for the Rural Broadband Access Loan and Loan Guarantee Program, and $ 13.406 million for the Community Connect Grant Program.

FY2010

The Obama Administration's FY2010 budget proposal requested a $38.495 million loan subsidy to support a loan level of $532 million for the Rural Broadband Access Loan and Loan Guarantee Program, and $ 13.406 million for the Community Connect Grant Program.

On June 18, 2009, the House Appropriations Committee approved the FY2010 agriculture appropriations bill (H.R. 2997; H.Rept. 111-181). The Committee approved a $28.96 million loan subsidy to support a loan level of $400 million for the Rural Broadband Access Loan and Loan Guarantee Program, and $ 17.976 million for the Community Connect Grant Program. The House passed H.R. 2997 on July 9, 2009.

On July 7, 2009, the Senate Appropriations Committee approved its version of the FY2010 agriculture appropriations bill (S. 1406; S.Rept. 111-39). Matching the Administration request, the Committee approved a $38.495 million loan subsidy to support a loan level of $532 million for the Rural Broadband Access Loan and Loan Guarantee Program, and $13.406 million for the Community Connect Grant Program. The Senate passed H.R. 2997 on August 4, 2009.

The Conference Report (H.Rept. 111-279) was filed on September 30, 2009, and adopted the House provisions: approved a $28.96 million loan subsidy to support a loan level of $400 million for the Rural Broadband Access Loan and Loan Guarantee Program, and $17.976 million for the Community Connect Grant Program.

THE AMERICAN RECOVERY AND REINVESTMENT ACT (P.L. 111-5)

On February 17, 2009, President Obama signed P.L. 111-5, the American Recovery and Reinvestment Act (ARRA). Broadband provisions of the ARRA provide a total of $7.2 billion, primarily for broadband grants. The total consists of $2.5 billion to RUS broadband loan, grant, and loan/grant combinations, and $4.7 billion to NTIA/DOC for a newly established Broadband Technology Opportunities Program.[58]

The ARRA does not specify how the $2.5 billion is to be divided between the RUS grant and loan programs. Regarding projects applying for funding, the ARRA states that:

- at least 75% of the area to be served by a project receiving these funds shall be in a rural area without sufficient access to high speed broadband service to facilitate economic development, as determined by the Secretary of Agriculture;
- priority shall be given to projects that will deliver end users a choice of more than one broadband service provider;
- priority shall be given to projects that provide service to the highest proportion of rural residents that do not have access to broadband service;
- priority shall be given to borrowers and former borrowers of rural telephone loans;
- priority shall be given to projects demonstrating that all project elements will be fully funded, that can commence promptly, and that can be completed; and
- no area of a project may receive funding to provide broadband service under the Broadband Technology Opportunities Program at NTIA/DOC.

According to the Administration's FY2010 budget proposal, USDA estimates that the $2.5 billion of budget authority will be apportioned as follows: $1.94 billion for grants and $485 million for loan subsidies to support $7.16 billion in direct loans. According to the budget summary, "the $2.5 billion in budget authority for broadband loans and grants provided by the Recovery Act will be administered under a series of notices to reflect the specific provisions of the Recovery Act, which differ from those of the 2008 Farm Bill."[59]

On July 1, 2009, RUS and NTIA jointly released the first Notice of Funds Availability (NOFA) detailing requirements, rules, and procedures for applying for ARRA grants, loans, and loan grant combinations.[60] The total amount available in this first funding round is $4 billion, including $2.4 billion (program level) under the RUS Broadband Initiatives Program (BIP), which will extend loans, grants, and loan/grant combinations to facilitate broadband deployment in rural areas; and $1.6 billion (budget authority) under the NTIA Broadband Technology Opportunities Program (BTOP), which will make available grants for deploying broadband infrastructure in unserved and underserved areas, enhance broadband capacity at public computer centers, and promote sustainable broadband adoption projects. **Table 4** compares the NOFA provisions for BIP and BTOP with similar provisions associated with the Community Connect Grant Program and the Rural Broadband Access Loan and Loan Guarantee Program.

First round applications for ARRA grants, loans, and grant/loan combinations were accepted between July 14 and August 20, 2009. Award decisions will be announced on or about November 7, 2009. The second funding round will likely open in late 2009 or early 2010; all awards must be made by September 30, 2010.

Table 4. Comparison of RUS and NTIA Broadband Programs

Community Connect Broadband Grants	Rural Broadband Access Loan and Loan Guarantee Program (2008 farm bill, section 6110 of P.L. 110-246)	Broadband Initiatives Program (BIP), Rural Utilities Service	Broadband Technology Opportunities Program (BTOP), National Telecommunications and Information Administration
Funding			
$13.4 million (FY2009)	$15.6 million (loan subsidy) $400 million (loan level) (FY2009)	$2.4 billion total for first round (program level), includes up to: • $1.2 billion for Last Mile projects ($400 million in grants for Remote Area projects, $800 million in loans and loan/grant combinations for Non-Remote projects); • $800 million in loans and loan/grant combinations for Middle Mile projects; and • $325 million for reserve fund.	$1.6 billion for first round (budget authority), includes up to: • $1.2 billion for broadband infrastructure grants (Last Mile and Middle Mile projects); • $50 million for Public Computer Center grants; • $150 million for Sustainable Broadband Adoption grants; and • $200 million for reserve fund.
Existing Broadband Providers			
Grants go to projects serving a single community with no existing broadband provider.	Prohibits the Secretary from making a loan in any area where there are more than 3 incumbent service providers unless the loan meets all of the following requirements: (1) the loan is to an incumbent service provider that is upgrading service in that provider's existing territory; (2) the loan proposes to serve an area where not less than 25% of the households are offered service by not more than 1 provider; and (3) the applicant is not eligible for funding under another provision of the Rural Electrification Act. Incumbent service provider is defined as an entity providing broadband service to not less than 5% of the households in the service territory proposed in the application.	Same as NTIA/BTOP definition. Additionally defines "Remote Area" as an unserved, rural area 50 miles from the limits of a non-rural area.	Eligible "unserved areas" defined as where at least 90% percent of households lack access to terrestrial broadband service. Eligible "underserved areas" for last mile projects if at least one of the following factors is met: (1) no more than 50% of the households in the proposed funded service area have access to facilitiesbased, terrestrial broadband service at greater than the minimum broadband transmission speed; (2) no broadband service provider advertises broadband transmission speeds of at least 3 megabits per second (Mbps) downstream; or (3) the rate of broadband subscribership for the proposed funded service area is 40% of households or less.

Table 4. (Continued)

Community Connect Broadband Grants	Rural Broadband Access Loan and Loan Guarantee Program (2008 farm bill, section 6110 of P.L. 110-246)	Broadband Initiatives Program (BIP), Rural Utilities Service	Broadband Technology Opportunities Program (BTOP), National Telecommunications and Information Administration
	Also prohibits the Secretary from making a loan in any area where not less than 25% of the households are offered broadband service by not more than 1 provider unless a prior loan has been made in the same area.		A proposed funded service area may qualify as underserved for middle mile projects if one interconnection point terminates in a proposed funded service area that qualifies as unserved or underserved for last mile projects.
Definition of Broadband			
Defines "broadband transmission service" as 200 kilobits per second (kbps) in both the downstream and upstream connection between provider and consumer.	Any technology identified by the Secretary as having the capacity to transmit data to enable a subscriber to the service to originate and receive highquality voice, data, graphics, and video. The Secretary shall, from time to time as advances in technology warrant, review and recommend modifications of rate-of data transmission criteria for purposes of the identification of broadband service technologies. The Secretary is prohibited from establishing requirements for bandwidth or speed that have the effect of precluding the use of evolving technologies appropriate for rural areas.	Same as NTIA/BTOP definition.	Two-way data transmission with advertised speeds of at least 768 kbps downstream and at least 200 kbps upstream to end users, or providing sufficient capacity in a middle mile project to support the provision of broadband service to end users.
Definition of Rural Area			
Any area of the United States not included within the boundaries of any incorporated or unincorporated city, village, or borough	Any area other than (1) a city or town that has a population of greater than 20,000 and (2) an urbanized area contiguous and adjacent to a city or town with a population greater than 50,000. The Secretary may, by regulation only, consider not to be rural an area that consists of any collection of	Any area, as confirmed by the latest decennial census of the Bureau of the Census, which is not located within: (1) a city, town, or incorporated area that has a population of greater than 20,000 inhabitants or (2) an urbanized area contiguous and adjacent to a city or town	Same definition as used by BIP/RUS. Applications to fund broadband infrastructure projects in areas which are at least 75% rural are required to be submitted to BIP. BTOP may make awards to such applications NTIA determines to be meritorious after RUS has reviewed the

Community Connect Broadband Grants	Rural Broadband Access Loan and Loan Guarantee Program (2008 farm bill, section 6110 of P.L. 110-246)	Broadband Initiatives Program (BIP), Rural Utilities Service	Broadband Technology Opportunities Program (BTOP), National Telecommunications and Information Administration
having a population in excess of 20,000 inhabitants.	census blocks contiguous to each other with a housing density of more than 200 housing units per square mile and that is contiguous with or adjacent to an existing boundary of a rural area.	that has a population of greater than 50,000 inhabitants. For purposes of the definition of rural area, an urbanized area means a densely populated territory as defined in the latest decennial census of the U.S. Census Bureau.	application and determined not to fund it. All other applications for Broadband Infrastructure projects, as well as applications for Public Computer Centers or Sustainable Broadband Adoption projects, must be submitted to BTOP.
Financial Obligation of Applicant			
Awardees must contribute a matching contribution equal to 15% of the requested grant amount. In-kind contributions for purposes that could have been financed with grant funds are allowed.	Allows the Secretary to require an entity to provide a cost-share in an amount not to exceed 10% of the amount of the loan or loan guarantee. Directs the Secretary to consider existing recurring revenues at the time of application in determining an adequate level of credit support. Requires the Secretary to ensure that the type, amount, and method of security used to secure a loan or loan guarantee is commensurate to the risk involved with the loan or loan guarantee, particularly when the loan or loan guarantee is issued to a financially healthy, strong, and stable entity. The Secretary is also required, in determining the amount and method of security, to consider reducing the security in areas that do not have broadband service.	For grants (Last Mile Remote Area projects): funding up to 100%. For grants receiving greater than 80% of eligible costs, the Administrator must determine that the awardee has a specific financial need that justifies funding greater than 80%; all applicants must be able to generate a minimum current ratioa of one by the end of the forecast period and demonstrate a positive cash balance for each year of the forecast period. For loans: the applicant must be able to generate sufficient revenues to cover expenses, have sufficient cash flow to service debts and obligations as they come due, and meet the minimum Times Interest Earned Ratio (TIER)[b] requirement of one by the end of the forecast period, as determined by RUS.	Required to provide matching funds of at least 20% toward the total eligible project cost.. Applicants must document their capacity to provide matching funds. NTIA will provide up to 80% of total eligible project costs, unless the applicant petitions the Assistant Secretary for a waiver of the matching requirement and that waiver is granted by the Assistant Secretary based on the applicant's demonstration of financial need. In-kind contributions, including third party inkind contributions, are non-cash donations to a project that may count toward satisfying the non-federal matching requirement of a project's total budget. In-kind contributions must be allowable project expenses.

a. "Current ratio" is defined as the applicant's current assets divided by the current liabilities.

b. TIER is defined as the ration of an applicant's net income (after taxes) plus (adding back) interest expense, all divided by interest expense (existing and any new interest expense including the interest expense associated with the proposed loan).

Endnotes

[1] Gillett, Sharon E., Massachusetts Institute of Technology, Measuring Broadband's Economic Impact, report prepared for the Economic Development Administration, U.S. Department of Commerce, February 28, 2006, p. 4. Available at http://www.eda.gov/ImageCache/EDAPublic/documents/pdfdocs2006/mitcmubbimpactreport_2epdf/v1/mitcmubbimpactreport.pdf.

[2] Crandall, Robert, William Lehr, and Robert Litan, *The Effects of Broadband Deployment on Output and Employment: A Cross-sectional Analysis of U.S. Data, June* 2007, 20 pp. Available at http://www3.brookings.edu/views/papers/ crandall/200706litan.pdf.

[3] Peter Stenberg, Mitchell Morehart, and Stephen Vogel, et al., *Broadband Internet's Value for Rural America*, United States Department of Agriculture, Economic Research Service, Economic Research Report Number 78, Washington, DC, August 2009, p. iii, http://www.ers.usda.gov/Publications/ERR78/ERR78.pdf.

[4] See for example: National Exchange Carrier Association (NECA), Trends 2006: *Making Progress With Broadband,* 2006, 26 p. Available at http://www.neca.org/media

[5] *Broadband Internet's Value for Rural America,* p. iii.

[6] Horrigan, John B., Pew Internet & American Life Project, *Home Broadband Adoption 2009,* June 2009, p. 14. Available at http://www.pewinternet.org/~/media

[7] FCC, *High-Speed Services for Internet Access: Status as of June 30, 2008,* Chart 6, available at http://hraunfoss.fcc.gov/edocs_public/attachmatch/DOC-292191A1.pdf .

[8] Rural Utilities Service, USDA, "Construction and Installation of Broadband Telecommunications Services in Rural America; Availability of Loan Funds," *Federal Register*, Vol. 65, No. 234, December 5, 2000, p. 75920.

[9] Rural Utilities Service, USDA, "Broadband Pilot Loan Program," *Federal Register*, Vol. 67, No. 15, January 23, 2002, p. 3140.

[10] Rural Utilities Service, USDA, "Broadband Pilot Grant Program," *Federal Register,* Vol. 67, No. 130, July 8, 2002, p. 45080.

[11] Title VI of the Rural Electrification Act of 1936 (7 U.S.C. 950bb).

[12] Section 772 of the FY2004 Consolidated Appropriations Act (P.L. 108-199) changed the definition of an "eligible rural community" to be defined as "any area of the United States that is not contained in an incorporated city or town with a population in excess of 20,000 inhabitants." Accordingly, the March 29, 2004 Notice of Funds Availability for the Rural Broadband Access Loans and Loan Guarantee Program defined "Eligible Rural Community" as follows:

> The definition of eligible rural community in Section 601(b)(2) of the Rural Electrification Act (7 U.S.C. 950bb)(b)(2), qualifying for financial assistance under the Rural Broadband Access Loan and Loan Guaranty Program, has been amended by provisions in the Consolidated Appropriations Act, 2004, to mean any area of the United States that is not contained in an incorporated city or town with a population in excess of 20,000 inhabitants. Therefore, an applicant no longer must demonstrate that it is not located in an area designated as a standard metropolitan statistical area. This change supersedes and nullifies contrary provisions in regulations implementing the broadband program found at 7 CFR part 1738.

[13] Rural Utilities Service, USDA, "Rural Broadband Access Loans and Loan Guarantees," *Federal Register*, Vol. 68, No. 20, January 30, 2003, pp. 4684-4692.

[14] The cash-on-hand requirement is waived for companies with two previous years of positive cash flow.

[15] USDA, Rural Utilities Service, "FCC/USDA Rural Broadband Educational Workshop," power point presentation, November 20, 2008. Available at http://www.usda.gov/rus/telecom/broadband FCC_USDABroadbandWorkshopNov20.pdf.

[16] USDA, News Release, "USDA Announces $267 Million Rural Broadband Loan," March 25, 2008. Available at http://www.rurdev.usda.gov/rd/newsroom/2008/RD_Broadband_Loans_3-25-2008.pdf.

[17] Rural Utilities Service, Department of Agriculture, "Rural Broadband Access Loans and Loan Guarantees," Proposed Rule, *Federal Register*, Vol. 72, No. 91, May 11, 2007, p. 26744.

[18] Rural Utilities Service, USDA, "Broadband Grant Program," 7 C.F.R. part 1739, *Federal Register*, Vol. 69, No. 144, July 28, 2004, pp. 44896-44903.

[19] A rural area is defined as "any area of the United States not included within the boundaries of any incorporated or unincorporated city, village, or borough having a population in excess of 20,000 inhabitants." (7 C.F.R. 1739.3)

[20] See CRS Report RL30719, *Broadband Internet Access and the Digital Divide: Federal Assistance Programs,* by Lennard G. Kruger and Angele A. Gilroy.

[21] In the Rural Electrification Loan Restructuring Act (P.L. 103-129, the 1993 farm bill), Congress amended the Rural Electrification Act to require that facilities financed under this program be capable of providing broadband service at the rate of 1 megabyte per second (7 U.S.C. 935(d)(3)(B)(iv)(I)(cc).

[22] USDA, Rural Utilities Service, Rural Development Telecommunications home page, http://www.usda.gov/rus/telecom/.

[23] Rural Utilities Service, *2010 Explanatory Notes,* available at http://www.obpa.usda.gov/25rus2010notes.pdf

[24] For more information on the Universal Service Fund, see CRS Report RL30719, *Broadband Internet Access and the Digital Divide: Federal Assistance Programs,* by Lennard G. Kruger and Angele A. Gilroy.

[25] Congressional Budget Office, Factors That May Increase Future Spending from the Universal Service Fund, CBO Paper, June 2006, p. 25. Available at http://www.cbo.gov/ftpdocs/72xx/doc7291/06-16-UniversalService.pdf.

[26] See *State Broadband Initiatives: A Summary of State Programs Designed to Stimulate Broadband Deployment and Adoption,* A Joint Report of the Alliance for Public Technology and the Communications Workers of America, July 2008, 54 pages. State program database available at http://www.speedmatters.org/statepolicy.

[27] California Broadband Task Force, *The State of Connectivity: Building Innovation Through Broadband,* Final Report of the California Broadband Task Force, January 2008, 83 p. Available at http://www.calink.ca.gov/ pdf/ CBTF_FINAL_Report.pdf.

[28] NGA Center for Best Practices, Issue Brief, "State Efforts to Expand Broadband Access," May 20, 2008, p. 1. Available at http://www.nga.org/Files/pdf/0805BROADBANDACCESS.PDF.

[29] For a summary of selected state broadband bills, see http://www.ncsl.org/programs/telecom/ .2008BroadbandBills.htm.

[30] Private communication, USDA, June 23, 2009.

[31] GAO, *Broadband Deployment is Extensive throughout the United States, but It Is Difficult to Assess the Extent of Deployment Gaps in Rural Areas*, p. 33.

[32] Testimony of Jim Andrew, Administrator, Rural Utilities Service, U.S. Department of Agriculture, "Broadband Program Administered by USDA's Rural Utilities Service," full committee hearing before the Senate Committee on Agriculture, Nutrition, and Forestry, 109th Congress, May 17, 2006.

[33] Rural Utilities Service, private communication, January 18, 2007.

[34] Rural Utilities Service, Department of Agriculture, "Rural Broadband Access Loans and Loan Guarantees," Proposed Rule, Federal Register, Vol. 72, No. 91, May 11, 2007, p. 26744.

[35] U.S. Department of Agriculture, Office of Inspector General, Southwest Region, *Audit Report: Rural Utilities Service Broadband Grant and Loan Programs,* Audit Report 0960 1-4-Te, September 2005, p. I. Available at http://www.usda.gov/oig/webdocs/09601-04-TE.pdf.

[36] Ibid., p. 6.

[37] Ibid., p. 8.

[38] Martinez, Michael, "Broadband: Loan Fund's Strict Rules Foil Small Municipalities," *National Journal's Technology Daily,* August 23, 2005.

[39] GAO, *Broadband Deployment is Extensive throughout the United States, but It Is Difficult to Assess the Extent of Deployment Gaps in Rural Areas,* p. 33-34.

[40] Ibid., p. 13.

[41] Ibid., p. 14.

[42] Ibid., p. 15

[43] Testimony of James Andrew, Administrator, Rural Utilities Service, U.S. Department of Agriculture, before the Subcommittee on Specialty Crops, Rural Development, and Foreign Agriculture, House Committee on Agriculture, May 1, 2007.

[44] Letter from Kyle McSlarrow, President and CEO, National Cable & Telecommunications Association to the Honorable Mike Johanns, Secretary of the U.S. Department of Agriculture, May 16, 2006.

[45] Testimony of Tom Simmons, Vice President for Public Policy, Midcontinent Communications, before Senate Committee on Agriculture, Nutrition, and Forestry, May 17, 2006.

[46] *Audit Report: Rural Utilities Service Broadband Grant and Loan Programs,* p. 17.

[47] Rural Utilities Service, private communication, January 18, 2007.

[48] U.S. Department of Agriculture, Office of Inspector General, Southwest Region, Audit Report Rural Utilities Service Broadband Loan and Loan Guarantee Program, Report No. 0960 1-8-Te, March 2009, p. 9.

[49] Ibid, p. 5.

[50] Ibid, p. 5-6.

[51] Ibid, p. 14.

[52] Rural Utilities Service, private communication, January 18, 2007.

[53] The statute (7 U.S.C. 950bb) allows States and local governments to be eligible for loans only if "no other eligible entity is already offering, or has committed to offer, broadband services to the eligible rural community."

[54] USDA, Rural Utilities Service, "FCC/USDA Rural Broadband Educational Workshop," power point presentation, November 20, 2008. Available at http://www.usda.gov/rus/telecom/broadband FCC_USDABroadbandWorkshopNov20.pdf.

[55] According to the GAO, satellite companies state that RUS's broadband loan program requirements "are not readily compatible with their business model or technology," and that "because the agency requires collateral for loans, the program is more suited for situations where the providers, rather than individual consumers, own the equipment being purchased through the loan. Yet, when consumers purchase satellite broadband, it is common for them to purchase the equipment needed to receive the satellite signal, such as the reception dish."

Satellite companies argue that in some rural areas, satellite broadband might be the most feasible and cost-effective solution. See GAO, *Broadband Deployment is Extensive throughout the United States, but It Is Difficult to Assess the Extent of Deployment Gaps in Rural Areas,* pp. 34-35.

[56] Michael J. Copps, Acting Chairman, Federal Communications Commission, *Bringing Broadband to Rural America: Report on a Rural Broadband Strategy,* May 22, 2009, 83 p.

[57] U.S. Dept. of Agriculture, *FY2008 Budget Summary and Performance Plan,* p. 44.

[58] For more information on ARRA broadband programs, see CRS Report R40436, Broadband Infrastructure Programs in the American Recovery and Reinvestment Act, by Lennard G. Kruger.

[59] U.S. Department of Agriculture, FY2010 Budget Summary and Annual Performance Plan, p. 50, available at http://www.obpa.usda.gov/budsum/FY10budsum.pdf.

[60] Available at http://www.ntia.doc.gov/frnotices/2009/FR_BBNOFA_090702.pdf.

In: Broadband and Rural America
Editor: Bethany J. Lenz
ISBN: 978-1-61668-637-6

Chapter 5

TESTIMONY OF JIM ANDREW TO THE HOUSE COMMITTEE ON AGRICULTURE, SUBCOMMITTEE ON SPECIALTY CROPS, RURAL DEVELOPMENT AND FOREIGN AGRICULTURE

Jim Andrew

Chairman McIntyre, Ranking Member Musgrave, and subcommittee members, thank you for the opportunity to testify today regarding broadband service in rural America. Some of you, who know me, know I live in rural America. I live outside of Millen Georgia, so far from town that I have to go back towards town to go hunting. I know what it means for a rural community to not have broadband service. I have been around long enough to have seen what has happened to too many of our small towns.

When the railroads passed some of our towns by, they dried up. When the interstate did not have an off ramp, the tourists and truckers did not stop to buy gas and meals, your town did not survive. Fifty-seven years ago when the telephone program was created in this very room, rural America did not have phones and it took a fourth of our population to produce the food and fiber needed for this Nation.

Rural America has changed but some things are still the same. We are still among the world's top producers of grains, soybeans, cotton, and other commodities. Down in Georgia, we still grow a few peanuts. But at the same time, rural America has diversified. The rural economy isn't just commodities anymore; it's manufacturing, services, and everything else as well.

So what does this have to do with broadband? Broadband is our railroad, our highway, our bridge to the rest of this country and to the world. With high-speed communications, we can bridge the barriers of time and distance to market our goods and services, have access to markets to both buy and sell.

Today's technology gives the student in the most rural school access to information and teaching that has long been available to more populated communities. With today's technology, advanced medical treatment and expertise is available to small rural hospitals.

Today's technology gives our communities and their citizens the ability to build businesses that provide jobs for our children.

My wife and I have two children, now adults. One lives in Baltimore and one in Atlanta. When they were younger, Millen, Georgia did not have the career opportunities that they were interested in pursuing. The loss of the next generation has been the problem for rural communities all over this country. With modern technology, we have an important new tool to plug the leak.

As I indicated before, we have administered the rural telephone loan program since it was created by Congress in 1949. In the mid 1990's, Congress amended the telecommunications section of the Rural Electrification Act to require new plants financed by us to be broadband capable at the speed of one meg. Placing this updated capability with USDA Rural Development made sense because we were already stringing the lines.

In addition to the Broadband Access Program, authorized by the 2002 Farm Bill, we have also had the pilot broadband loan program that was a product of the appropriations process for fiscal years 2001 and 2002 and the Broadband Community Connect Grant program.

A third program -- the Distance Learning and Telemedicine Loan and Grant program (DLT) -- was created in the 1990 Farm Bill. The first DLT grants were made in 1994. While it is not a broadband program, it makes use of broadband connectivity to improve educational opportunities and improve the quality of health care for rural citizens.

THE PRESENT

We have made important accomplishments with the Broadband Access Program. To date, we have approved 69 loans for $1.21 billion to finance facilities in 40 states. These projects, when fully functional, will serve a projected 520,000 households in 1,081 communities. Eight projects have been completed in Kansas, Louisiana, Michigan, North Dakota, Nebraska, South Dakota, Texas and Washington State.

Of these loans, approximately 40 percent of the communities approved for funding were unserved at the time of loan approval. An additional 15 percent had only one broadband provider at the time of loan approval.

To date, the program has received 198 applications totaling $4.3 billion. As of today, we have 21 applications under review for more than $950 million. We anticipate another $400 million in applications this year. We have administered this program with no increase in personnel. In fact, we have fewer employees than we had when the program began.

From the beginning the loan applications have been larger and more complicated than anything else the Telecom Program staff has experienced. In 2003, the average loan size was $11.2 million. In 2006, the average was $44 million.

During the past six years, we have made 129 Community Connect grants for a total of $57.8 million. Under the Pilot Broadband Program, we made 28 loans totaling $180 million. Since 2002, USDA has invested more than $166 million in its Distance Learning and Telemedicine (DLT) Program, allowing 3,796 rural educational facilities to expand their access to modern telecommunications technology, and 2,226 health care institutions to develop technologies needed to enhance local medical care.

Let me make one very important point….no one else is making loans for broadband service in rural America. We are committed to making this work.

Success Stories

We have a number of success stories that you can review on our web site. I want to mention some of them here today. Nex-Tech, Inc. of Lenora, Kansas has received two loans totaling just under $12 million. This company has used a combination of fiber/coaxial cable and DSL to build a system that provides broadband data services, voice and video at speeds over six Mbps to Plainville, Smith Center, and Stockton, Kansas. These areas had poor or non-existent data services. The system will serve 2,251 voice lines, 1,855 video subscribers, 768 high-speed data and 528 dial-up subscribers with new or improved service.

What do these loans mean? What is the impact? Osborne Industries in Osborne, Kansas is only one example. The company has not only remained in town; it's more than doubled in number of employees. It has become the largest employer in the community, thanks to broadband capability that has allowed it to reach not only across the nation but to international markets. Just this year, the employees have been able to purchase the company. They are working for themselves because they took the tool of broadband and have been able to make the concept of "living locally and competing globally" a fact.

And just down the road, in Phillipsburg, Kansas, a new ethanol plant was built and depends on Nex-Tex for its broadband service. This $54 million business created 34 jobs and is expanding.

A second example is a small company that is headquartered in Frankenmuth, Michigan. Air Advantage began with a Community Connect grant and with that start was able to graduate to two loans totaling $1.5 million. They have used a fixed wireless system that provides broadband service to 37 rural communities in Michigan. This company and its parent, Zimco, were founded by local businessman Scott Zimmer and his wife with a mission to improve the education and communications in their community.

Air Advantage now serves 1,100 customers with another 1000 projected in the near future. Their subscriber rates exceed their original projections. In addition, Scott and his wife successfully assisted local schools with applications for distance learning grants.

A typical success story from the Pilot Broadband loan program is Consolidated Enterprises, Inc. With a loan of just under $20 million, this project brought broadband services to the town of Dickenson, North Dakota. With the broadband loan, this company was able to purchase and upgrade an existing cable company and replace existing lines with new hybrid fiber/coax facilities in and around Dickenson. Last year, Consolidated reported more than 3,000 broadband customers receiving service. In addition, Consolidated's parent company, Consolidated Telecom, invested significant general company funds to ensure that the outlying rural areas were also upgraded and now offers advanced broadband services to those residents as well.

THE PAST

Now let's talk about some of the bumps in the road this program has incurred. The law creating the program instructed that regulations be issued without public comment. That meant we got them out faster, but it also meant that we did not have the amount of input that comes with proposed regulations published for comment.

We are now developing revised regulations, which are being reviewed by OMB and will be published for public comment very soon. We expect that the proposed regulations will receive extensive comments. We also believe the process will be of help to this subcommittee as you begin drafting the legislative language on the next farm bill.

In addition to the lack of public comment, the volume of loan applications has been an issue. In July and August of 2003, we received applications totaling over one billion dollars. Needless to say, we were overwhelmed. We had to rework how we handled loan applications. These were not our traditional telephone loan applications.

We were concerned from the beginning that the loans we made had to be good loans. First, this is not just the government's money. This is the taxpayers' money. It needs to be spent in the way that Congress intended it be spent and in a way that accomplishes the goals of Congress. Second, no one benefits from a bad loan. Not the applicant, not the taxpayer, and not the people who are looking for broadband service.

This agency has always been concerned about security of a loan. For utility programs, it is often more important to have an accurate projection of the revenue stream than it is to tie up the hard assets of a borrower, although we try to do both.

We have returned 108 loan applications for a variety of reasons. In some cases, review indicated that they did not have the credit support or equity needed. In others the applications were incomplete, had unsubstantiated or missing financial or technical information, or did not meet other criteria in the statute or regulations. It takes just as long, and sometimes longer, to review an application that is not approved as it does one that is approved.

A great deal of concern has been expressed from applicants, private industry, and Congress, including the House Committee on Agriculture, that we were not getting a sufficient number of loans approved fast enough. In any business, time is money and in the broadband industry, the changes occur almost daily.

We understand those concerns and are trying to respond appropriately. We have worked to streamline our review process, make the applications more "user friendly" and opened up our process to the extent possible.

Another frequently raised concern is that we are lending into existing service. In many cases we are doing just that, lending into some level of service. Let me go over some of the questions we face as we approach each application.

The statute requires us to give first priority to applicants that will provide service where there is no service. We do that. We want to see more applications that serve an unserved area. If you want rural citizens to have access to broadband, you are going to lend into competition.

For example, we had an application where we had two servers saying they were in one of the towns in question. Our review found that they were providing service in that community. But at the same time, I had a letter on my desk from the mayor of the town in question stating that they had lost two businesses because they could not get access to broadband and the college in the community could not get broadband service on the campus. We made the loan.

On the other hand, we hear and understand the concern. This is the first program in the utilities part of Rural Development where we lend into competitive areas. It raises new issues. We have met with every group that has asked to meet with us on this issue. The Assistant Administrator for Telecommunications Programs, Ms. Jacki Ponti, has made a concerted effort to improve the openness of the process:

- New USDA/RD web link on the Federal Communications Commission (FCC) Rural Home Page.
- New and improved web page design that is easier to use and easier to find information.
- New and improved weekly broadband report indicating companies with applications pending and the communities proposed to be served if the application is funded.
- Under construction is a new Community Search Function that will help everyone with an interest find out what communities are being served, which ones have applications pending, and which companies are providing service.

Where Do We Go Tomorrow?

Congress has a challenging task on hand to write a Farm Bill for 2007 that will cover commodities, conservation, food and nutrition, and of rural development. We are committed to working with you on this task. My area of responsibility is small, but I think it is important. Here is what we are doing.

We have proposed changes in the regulations that we hope will address some of the concerns that have been expressed. As these are published and comments come in, I expect I will hear from a number of members of Congress. We will have an open discussion as to how this program should work.

In addition, we are working to improve program delivery in the field. This begins with strong leadership from the Rural Development State offices with 47 state offices covering all 50 states. These State Directors with their staffs provide the eyes and ears of where the needs are in rural America.

Under the supervision of the State Directors, Telecom Program Staff provide Financial and Engineering support and guidance in all 50 states with 28 General Field Representatives monitoring more than 1000 telecom and broadband loan and grant projects across the country.

Finally, twenty three broadband workshops and outreach events have been held to inform potential applicants and community leaders about the program, as well as assist existing program borrowers with compliance requirements.

In closing, USDA can successfully operate a rural Broadband loan program because:

- We have expertise with the program and a long history in administering loan and loan guarantee programs.
- There is an established distribution system with field representatives and state offices in the heart of rural America

- We have developed a qualified and dedicated headquarters staff of 16 in the national office comprised of Engineers, loan specialists, and economists to provide broadband expertise and detailed evaluation of proposals.

There is a demand for broadband in rural America. We see it and hear it every day. We want to do a better job of identifying where the holes in service exist. We can only make loans to those who submit applications. Local leadership is vital to every rural development program.

The technology is changing and so are the business plans. We cannot be much leaner, but we can be smarter in how we manage this program. I look forward to your questions.

In: Broadband and Rural America
Editor: Bethany J. Lenz

ISBN: 978-1-61668-637-6

Chapter 6

TESTIMONY OF MR. KEVIN FELTY, GENERAL MANAGER, PLAINS COOPERATIVE TELEPHONE ASSOCIATION BEFORE THE SUBCOMMITTEE ON SPECIALTY CROPS, RURAL DEVELOPMENT AND FOREIGN AGRICULTURE

Kevin Felty

Good Afternoon Chairman McIntyre, Ranking Member Musgrave and committee members.

My name is Kevin Felty I am the General Manager of Plains Cooperative Telephone Association and its wholly owned, deregulated subsidiary, Plains Communications Services. Our headquarters is located in Joes, Colorado on the extreme eastern plains of our state. Plains Telephone provides voice and data services to 1350 access lines in 7 exchanges covering 2,000 sq. miles, with a density of .7 access lines per sq. mile. Despite the vast expanse we have 100% availability of broadband services via our telephone plant with a penetration rate of 32%. We have been referred to as "the poster child for rural high cost companies"

As a 45 year traditional borrower of RUS funds we have an inherent understanding of the loan process and requirements associated with doing business with the RUS. Programs like the Broadband Loan and Grant are key components of most company's business plan if it intends to serve rural America. Traditional capital sources do not understand the costs associated with serving these sparely populated areas and they do not like the slim margins associated with this product line, furthermore most large companies simply don't play where we chose to live. That leaves the deployment of advanced service up to companies that understand this type of business model.

My prior position before coming General Manager at Plains Cooperative Telephone was with Sunflower Telephone Company (A FairPoint Communications Company.) While serving as Manager of Sunflower Telephone I actively pursued and was awarded a Community Connect Grant. This grant allowed Sunflower to bring broadband services to 3 exchanges that had a total of 307 access lines in a 900 sq. mile service area. Today 20% of the customers of Towner, Sheridan Lake and Hartman Colorado have subscribed to broadband services and are using the service as an everyday tool in their business. During the scoring process of the grant we received a 70% out of a possible 70% in the demographic portion of the application. (Density, mean family income and other quantifiable measurements) By all accounts the $300,000.00 capital investment needed to complete this project would not have even come close to a satisfactory rate of return on the investment for any company. Without this grant from the RUS these committees would still be on the wrong side of the digital divide.

While I set here and sing the praises of these programs they are not without systemic problems. In an effort to secure financing for Plains Telephone subsidiary's 700 MHZ broadband wireless project we did a review of Rural Broadband programs administered by the Rural Utilities Service. We found the process for a new small start up company administratively cumbersome as evident by the 38 page application guide and the 53 page actual loan application. We did manage to fund the project internally and ended up delaying the rollout.

In light of recent and past criticism of the Broadband Loan and Grant Program a few inherent questions should be ask.

1. Does the staff at RUS have a clear understanding of the legislative intent?
2. A simple redefinition of what is broadband may be considered? (currently the FCC and RUS defines broadband as 200 kbps)
3. Quantify served vs. underserved areas? (How many RUS funded competitors in a rural market constitute fully served?)
4. In today world should broadband services be considered a basic service?
5. Is current staff and technology at the RUS adequate to fulfill the wishes of congress?
6. Can the process be streamlined?

I have included excerpts in my testimony from a September 2005 Inspector General Audit of the RUS Broadband programs:

> "Congress's current language was meant to serve only as a broad definition for bringing broadband service to rural citizens, and that the clear purpose behind the program was to fund service to the truly rural. We do not believe Congress foresaw that this broad definition might be used to justify funding loans to affluent suburban communities while other more rural communities remained underserved."

The audit went on to state:

> "Based on this review, we found that RUS has not maintained its focus on rural communities without preexisting service. Although the language of the law specifies that these Federal loans and grants are for rural communities, RUS has codified and implemented a definition that cannot reliably distinguish between rural and suburban areas. Due to this ambiguous definition, the agency has issued over $103.4 million in loans to 64 communities near large cities, including $45.6 million in loans to 19 planned subdivisions near Houston, Texas."

While these instances were splashed all over the press and caused embarrassment to the RUS and the USDA, I still believe the program is viable and with a some modification can be accessed by the parties it was intended for as well as fulfill the legislative directive. The establishment of a Rural Broadband Initiative could provide direction for all parties involved in bringing parity to these type services.

In closing I would ask congress and this committee to continue to support the long standing national social policy of quality, affordable telecommunications services for all of American no matter the remoteness or cost of supplying these services.

I would be happy to answer any questions the committee has.

Thank you.

In: Broadband and Rural America
Editor: Bethany J. Lenz

ISBN: 978-1-61668-637-6

Chapter 7

WRITTEN STATEMENT OF JACK M. GELLER, PHD., PRESIDENT, CENTER FOR RURAL POLICY & DEVELOPMENT, SAINT PETER, MINNESOTA, BEFORE THE U.S. HOUSE OF REPRESENTATIVES COMMITTEE ON AGRICULTURE

Jack M. Geller*

Chairman McIntyre, members of the subcommittee, it is indeed an honor to have the opportunity to provide input as you and your colleagues take up the important task of crafting a new farm bill. My name is Jack M. Geller, president of the Center for Rural Policy & Development in Saint Peter, Minnesota. The Center is Minnesota's only nonprofit, nonpartisan rural policy research center. As such, we are dedicated to researching, understanding and advising Minnesota's policymakers on the forces that impact rural Minnesotans and the communities they reside in.

My comments this afternoon will be limited in scope and will focus on the importance of having accessible and affordable high-speed telecommunications infrastructure throughout rural America. In that discussion I will touch upon the adoption and deployment of such infrastructure throughout rural Minnesota and what we have learned about the value of having good objective data on the utilization and consequences of broadband technology to Minnesota policymakers, industry executives and rural organizations.

Since the turn of the 21st century, added emphasis has been placed at both state and national levels on the status of the telecommunications infrastructure throughout rural America. As our nation's economy continues to rapidly transform into a technology-driven, information-based economy, the ability to access advanced telecommunications services has become vital. As FCC Commissioner Michael Copps noted last month after learning of a new OECD report that dropped the United States in its ranking from 12th to 15th among the OECD

* Corresponding author: 600 South Fifth Street, Suite 211, Saint Peter, Minn. 56082 – (507) 934-7700, Email: jgeller@ruralmn.org, On the web: www.ruralmn.org

nations in broadband penetration, "These rankings are not a beauty contest. They're about our competitiveness as a country and creating economic opportunity or all of our people." Accordingly, it has become clear that without the advanced telecommunications tools and resources to adequately compete in the national and global marketplace, businesses and industries throughout rural America will continue to be at a serious disadvantage.

But as you know, business applications and economic competitiveness are not the only factors behind the need for these high-speed connections. Unlike the static Internet applications of yesterday, today's more modern and sophisticated applications are designed with the assumption that the end user has a high-speed or broadband connection. Whether it's for social networking, personal entertainment, research, or the delivery of public services, modern Internet applications utilize the transfer of extremely large audio, video, data and graphics files that at best will frustrate the dial-up user and at worst, completely overwhelm a dial-up connection and render it useless.

Further, please understand that the need for broadband is not limited to popular video sites as YouTube or audio sites such as iTunes. One area often left out of such discussions are the efforts that state, local and federal governments expend to find efficiencies in the delivery and processing of public services on the Internet. Today we know that a significant percentage of Americans file their incomes taxes online, but we often do not consider that the majority of businesses routinely pay quarterly sales taxes, unemployment taxes and other business taxes on line. In fact, the Internet is quickly becoming the primary method by which businesses interact with their government. Further, today in Minnesota you can purchase your car license tabs, local building and other permits, pay your municipal utility bills and even file for unemployment benefits online. Undoubtedly, our federal, state and local governments will continue to aggressively accelerate their efforts to seek efficiencies by pushing more and more public services onto the Internet. Further, now that we officially have electronic voting machines in many states, don't be surprised we start hearing about pilot tests of actual online voting before the expiration of this next farm bill. Imagine -- the ability to participate in our democratic process with an Internet connection.

And it is for these very reasons that federal agencies such as the USDA Rural Utilities Service have put funding programs in place to assist rural America in remaining viable, economically competitive – and connected.

As a result of the Internet's rapid and functional integration into our daily lives, the Center for Rural Policy and Development decided back in the year 2000 to begin to annually monitor the adoption, diffusion and utilization of these digital technologies throughout rural Minnesota. Working with both government and telecommunications industry groups, we have amassed over the years a valuable longitudinal database which allows both policy makers and industry groups to better understand how rural Minnesotans are engaging these newer digital tools.

Accordingly, as once can see from Figure 1, while there has been a modest increase in home computer ownership and slightly higher gains in Internet connectivity throughout rural Minnesota households since 2001, there has been an unmistakable linear increase in broadband penetration, as rural Minnesotans switch from their dial-up connections to broadband connections. In fact, at the end of 2006 we reported that 39.7% of rural Minnesota households subscribed to a broadband service. This is just a few a percentage points below the national subscription rate of 42% as reported by the PEW Internet and American Life Project.

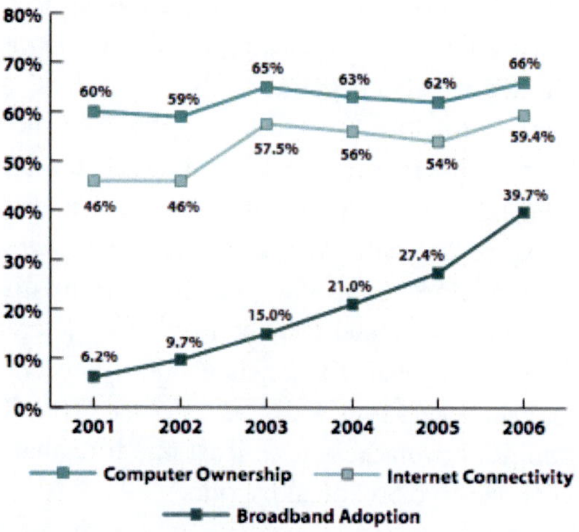

Figure 1. Computer ownership, internet connectivity and broadband adoption in rural Minnesota 2001-2006

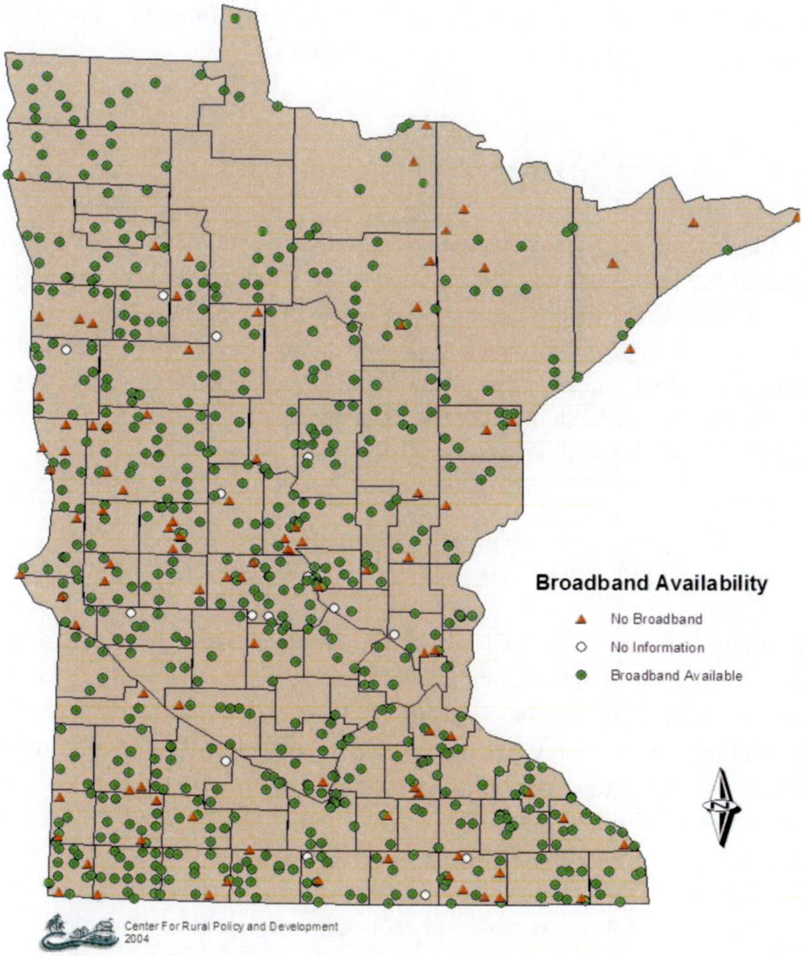

Figure 2. Broadband availability in Minnesota

Of course, the adoption of high-speed telecommunications services is simply not possible if such broadband services are not accessible and available throughout rural America. And it was for that simple reason that we also began to work with Minnesota's broadband providers to monitor the deployment of broadband services as well. The outcomes of these monitoring efforts have been valuable to policymakers, broadband advocates and industry officials, as we no longer need to make generalizations based upon case-specific or anecdotal events.

As one can see from Figure 2, access to broadband services, while not ubiquitous, is quite widespread throughout rural Minnesota; as the green dots on the map designate communities that have access to at least one broadband provider, red triangles represent communities that do not have a broadband provider and a few white dots represent rural markets where we were simply unable to confirm access to broadband services. Overall, more than 85 percent of Minnesota's rural communities have access to at least one broadband provider and more than one-third have access to competitive broadband services.

Essentially, today in rural Minnesota if you live within the municipal boundaries of our more than 800 cities and towns, there is a very high probability that you will have access to at least one broadband provider. However, we cannot sound so optimistic if you reside outside these municipal boundaries. Across Minnesota's countryside, accessibility and availability of broadband services is considerably less consistent. And in Minnesota, more than 900,000 residents currently live outside these municipal boundaries.

USDA as the Federal Lead

Among the federal agencies designated to assist rural communities with their advanced telecommunications infrastructure, the USDA Rural Utilities Service is clearly the most comprehensive. With more than $1 billion appropriated, the Rural Utilities Service enables communities through its Rural Broadband Access Loan Program, its Community Connect grant program and its Distance Learning and Telemedicine grant program. Clearly, without the Rural Utilities Service funding, access to advanced telecommunications technologies throughout rural America would be even further behind.

The Need of Better Information

But the reality is that in spite of all this added emphasis on advanced telecommunications services, there is embarrassingly poor information and data on the adoption, diffusion, deployment and utilization of these advanced technologies throughout rural America. Broadband adoption rates and utilization patterns are poorly assessed throughout rural America. This is especially true at the state-level. Agencies such as the National Telecommunications Information Administration occasionally conduct such technology assessments, but they do not differentiate rural areas from urban areas. And even the nationally-recognized PEW Foundation's Internet and American Life project only occasionally conducts rural assessments. In fact today, the federal government still has no credible list simply identifying which communities do, or do not have access to broadband technology throughout rural America.

There is little doubt that if Rural America is going to successfully compete in this emerging information economy and create vibrant and connected communities, programs such as the USDA/RUS Telecommunications Programs will be essential. Yet, at the same time, simply funding rural infrastructure projects without fully understanding the need, the impact and the consequences will inevitably lead to untargeted and sometimes unnecessary spending. Simply put, when Congress targets federal funds for rural telecommunications infrastructure, it is generally done with some assumptions about the value of such public investments. These assumptions include:

- **That Rural Businesses** will be more economically competitive with access to advanced telecommunications technology.

- **That Rural Communities** will be better able to recruit new and retain existing businesses with access to advanced telecommunications technology.

- **That Rural Residents** will be more connected; enhance quality-of-life opportunities; be able to telecommute; start-up home-based businesses; and more easily access online educational opportunities with these technologies.

- **And That Rural Governments** will be better able to provide access to public services and serve their constituencies more effectively by enhancing their online services.

Accurate data and information on the availability and consequence of this significant public investment is essential. Such information will help target future investments, establish "best practices" in broadband adoption strategies, and provide the type of information that policymakers at both the state and national levels need to make sound telecommunications policy decisions.

So while the Rural Utilities Service may have accurate information about the number of borrowers and grantees, or loan repayment schedules and loan default rates, more substantive questions go unanswered:

- Are rural communities better able to retain and recruit businesses as a result of these telecommunications investments?
- Are such investments better enabling rural entrepreneurs to start-up new businesses or increase their market share?
- Is the rural regional economy more competitive as a result of USDA's public investments in telecommunications infrastructure?
- Are rural residents enhancing their online educational opportunities as a result of these public investments?
- Are rural governments better able to offer citizens enhanced choices in the delivery of public services as a result of these public investments?

The answers to questions such as these that address the substantive consequences of public investments in the rural telecommunications infrastructure would be invaluable to policymakers and agency officials in help target future investments.

Recently the Federal Communications Commission validated the importance of such information when they announced their effort to determine whether high-speed Internet access is being made available fast enough, calling it "critical" to the nation's economy. Specifically, the FCC emphasized the need to assess the availability of broadband service in rural and other under-served areas, along with the need to improve data collection methodologies on wireless broadband services.

Equally, if not most importantly, agencies such as the Rural Utilities Service need such data to provide objective feedback on the impact of their programs, to make the necessary adjustments and modifications that all program managers must make from time to time. And USDA Rural Development state directors also need to better understand the consequence of these telecommunications investments as they coordinate them with their portfolio of housing, infrastructure and business development programs to further advance their important mission throughout rural America.

In closing, allow me to offer the observation that if the USDA Rural Utilities Service were a technology company making such sizeable technology investments, it would be unheard of that it were not spending a sizeable percentage of its revenues on research and development. The need to understand the consequences and outcomes of these public investments is evident. Allow me to suggest that in the case of the USDA Rural Utilities Service, a simple 2-5 percent set-aside of program and loan funds for such research & development would be one the best investment strategies that Congress could make.

Chairman McIntyre, members of the subcommittee, thank you again for this opportunity to address you on this important matter.

Committee on Agriculture U.S. House of Representatives Information Required from Non-governmental Witnesses

Jack M. Geller Ph.D. is the founding President of the Center for Rural Policy and Development in St. Peter, Minnesota. The Center is Minnesota's only nonprofit, non-partisan, rural policy research center; and as such, is dedicated to the study of the economic, social and political forces that impact rural Minnesotans and the communities they reside in.

Dr. Geller received his doctorate in rural sociology from Iowa State University in 1986. Since 2001 he has annually conducted statewide studies and has written extensively on the adoption, diffusion and deployment of high-speed Internet (or broadband) services throughout rural Minnesota.

Prior to his appoint at the Center, Dr. Geller served as Director of Health Services Research at the Marshfield Medical Research Foundation (Marshfield, WI), where he assisted in the development of a system-wide telemedicine project that is still in use today.

From 1987-1996 Dr. Geller served as Professor of Community Medicine and Rural Health; and from 1991-1996 he served as Director of the Center for Rural Health at the University of North Dakota School of Medicine & Health Sciences. During his tenure at the University of North Dakota, Dr. Geller helped create and was a governing board member of the "Rural Health Telecommunication Project"; a five-year $4 million federally-funded demonstration project designed as a statewide interactive video system, linking eleven state-supported institutions of higher education throughout the state of North Dakota.

Jack, his wife Diane and their four children reside in rural Mankato, Minn. Township.

Committee on Agriculture
U.S. House of Representatives
Required Witness Disclosure Form

House Rules* require nongovernmental witnesses to disclose the amount and source of Federal grants received since October 1, 2004.

Name: Jack Mitchell Geller

Address: 129 Ironwood Court, Mankato, MN 56001

Telephone: 507-934-7700

Organization you represent (if any): Center for Rural Policy & Development

1. **Please list any federal grants or contracts (including subgrants and subcontracts) <u>you</u> have received since October 1, 2004, as well as the source and the amount of each grant or contract. House Rules do <u>NOT</u> require disclosure of federal payments to individuals, such as Social Security or Medicare benefits, farm program payments, or assistance to agricultural producers:**

Source: None **Amount:** 0

Source: ______ **Amount:** ______

2. **If you are appearing on behalf of an organization, please list any federal grants or contracts (including subgrants and subcontracts) <u>the organization</u> has received since October 1, 2004, as well as the source and the amount of each grant or contract:**

Source: None **Amount:** 0

Source: ______ **Amount:** ______

Please check here if this form is NOT applicable to you: ______

Signature: ______

* Rule XI, clause 2(g)(4) of the U.S. House of Representatives provides: *Each committee shall, to the greatest extent practicable, require witnesses who appear before it to submit in advance written statements of proposed testimony and to limit their initial presentations to the committee to brief summaries thereof. In the case of a witness appearing in a nongovernmental capacity, a written statement of proposed testimony shall include a curriculum vitae and a disclosure of the amount and source (by agency and program) of each Federal grant (or subgrant thereof) or contract (or subcontract thereof) received during the current fiscal year or either of the two previous fiscal years by the witness or by any entity represented by the witness.*

In: Broadband and Rural America
Editor: Bethany J. Lenz

ISBN: 978-1-61668-637-6

Chapter 8

TESTIMONY OF MR. DENNY LAW, EASTERN REGIONAL MANAGER, GOLDEN WEST TELECOMMUNICATIONS, WALL, SOUTH DAKOTA; BEFORE THE HOUSE COMMITTEE ON SPECIALTY CROPS, RURAL DEVELOPMENT AND FOREIGN AGRICULTURE

Denny Law

Chairman McIntyre, Ranking member Musgrave, members of the subcommittee, I would like to that you for the opportunity to be here today. My name is Denny Law and I am here today on behalf of the National Telecommunications Cooperative Association and Golden West Telecommunications Cooperative in Wall, South Dakota where I serve as Eastern Regional Manager. Golden West Telephone Company was incorporated in 1916 as an effort to provide telephone service between the towns of Interior and Quinn, SD. During the Great Depression, Golden West suffered setbacks and the assets were sold by the county sheriff to pay taxes. After President Truman signed the telephone amendments to the Rural Electrification Act in 1949, residents of the community in Quinn met to form Golden West Telephone Cooperative and soon applied for a loan from the REA. Beginning with telephone line strung along fence posts to farms and ranches, Golden West Telecommunications and its subsidiaries now provide service to over 43,000 telephone customers, 15,000 internet subscribers, and 10,000 cable television customers. Golden West serves customers across 63 telephone exchanges and 24,500 square miles — an area larger than the states of New Hampshire, Massachusetts and New Jersey combined.

While Golden West serves a large geographical area, our resident population is very small. The largest community Golden West serves is Hot Springs, South Dakota with a population of approximately 4,000 people. At the other end of the spectrum, Golden West provides telecommunications services to the small community of Hayes. The entire Hayes telephone exchange encompasses 1,119 square miles, but has only 192 access lines, which equates to 0.17 access lines per square mile. Residents in the Hot Springs exchange as well as the Hayes exchange have access to broadband services due, in part, to the RUS loan programs that have provided affordable financing to construct these telecommunications facilities.

Golden West also provides telecommunications service on four Native American tribal reservations in South Dakota, including the Pine Ridge Indian Reservation. Bringing service to parts of the Pine Ridge Indian Reservation in 1973 when no other company would provide service, Golden West formally purchased the entire Pine Ridge telephone exchange in 1981. Of the almost 1,700 access lines comprising the Pine Ridge exchange, close to half receive service through the FCC's Lifeline or Link-up programs which provides discounted telephone service to customers on public assistance. According to information compiled in the 2000 Census, landline telephone penetration on the Pine Ridge Reservation was measured at 75.2 %. This figure has also been confirmed by the FCC's Indian Telecommunications Initiative Study on Telephone Subscribership on American Indian Reservations and Off-Reservation Trust Lands and a January, 2006 GAO Report on the challenges of telecommunications for Native Americans on Tribal Lands.

Throughout Golden West's history, we have been borrowers through the REA and since USDA's most recent reorganization, the Rural Utilities Service. Due to RUS funding, many communities served by independent telephone cooperatives throughout the United States have significantly higher broadband deployment than neighboring communities served by regional Bell operating companies. While the mantra from Congress is there is a "digital divide", that is far from the case in many areas of rural America. What has been overlooked is that since 1995 RUS telecommunications infrastructure loans have been required to be in harmony with state modernization plans and be broadband capable. RUS telecommunications lending has stimulated billions of dollars in private capital investment in rural communications infrastructure.

In recent years, on average, less than a few million in federal subsidy has effectively generated $690 million in federal loans and guarantees. For every $1 in federal funds invested in rural communications infrastructure, $4.50 in private funding has been invested. By the early part of this decade, RUS telecommunications borrowers had used more than $12 billion in loans to build a network of more than one million miles of infrastructure to provide advanced communications services across rural America. NTCA members' tremendous record of broadband deployment in some of the most rural and remote areas of rural America is largely due to the RUS telecommunications loan program. NTCA's 2006 annual member survey shows more than 96% of it's members are offering broadband (200 Kbps) to some portion of their customer base and 88% of it's members are providing broadband at speeds of more than 1 Megabyte.

Since the initiation of the RUS broadband loan program five years ago, RUS has made over $1 billion in loans to 68 entities, including 6 to South Dakota telcos. RUS Broadband loans have been used by some rural independent companies to increase broadband capabilities within their traditional service territories.

Other independent telecommunications providers in South Dakota have used the broadband loan program to overbuild or compete in communities outside their traditional ILEC territories where broadband deployment may not be as advanced. However, there are also stories in many parts of the country where broadband loans have been approved where broadband already services existed through two or three competitors. In the case of a loan in Oregon, a loan was made in communities where broadband deployment was at 99% or greater. This loan was also made over the objections of the RUS Field Representative. While overbuilding and competition is allowed under the regulations, I do not believe this is what Congress has intended when the Farm Bill was signed into law.

The Rural Broadband Improvement Act (11R2035) was recently introduced by Reps. Stephanie Herseth Sandlin (D., S.D.) and Jerry Moran (R., Kan.). Golden West believes the Rural Broadband Improvement Act would provide better definitions between rural and urban as well as limit loan fund availability for applicants proposing to serve areas that already have a broadband provider. Golden West supports the principles outlined in HR2035. NTCA also agrees with the direction of this legislation and is committed to continuing to refine it to the degree necessary to ensure the programs integrity, effectiveness, and sustainability as well as ensure the areas that are most in need of such financing receive it.

As a telco provider, I am well aware of the challenges faced by a federal agency trying to determine where broadband currently exists. Under the FCC's Form 477, broadband deployment is reported only by a zip code and does not paint an accurate picture of what is available in rural America. I believe that RUS should be able to better determine the level of broadband deployed based on their own lending portfolio inventory. As Golden West has just been approved for our "Y" loan--- representing our 25th telecommunications infrastructure loan through RUS, I am well aware of the high standards required for borrowing from the American taxpayers. I can testify that the level of information required by RUS for approval of an infrastructure loan under its longer standing loan programs is extensive. A similar level of information should be required under the broadband program. Whether an infrastructure loan or a broadband loan; the painstaking process of an RUS loan application is designed to protect the American taxpayers. This principle should he reinforced as Congress looks to reauthorize the program.

One possible change that others have addressed is reforming the current public notice process that is utilized under the broadband loan program. At the federal, state and municipal level, the "legal notice" is often accomplished through the publication of a newspaper notice or advertisement. Some broadband loan applicants have taken advantage of non-specific public notice requirements and merely placed ads in larger statewide papers and purposely avoiding local newspapers published in the communities that are proposed to be served. By doing this applicants are able to prevent comment by incumbents already providing broadband. NTCA has suggested to USDA and to Congress that this process should be reformed by establishing a system that requires notification by certified letter to all service providers in the affected area. While the RUS field representative is supposed to verify the level of broadband deployment in the area that is to be served, the awarding of some RUS loans indicates that this is not occurring. A thorough verification process needs to occur and would end the gaming of the system by all potential competitors.

Given Golden West's success in connecting farms, ranches, small communities and tribal areas throughout South Dakota illustrates well how vital RUS Loan programs ensure that funds address the needs of those residents living in rural America. In many communities in South Dakota, a number of rural independent telecommunications companies provide broadband in areas outside the city limits where a cable company or an RBOC has no interest in providing service. These rural residents deserve the same consideration as those living within the city limits. Given the tremendous costs associated with connecting such long loops, Congress and USDA should examine how to better provide funds to these "last mile" customers. What's good enough for residents living just miles away from the Houston skyline should be even better for residents living outside of Wall, South Dakota.

Golden West and other rural independent telcos have an extensive history of providing voice and advanced telecommunications services to some of the most remote, high cost areas of rural America. Since the 1950's, RUS infrastructure loans have provided a solid foundation for the deployment of these basic and advanced telecommunications services in rural areas throughout this county. The RUS Broadband loan program is a more recent application and there are many examples of loans being awarded to provide broadband to communities or regions where broadband did not exist, but it is obvious the program could he working better. I welcome the opportunity to answer any questions you might have.

In: Broadband and Rural America
Editor: Bethany J. Lenz

ISBN: 978-1-61668-637-6

Chapter 9

STATEMENT OF WALTER B. MCCORMICK, JR., PRESIDENT AND CEO, U.S. TELECOM ASSOCIATION TO THE HOUSE COMMITTEE ON AGRICULTURE, SUBCOMMITTEE ON SPECIALTY CROPS, RURAL DEVELOPMENT AND FOREIGN ASSISTANCE

Walter B. McCormick

Chairman McIntyre, Ranking Member Musgrave, members of the subcommittee: Thank you for this opportunity to appear before you today. I am Walter McCormick, president and CEO of the USTelecom Association. This committee has been at the forefront of helping advance the development of rural America, from bringing electricity and safe, running water to communities that never had it before, to connecting the country via the telephone and now, we all hope, via the high-speed Internet. USTelecom and its member companies are proud of the role we play connecting the country, and we wholeheartedly support the objective of ubiquitous, nationwide broadband. We were pleased to see "affordable broadband access for all Americans" as a component of Speaker Pelosi's Innovation Agenda. We believe this should be a *non*partisan objective of our nation's leaders, and we believe that the Rural Utilities Service (RUS) has a critical role to play in helping bring broadband to rural areas currently not served.

USTelecom represents innovative companies ranging from the smallest rural telecoms in the nation to some of the largest corporations in the U.S. economy. Our member companies offer a wide range of services across the communications landscape, including voice, video and data over local exchange, long distance, Internet and cable networks. USTelecom is the nation's oldest – and largest – association representing rural telecom providers. The vast majority of our member companies are rural providers. They are small businesses serving small communities. They are proud members of these communities and deeply committed to their future development. What unites our diverse membership is our shared determination to deliver innovative voice, video and data services to the consumer—a commitment we know is shared by this subcommittee. So we appreciate the opportunity to be here today.

REGULATORY CHANGES HAVE SPURRED BROADBAND DEPLOYMENT

The Federal Communication Commission's decisions that oriented the communications marketplace away from government-managed to market-based competition has resulted in an explosion of broadband coverage across the nation. In March 2002, the FCC clarified that highspeed cable-modem service is an information service not subject to unbundling and other Title II regulations of the Communications Act. In August 2003, the FCC exempted wireline fiber facilities from the Commission's unbundling requirements. In September 2005, the FCC clarified that wireline broadband Internet access service is also an information service not subject to unbundling and other Title II regulations of the Communications Act. These actions have accelerated broadband deployment in the United States from just over 4 million broadband lines in 2000 to just under 16 million broadband lines in 2002 to approximately 32 million lines in 2004 to almost 65 million lines in 2006. This demonstrates a direct correlation between the FCC's market-based policies and the explosion of broadband subscribers in the United States.

The lack of regulation on wireless services also has permitted wireless broadband services to explode as well. In June of 2005, there were almost 380,000 wireless broadband subscribers; in June of 2006, there were more than 11 million.

Internet access is available through DSL, or cable modem, or wireless, or satellite – and, increasingly, over power lines and municipal wi-fi systems. In fact, there are more than 1,270 broadband service providers in the U.S. today.

Against this competitive backdrop, North American telecommunications companies are projected to spend $70 billion on new infrastructure this year. The next wave of broadband innovation holds the promise of significant, life-enhancing advances from health care to the environment to education and to our economy. It is critical, as you know, that these opportunities be accessible in rural America, as well. Mr. Chairman, much has been made recently of new international broadband penetration rankings from the Organization of Economic Cooperation and Development (OECD). We have some issues of our own with our country's current ranking of 15th in the world. We feel it significantly undercounts, for example, connections in the U.S. business market. It certainly also under-values the markedly more intense facilities-based competition we have here in the U.S. But the most striking dissimilarity is that 10 of the 11 countries allegedly in front of us are significantly smaller than the U.S.—as diminutive as Norway, which is comparable in geographic size to New Mexico. A majority also have much smaller *populations*, including Iceland, an entire country that is comparable to the metro area of Naples, Florida. The exception is Canada, which is a country of vast geographic expanse. However, 80% of the population is clustered along the U.S. border. So the true broadband challenge before our country is precisely the challenge we are here today to discuss. How can we most efficiently work together to connect parts of the country where the marketplace alone is incapable of attracting the significant investment necessary to truly build a broadband nation?

The RUS Broadband Program -- Modest Changes Could Produce Dramatic Results

Our member companies want to work with the subcommittee to completely close the gap in broadband coverage. In its relatively brief history, the RUS broadband loan program has achieved some successes. But we believe with modest changes, largely based on the successful RUS telephone program, the program could accomplish even more.

As the subcommittee begins to write the rural development title of the Farm Bill, USTelecom would make the following recommendations to advance our collective goal of helping the nation achieve universal broadband penetration:

(1) Better target areas currently not served;
(2) Enhance incentives for investment in the areas not served;
(3) Expand program eligibility;
(4) Improve processing at USDA, and
(5) Explore public-private partnerships.

Revise the Eligibility Rules to Better Target Areas Not Served

We believe the primary weakness of the current program is that it does too little for areas with no access to broadband. Although the nation is dotted with areas currently not served, the USDA Inspector General concluded the program's focus has shifted away from rural communities that would not, without government assistance, have access to broadband technology.

In revising eligibility rules, we believe the subcommittee need look no further than the RUS telephone program. This program has a 60-year record of success, and we believe it holds important lessons for broadband. In the telephone program, initial loans to areas with adequate, existing service are discouraged. In fact, the RUS administrator must issue a non-duplication finding prior to making a loan. In the broadband program, such a requirement would help direct funds to where they are most needed – those areas with no existing broadband service. Making loans for duplicative facilities and service, when other citizens in rural America reside in areas with no service at all, is a waste of scarce government resources. In addition, the telephony program requires that service be extended to the widest practical number of users in the service area, avoiding a problem that has sometimes arisen in the broadband program, where service is only provided within town limits, but not to the surrounding county.

Enhance Incentives for Investment in Areas Not Served

Providing broadband service in rural and remote areas is a challenging proposition. While the current practice of offering cost-of-money loans makes projects financially viable in some areas, other higher cost areas will require below-cost loans or a combination of loans and grants to make a costly infrastructure build feasible. This will become increasingly important as the program narrows to focus on areas with truly no access. Congress should encourage RUS to look at the unique needs of these areas and to enhance incentives for the private sector to act. Taxpayers will reap the benefits through loan repayments and tax revenues generated by broadband-driven economic development. We believe that taking these basic steps would increase the number of loan applications to areas with no service facing significant economic barriers to investment, such as low population densities or difficult terrain.

Expand Eligibility to More Applicants

We also believe steps should be taken to expand the number of companies eligible for broadband loans. When the broadband program was established, a provision was adopted prohibiting loans to telephone companies with more than 2% of the nation's access lines. This is counterproductive. Some USTelecom members serve rural areas that would otherwise qualify for broadband loans. For example, the FCC classifies Embarq as a rural carrier in 17 of the 18 states it serves, yet it is prohibited from applying for RUS broadband funds. Meanwhile, RUS is searching for more applications from carriers seeking to serve untouched areas. Again, if I might refer you to the successful, 60-year-old telephony program – it has never had a 2% restriction, and it has never suffered as a result. The emphasis in our view should be on the infrastructure needs of a community, not on the company willing to serve it.

Improve Processing at USDA

USTelecom also advocates that steps be taken to improve processing of loan applications at USDA. At present, the broadband and telephony programs have access to a small number of attorneys in the Agriculture Department's general counsel office. This has created a bottleneck when legal decisions are needed and caused delays in processing loan applications—delays that too often put broadband deployment on hold in communities with no service.

Explore Public-Private Partnerships

Finally, I point the subcommittee's attention to the successful public-private partnership in Kentucky, driven by a non-profit organization called Connect Kentucky. Connect Kentucky has worked with the RUS broadband program, but has gone much farther than would have been possible with RUS alone. Its first objective was to map broadband availability in the whole state, something that no other state has done. Then it created technology teams in each community that lacked broadband. These teams looked at computer ownership, technological literacy, and other factors to increase demand for broadband. At the same time, the teams worked with broadband providers to match up new demand with new broadband deployments. By the end of 2007, Kentucky will go from having one of the lowest broadband subscription rates in the country to having broadband available to 100% of its households. That's impressive progress, and we think Congress might look to Connect Kentucky as a model for what works. In fact, we understand that Senator Durbin has recently introduced legislation that would create a national program based on the Connect Kentucky model.

Mr. Chairman, in closing, let me reiterate that it is critically important that rural areas be included in the nationwide drive for greater bandwidth capacity. This modernization of the nation's communications infrastructure will seed economic growth and expand opportunities ranging from telecommuting to distance learning to telemedicine. Mr. Chairman, nowhere in the nation do these advances hold more potential than in rural America.

After 60 years of success, the RUS loan programs remain an essential public-private partnership conceived with the best of intentions—spreading opportunity throughout the country—helping the private sector overcome the often significant economic barriers associated with our nation's vast geography. The results have been impressive: RUS generates more revenue than it costs. It provides incentives where the market does not so private companies can invest in infrastructure that promotes rural economic development. And, it expands our citizens' access to services that can vastly enhance their quality of life and the economic opportunities available to them in their own communities. I'd also like to add that RUS has never lost a dime of taxpayer money because of a telecom carrier default.

We thank you for your invitation to appear today. USTelecom and its member companies look forward to working with the subcommittee and this Congress to achieve our shared objective of making broadband as ubiquitous today as electricity, water and telephone service. Broadband is an essential building block of every modern American community. We look forward to working with you to make its many opportunities accessible to all Americans. Thank you.

WALTER B. MCCORMICK

President & CEO

Walter B. McCormick Jr. is the President & CEO of the United States Telecom Association (USTelecom), the nation's premier trade association representing service providers and suppliers in the new telecommunications marketplace.

A respected Washington veteran with more than 20 years experience in telecommunications, McCormick joined USTelecom in 2001 and has guided the organization's evolution into one of the top trade associations in the nation's capital. McCormick united a broad and diverse industry to pursue two critical advocacy goals: (1) securing market-based competition that allows consumers, rather than the government, to determine which communications services and technologies succeed; and (2) preserving the future of universal service, which ensures all Americans have affordable, reliable basic phone service. USTelecom has mounted a compelling case for reform, as evidenced by the fact that Congress is now preparing for a historic update to the nation's telecom laws--an effort few in Washington believed was imminent just one year ago.

McCormick first established himself as one of the nation's leading trade association executives while serving at the helm of the American Trucking Associations. There he led a broad corporate restructuring which merged competing membership organizations and refocused the enterprise on its core missions of lobbying, litigation and public relations. The results were extraordinary: record dues revenues, a reinvigorated membership and significant legislative victories. *Washingtonian* magazine has profiled McCormick as a trade association executive with 'real clout.' The Capitol Hill newspaper *The Hill* has consistently named McCormick as a 'rainmaker' and one of 12 'top trade association lobbyists' in Washington.

Prior to his trade association work, McCormick headed the practice group on Regulatory Affairs, Public Policy, and Legislation at Bryan Cave LLP, an international law firm of more than 500 lawyers. Before practicing law, he had a distinguished career in government. McCormick was nominated by President George Bush, and confirmed by the U.S. Senate, as General Counsel of the U.S. Department of Transportation, where he served under then-Secretary Andrew Card. He also served for more than a decade on the U.S. Senate staff. McCormick was General Counsel of the U.S. Senate Committee on Commerce Science and Transportation in the 99th Congress, and was Republican Chief Counsel and Staff Director of the Committee in the 100th, 101st, and 102nd Congresses. During his tenure, he was responsible for drafting many telecom and transportation laws, and was identified by *Roll Call* magazine as one of the 50 most influential staffers on Capitol Hill.

McCormick holds degrees in journalism and law from the University of Missouri. He has studied international economics and political science at Georgetown University, and has completed the program for senior managers in government at Harvard University's John F. Kennedy School of Government. He is a member of the Board of Directors of Rockhurst University and a member of the District of Columbia Bar, the Missouri Bar and the Federal Communications Bar Association.

Committee on Agriculture
U.S. House of Representatives
Required Witness Disclosure Form

House Rules* require nongovernmental witnesses to disclose the amount and source of Federal grants received since October 1, 2004.

Name: Walter B. McCormick, Sr.

Address: 607 14th St. NW, Suite 400

Telephone: 202-326-7300

Organization you represent (if any): UNITED STATES TELECOM ASSOCIATION

1. **Please list any federal grants or contracts (including subgrants and subcontracts) you have received since October 1, 2004, as well as the source and the amount of each grant or contract. House Rules do NOT require disclosure of federal payments to individuals, such as Social Security or Medicare benefits, farm program payments, or assistance to agricultural producers:**

Source: N/A **Amount:** ______

Source: ______ **Amount:** ______

2. **If you are appearing on behalf of an organization, please list any federal grants or contracts (including subgrants and subcontracts) the organization has received since October 1, 2004, as well as the source and the amount of each grant or contract:**

Source: U.S. Dept. of Energy **Amount:** $150,000

Source: ______ **Amount:** ______

Please check here if this form is NOT applicable to you: ______

Signature: [signature]

* Rule XI, clause 2(g)(4) of the U.S. House of Representatives provides: *Each committee shall, to the greatest extent practicable, require witnesses who appear before it to submit in advance written statements of proposed testimony and to limit their initial presentations to the committee to brief summaries thereof. In the case of a witness appearing in a nongovernmental capacity, a written statement of proposed testimony shall include a curriculum vitae and a disclosure of the amount and source (by agency and program) of each Federal grant (or subgrant thereof) or contract (or subcontract thereof) received during the current fiscal year or either of the two previous fiscal years by the witness or by any entity represented by the witness.*

PLEASE ATTACH DISCLOSURE FORM TO EACH COPY OF TESTIMONY.

In: Broadband and Rural America
Editor: Bethany J. Lenz

ISBN: 978-1-61668-637-6

Chapter 10

TESTIMONY OF TOM SIMMONS, SENIOR VICE PRESIDENT OF PUBLIC POLICY, MIDCONTINENT COMMUNICATIONS, BEFORE THE HOUSE AGRICULTURE SUBCOMMITTEE ON SPECIALTY CROPS, RURAL DEVELOPMENT AND FOREIGN AGRICULTURE

Tom Simmons

Chairman McIntyre, Ranking Member Musgrave, and Members of the Committee, thank you for inviting me to testify today. My name is Tom Simmons and I am the Senior Vice President of Public Policy for Midcontinent Communications, a leading provider of cable telecommunications services in rural America, including analog and digital cable television, broadband Internet, and local and long distance telephone services. We serve over 200,000 customers in approximately 200 communities in North and South Dakota, Western Minnesota, and Northern Nebraska, generally classified as small or rural. The size of our communities ranges from densities of 5 to 116 homes per mile of cable plant and populations range from less than 30 in Barlow, North Dakota to our largest community, Sioux Falls, South Dakota, which has a population of more than 150,000.

Midcontinent launched its broadband Internet service nearly ten years ago, on April 15, 1996 in Aberdeen, South Dakota, and made a pledge then to bring advanced broadband services to as many customers as possible regardless of the size of the community. At the end of 2005, we completed a project to rebuild our cable plant to 750 MHz or better in 50 more Midcontinent communities, bringing our total of upgraded systems to 152, serving over 95% of Midcontinent's customers.

Customers in these communities now enjoy over 150 channels of analog and digital video programming, broadband Internet service, high definition television, digital video recording capability, and in many communities, video on demand. Midcontinent is also a certificated local exchange telephone service provider in North Dakota, South Dakota, and Minnesota.

Midcontinent first launched facilities-based circuit switched telephony in 2000, and is converting previously resold phone service to its own facilities-based digital phone product. By the end of this year, most of its phone customers will receive service via Midcontinent's facilities-based network.

All of this has required Midcontinent, a privately held company, to invest well over \$100,000,000 in private risk capital to bring advanced services to our customers in rural America without the assistance of public funds, and we hope to continue doing so. We're proud of our ability to deliver the services our customers in relatively small towns demand, which are no less than those desired and expected in suburban and major metropolitan areas.

And Midcontinent is not alone. The cable industry has invested over \$110 billion during the last ten years to become the largest provider of broadband service in America. According to Kagan Research, cable Internet service is now available to 94 percent of *all* U.S. households including hundreds of small towns and communities in rural areas all across the country. Nearly 30 million American households subscribe to cable's broadband service and according to a report just released by the OECD, 58 million American households subscribe to a high-speed Internet service. As broadband availability and penetration continue to increase, so do download speeds. While the cable industry began by offering 1-1.5 Mbps ten years ago, most cable operators today offer speeds above 5 Mbps, and while some are already offering speeds of up to 50 Mbps, a new technology will soon allow speeds above 100 Mbps. As broadband availability and speed has increased, the price-per-megabit of service has dropped due to a highly competitive marketplace in which the telephone companies' Digital Subscriber Line (DSL) service is available to 70 percent of all US households that can access ILEC telephone service. At the same time, the number of "alternate" broadband technology subscribers (such as WiMax and broadband-over-power line subscribers) also continues to increase.

However, despite the widespread availability of broadband service, there are still remote, sparsely populated areas of the country that are not served. Recognizing this, the cable industry supports legislative initiatives and government programs designed to promote broadband deployment in these unserved, rural areas. These include:

- Tax credits or other tax incentives to providers that build out in rural areas that are unserved by an existing broadband provider.
- Any use of USF money to support broadband deployment should be targeted to unserved areas.
- Expansion of the FCC's Lifeline and Link-Up Programs to help ensure that broadband access is extended to low-income households.
- Public-private partnerships to provide broadband in unserved areas.
- Passage of H.R. 743 and S. 156 that would make permanent the current moratorium on Internet access taxes and unfair taxes on electronic commerce.
- Reform of the Rural Utilities Service Broadband Loan Program so that funding is targeted specifically to unserved areas.

The Rural Utilities Service Broadband Loan Program is a prime example of a program that was intended to make it economically feasible for the private sector to serve rural communities that lack broadband access. Though we support the goals of the RUS Broadband

Loan Program, we are very concerned about how this program has been implemented and managed. Our concerns focus on two areas:

> First, that RUS loans are largely being used to subsidize broadband deployment in areas *already served* by companies that deployed broadband service without a government subsidy, using private risk capital.
>
> Second, that RUS rules make it difficult for *anyone*–existing providers, the public, and even RUS staff to assess the accuracy of the claims made by an applicant regarding existing broadband service in the area it proposes to serve. As such, the RUS is often unable to determine whether the grant of a loan will undermine private sector competition. More importantly, because the process lacks transparency, taxpayer funds are being misspent on projects that are not extending broadband service to unserved rural communities.

In September 2005, the U.S. Department of Agriculture Inspector General's Audit Report on the RUS' Broadband Grant and Loan Programs (OIG Report) found that this program has "not maintained its focus on rural communities without preexisting service" (OIG Report at ii). Instead, it is largely being used to subsidize competition in areas where one, and in many cases, multiple providers of broadband service, exist. To this extent, the private entrepreneurs' reward for being the first risk takers in rural America is to face a government-subsidized competitor.

The RUS itself recognizes the difficulties presented by subsidized competition in rural America. Its own regulations *prohibit* the granting of a loan in a market where an RUS borrower already exists. The RUS doesn't want to put its borrowers at risk by subsidizing competition. That same principal should apply with regard to private sector competition.

Providing broadband service in high cost rural areas is economically risky at best. Midcontinent and other cable operators in rural communities all across America have taken, and continue to take, that risk. However, that risk could become unbearable if we are faced with a competitor subsidized by the government. Subsidizing a company to overbuild an existing provider could have the perverse effect of making it increasingly difficult, if not impossible, for a company that entered the market first using private risk capital to continue to provide quality service in that market. Additionally, the threat of a government subsidized competitor in rural markets also creates a disincentive for a company that does not receive federal support to extend service to otherwise unserved rural communities. Why should we build out to a rural area if we know our reward will be to face a government subsidized competitor? Finally, and most importantly, subsidizing competition is a waste of the finite loans funds available that should instead be targeted to unserved areas where a market-based solution has not developed.

There are numerous examples of loans being granted in areas already served by one or more providers. The case closest to Midcontinent involves Mitchell, South Dakota, a small city of a little over 14,500 residents. When the RUS granted a loan in excess of $20 million to Sancom, Inc. to overbuild us in Mitchell, we were already competing with Qwest for telephone customers, as well as with two DBS companies for each and every video customer. Midcontinent's investment in private risk capital to upgrade our system in Mitchell allowed us to offer our customers a variety of advanced services, including high speed Internet access at speeds of 3 mbps downstream and 256k upstream, which has been further upgraded to 8 mbps downstream. Midcontinent also provides high definition television service and telephony. Digital phone service is now also available. And Mitchell was not the only town overbuilt in South Dakota. As an RUS official reported at a 2004 South Dakota Public Utilities Commission Wireless Conference in Spearfish, South Dakota, the RUS had approved $37 million in loans to South Dakota companies by that time, but none of that money was targeted to provide broadband service to any of the more than 70 communities in that state that had no access to broadband service.

Many other companies have faced similar situations. In Fairfield, Iowa, the RUS granted a $9.475 million loan to an applicant to compete against two existing broadband providers. The cable operator in that case had invested millions of dollars to bring this town of approximately 9,500 people a state-of-the-art 860 MHz system, with capacity that exceeds or equals the capacity of major metropolitan areas. Services offered by this cable operator today include high speed Internet access at speeds of up to 15 mbps downstream (at the customer's option) and 512k or higher upstream, 194 video channels, high definition video service, and video on demand. Digital phone service will also soon be launched.

As the OIG Report makes clear, loans to fund competition have been granted time and again, and, as a result, "[the] RUS may be setting its own loans up to fail by encouraging competitive service; it may also be creating an uneven playing field for preexisting providers operating without Government assistance" (OIG Report at ii). To prevent such a scenario, the OIG recommended that before approving such loans, the RUS should conduct objective market research. Without such research, it believed the RUS could issue loans to companies with little chance of survival, which "would not appear to be a suitable use of Federal funds" (OIG Report at 16). The OIG Report illustrates that the RUS broadband loan program, as currently structured, unnecessarily places taxpayer dollars at risk, creates unfair subsidized competition, and does little to promote the goals of Congress and the Administration to bring broadband to every American.

Last week, Rep. Stephanie Herseth Sandlin (D-SD), along with lead Republican co-sponsor Rep. Jerry Moran (R-KA), introduced legislation to reform the RUS broadband loan program. First, HR 2035 would redefine "eligible rural community" so that scarce federal resources are not squandered on projects that serve densely populated suburban communities where broadband competition already thrives. By excluding these larger and densely populated regions, the program can focus on granting loans to the communities that most need it--rural areas that do not have access to a broadband service.

Second, the bill would also create an incentive for applicants to deploy broadband service to as many unserved households as possible. Specifically, applicants proposing to serve areas where more than 50% of the households do not have broadband service available at the time of the application may receive a loan that covers the full amount of the project, despite the presence of some competition. However, applicants proposing to serve areas where fewer than 50 percent of the households are unserved will only receive funding for the proportional amount of unserved households. This important distinction will encourage loan applicants to focus on finding and bringing broadband service to a significant number of unserved households. The Committee may also want to consider giving priority to loans that would result in bringing broadband service to the greatest number of unserved households. Another option would be to require that all projects that receive an RUS loan would be required to serve a certain minimum percentage of unserved households.

Mr. Chairman, the legislative changes prescribed in HR 2035 would do much to ensure that the RUS refocuses the broadband loan program on its original mission of bringing broadband service to unserved rural areas. The cable industry also believes that considerable changes have to be made to the RUS's broadband loan application assessment and approval process, since that process is in part responsible for the approval of many loans in areas that are already served, in many cases by a significant number of providers, or in areas which this program was not designed to reach. As Under Secretary of Agriculture Thomas Dorr acknowledged, at a Senate Commerce Committee hearing on March 6, 2006, "[a]s good stewards of the taxpayers' money, [the RUS] must make loans that are likely to be repaid" and that one of the agency's challenges "in determining whether a proposed project has a reasonable chance of success is validating the market analysis of the proposed territory…" (Dorr Testimony at 6). The RUS' own processes, however, make it difficult for the RUS to meet this challenge.

Specifically, the RUS should improve the process in place to solicit the data it needs to validate the applicant's market analysis. The public, including existing providers, receives little notice when RUS applications are filed. Current RUS rules require only a one time legal notice by the applicant in a newspaper of its choice at some time prior to the filing of its loan application. There is no guarantee that an existing provider will see the notice, thereby denying the RUS an accurate picture of the broadband services available in that market.

Furthermore, the RUS does not disclose when an application has been filed, the amount of the loan the applicant is requesting, or any assertions made about existing broadband service in a community, including whether the community is already served by one or more broadband providers. It does periodically update a relatively new online list of towns and unincorporated rural areas that are covered by pending or approved applications, along with information about the company applying for the loan. There is, however, no true public notice and comment period that ensures the public is heard, and its information is taken into account, prior to a loan being granted.

In fact, the only way anyone can find out how an applicant is characterizing a market is through a Freedom of Information Act (FOIA) request, and in our experience, such requests are usually fulfilled after the application is approved. When Midcontinent finally received a response to its FOIA request on the Mitchell, South Dakota application, we were more than a bit surprised to see that the applicant had blacked out its assertions about our company, apparently based on a claim that this information was proprietary or confidential. Other companies have reported that when they received FOIA responses, they found that applicants had made misstatements about, for example, their Internet speeds, system capacity, quality of service, and the number of video channels being offered.

We believe that the current notice and disclosure rules and practices limit the ability of the RUS to properly assess a loan's likelihood of repayment given the potential lack of data needed to evaluate an applicant's claims regarding the level of broadband service in a market. A more transparent, open process allowing for disclosure of non-proprietary, non-confidential information to the public would assist RUS staff evaluating loans and benefit the public, whose tax money supports this program. We acknowledge the need to protect proprietary information; but we believe it is essential to make available information being provided in a loan application about third parties and the state of broadband in a market. The RUS should know that information before it fully and fairly represents the market it is evaluating to determine the feasibility of a proposed loan and whether granting the loan is appropriate. Absent such information, the RUS had experienced $30.4 million in loan defaults as of Fall 2005 (OIG Report at i).

As part of your legislative efforts to reform and improve the RUS Broadband Loan Program, we recommend that Congress also direct the RUS to implement the following regulatory changes:

1. *Define "unserved":* The rules should be modified to simply define an unserved household as one where broadband service is not being provided and no entity has begun construction of the facilities to provide such service as of the date the application for a loan is filed with the RUS.

2. *Require an independent verification of "unserved households":* As there has been incorrect market data supplied to the RUS in past broadband loan applications, the RUS itself should make an independent determination regarding the number of unserved households in a proposed service area, based on a combination of information gathered independently by the agency, and information from the applicant and the public.

3. *Adopt a notice and comment period:* By adopting a formal notice and comment period for each filed application, the RUS will have access to additional information in making its determination of whether an area is served or unserved

In closing, let me reiterate that Midcontinent supports the federal government's goal of ensuring that all Americans have access to broadband services. We have invested millions of dollars to help that goal become a reality. We recognize that government subsidies may be the only answer in some rural areas. However, any government program designed to promote broadband deployment must be carefully defined and targeted at those areas that lack broadband service. We believe that Rep. Herseth Sandlin's legislative changes to the RUS broadband loan program coupled with regulatory enhancements and vigilant oversight will help this program accomplish its goals and increase broadband access in rural America.

Mr. Chairman, thank you for inviting me to testify today. I would be happy to answer any questions you or the Members of the Committee may have.

CHAPTER SOURCES

The following chapters have been previously published:

Chapter 1 – This is an edited, reformatted and augmented version of a United States Federal Communications Commission publication, dated May 22, 2009.

Chapter 2 – This is an edited, reformatted and augmented version of a United States Government Accountability Office, Report to Congressional Committees. Publication GAO-06-426, dated May 2006.

Chapter 3 – This is an edited, reformatted and augmented version of a United States Department of Agriculture, Economic Research Report, Report Number 78, dated August 2009.

Chapter 4 – This is an edited, reformatted and augmented version of a United States Congressional Research Service publication, Report Order Code RL33816, dated October 6, 2009.

Chapter 5 – These remarks were delivered as Statement of Jim Andrew, before the Subcommittee on Specialty Crops, Rural Development and Foreign Agriculture, U.S. House of Representatives, dated May 1, 2007.

Chapter 6 – These remarks were delivered as Statement of Kevin Felty, General Manager, Plains Cooperative Telephone Association, before the Subcommittee on Specialty Crops, Rural Development and Foreign Agriculture, U.S. House of Representatives ,dated May 1, 2007.

Chapter 7 – These remarks were delivered as Statement of Jack M. Geller, Ph.D., President, Center for Rural Policy and Development, before the Subcommittee on Specialty Crops, Rural Development and Foreign Agriculture, U.S. House of Representatives, dated May 1, 2007.

Chapter 8 – These remarks were delivered as Statement of Denny Law, Eastern Regional Manager, Golden West Telecommunications, before the Subcommittee on Specialty Crops, Rural Development and Foreign Agriculture, U.S. House of Representatives, dated May 1, 2007.

Chapter 9 – These remarks were delivered as Statement of Walter B. McCormick, Jr., President and CEO, U.S. Telecom Association, before the Subcommittee on Specialty Crops, Rural Development and Foreign Agriculture, U.S. House of Representatives, dated May 1, 2007.

Chapter 10 – These remarks were delivered as Statement of Tom Simmons, Senior Vice President of Public Policy, Midcontinent Communications, before the Subcommittee on Specialty Crops, Rural Development and Foreign Agriculture, U.S. House of Representatives, dated May 1, 2007.

INDEX

B

C

D

E

F

G

H

N

O

P

S

T

U

V

W

Y

Z